USING
IBM® SPSS®
STATISTICS

SECOND EDITION

I dedicate this textbook to my
three children, Sally, James (1965–1996),
and Wendy. The encouragement and support for their
father and his educational pursuits was (and is) above the call of duty.

—*James O. Aldrich*

I dedicate this book to my son,
Randy Cunningham, and my friend, Glenn Bailey.

—*James B. Cunningham*

USING

IBM®

SPSS®

STATISTICS

An Interactive Hands-On Approach

SECOND EDITION

James O. Aldrich ■ James B. Cunningham

California State University, Northridge

Los Angeles | London | New Delhi
Singapore | Washington DC | Boston

Los Angeles | London | New Delhi
Singapore | Washington DC | Boston

FOR INFORMATION:

SAGE Publications, Inc.
2455 Teller Road
Thousand Oaks, California 91320
E-mail: order@sagepub.com

SAGE Publications Ltd.
1 Oliver's Yard
55 City Road
London EC1Y 1SP
United Kingdom

SAGE Publications India Pvt. Ltd.
B 1/I 1 Mohan Cooperative Industrial Area
Mathura Road, New Delhi 110 044
India

SAGE Publications Asia-Pacific Pte. Ltd.
3 Church Street
#10-04 Samsung Hub
Singapore 049483

Acquisitions Editor: Vicki Knight
Editorial Assistant: Yvonne McDuffee
Production Editor: Bennie Clark Allen
Copy Editor: QuADS Prepress (P) Ltd.
Typesetter: C&M Digitals (P) Ltd.
Proofreader: Gretchen Treadwell
Indexer: Wendy Allex
Cover Designer: Janet Kiesel
Marketing Manager: Nicole Elliott

Printed in the United States of America.

ISBN 978-1-4833-8357-6

This book is printed on acid-free paper.

Certified Chain of Custody
Promoting Sustainable Forestry
www.sfiprogram.org
SFI-01268

SFI label applies to text stock

15 16 17 18 19 10 9 8 7 6 5 4 3 2 1

BRIEF CONTENTS

Detailed Contents

SAGE was founded in 1965 by Sara Miller McCune to support the dissemination of usable knowledge by publishing innovative and high-quality research and teaching content. Today, we publish more than 750 journals, including those of more than 300 learned societies, more than 800 new books per year, and a growing range of library products including archives, data, case studies, reports, conference highlights, and video. SAGE remains majority-owned by our founder, and after Sara's lifetime will become owned by a charitable trust that secures our continued independence.

Los Angeles | London | New Delhi | Singapore | Washington DC | Boston

PREFACE TO THE SECOND EDITION

This second edition was written while using IBM® SPSS® Statistics* Version 22. The first edition was written while using Versions 18 and 20. Although Version 22 is the most recent version available, it is certainly compatible with the earlier releases.

As in the first edition, this book can be used in conjunction with an instructor or as a self-instructional guide. It retains the well-received bulleted points, which inform the reader in exacting terms what has to be done to accomplish certain statistical operations while using the SPSS program. We have improved the self-instructional aspect of the book by adding more SPSS screenshots. The screenshots are complemented with a generous supply of callouts that are used to direct the reader's attention to specific control points.

REASONS FOR WRITING THIS BOOK △

One of the motivating factors in writing this book was to provide readers with the knowledge to effectively use the power of the SPSS program to analyze data of their choosing. It is the ability to analyze one's own data, see them come to life, that makes data analysis an exciting adventure into the unknown. We felt that many (or most) of the SPSS instructional textbooks utilize existing databases and provide minimal, if any, guidance on how to structure and enter data. In this second edition, we continue with the philosophy that it is wise to know how to enter data into the SPSS

---

*SPSS is a registered trademark of International Business Machines Corporation.

program. On leaving the academy and finding work in the real world, the ability to analyze data using SPSS can prove extremely useful in advancing one's career. In this edition, we continue to provide the reader with many opportunities for actually entering data, not just opening existing databases. We encourage readers to enter their own personal data as this makes the discovery process that much more exciting. There are few things in research that are more rewarding than making that final click on the mouse and watching your mass of numbers come to life with new meaning and purpose. Whether it's a graph, a prediction equation, or perhaps a statistical test showing a significant difference between groups, the discovery of the unknown that was hidden within the data can be extremely gratifying. The rewards of data analysis can give, and often have given, new meaning to the lives of researchers and to entire societies that benefit from discovery.

Δ   NEW FEATURES FOR THIS EDITION

Perhaps one of the most important additions to this second edition are the practice exercises at the end of each chapter. Detailed answers and explanations for these *review exercises* are provided in Appendix C at the end of the book. In many cases, these detailed answers (including relevant output screenshots) actually qualify as additional examples in each chapter.

Also new to this edition are the completely revised Chapters 8 and 9 on data graphing. These updated chapters present more complex graphing challenges than those given in the first edition. We feel that the detailed instruction in these new chapters will give the student the ability to produce and edit graphs having a truly professional appearance. These revised chapters present charts showing both *descriptive* univariate and *exploratory* bivariate graphing examples. This edition gives the reader hands-on experience in producing quality graphs by using the SPSS feature known as the *Chart Builder*. Knowledge of the *Chart Builder* will surely enhance one's ability to better understand data through graphing and visualization of summarized databases.

Although our original intent in writing this book was to publish a short "primer," it did grow a little beyond our expectations. Our readers commented positively, and some wanted more depth. Some readers wanted us to expand the coverage in order to make it suitable for graduate students. With that in mind, we added two new chapters to this second edition. Chapter 23 presents *logistic regression*, which serves as a natural extension of our *single* and *multiple regression* chapters. We chose to present the *binary logistic regression* method, which is easily understood and nicely handled by SPSS. Chapter 24, also new, is on *factor analysis*. We chose the popular method of *principal component factor analysis* as a way to introduce students to this type of analysis. This particular method of analysis will give the reader new insight into statistical tools that don't fall within the scope of testing for significance or prediction. Furthermore, we have found that the *principal component* approach to factor analysis can be an exciting *descriptive/exploratory* method for the new student/statistician. Discovering new *latent variables* can provide openings for creativity and can actually be fun! Such creativity and fun will be within the reach of anyone reading and practicing our factor analysis chapter.

We have also expanded and completely revised our two chapters on *chi-square analysis* with the idea of adding depth to our illustrations. We also did this to illustrate the different ways to input the frequency and proportional data to get SPSS to successfully do the chi-square test. We retained the two separate chapters for *goodness of fit* and *test of independence*, but each chapter now shows multiple ways for structuring and entering data for the chi-square analysis

Some minor changes that should prove useful include a new section in Chapter 4 that shows how SPSS can provide assistance by suggesting the *level of measurement* for your variables. The *data transformation* information was moved from an appendix to Chapter 6. Also added to Chapter 6 is a handy feature that allows one to split cases into groups for independent analysis. Chapter 18, on *analysis of covariance*, was also revised to directly include the test for the *homogeneity of regression slopes* (moved from the appendix).

DATA USED IN THIS BOOK   Δ

As in any book concerned with data analysis, a large amount of data and many databases are required. In some cases, we used real data, such as the database listed in Appendix A1 and Appendix A2, called *class survey*. However, in many instances, especially in the Review Exercises, the data were manufactured for the purpose of demonstrating a particular statistical technique. The results of the demonstrated analysis should be considered as only a demonstration of a statistical process—not as research facts. You will also notice that many databases from the SPSS samples files are used—these are also the result of data manufactured by SPSS for instructional purposes. We encourage readers to use their own data to duplicate some of the techniques illustrated in this book.

Δ OVERVIEW OF THE BOOK'S STRUCTURE

The book is unique in that it encourages the reader to interact with SPSS on the computer as he or she works through the examples in each chapter. This approach to learning may be novel to the reader, but we feel that the best way to learn a subject is to interact with it in a meaningful manner. We have made every effort to ensure that the book is "user-friendly" as we guide the reader through the interactive learning process. Bulleted phrases provide step-by-step procedures to be followed by the reader when completing the exercises.

Another novel approach taken in this book is the inclusion of parametric and nonparametric statistical tests in the same chapters. Other books describe parametric and nonparametric tests in separate chapters, which we feel is inefficient because it forces the reader to continually move from one section of a book to another in search of the rationale justifying the use of either type of test.

This second edition of *Using IBM® SPSS® Statistics: An Interactive Hands-On Approach* not only can be a useful resource for readers who may have some background in statistics but will also provide basic information to those individuals who know little or nothing about statistics. The book is for those who want SPSS to do the actual statistical and analytical work for them. They want to know how to get their data into SPSS and how to organize and code the data so SPSS can make sense of them. Once this is accomplished, they want to know how to ask SPSS to analyze the data and report out with tables and charts in a manner understood by the user. In short, they want SPSS to do the tedious work!

Δ OVERVIEW OF THE BOOK'S CHAPTER AND APPENDIX CONTENT

All chapters include screenshots showing the reader exactly how and where to enter data. The material covered in Chapters 1 through 4 provides basic but essential information regarding navigating in SPSS, getting data in and out of SPSS, and determining the appropriate level of measurement required for a statistical test. Chapters 5 and 6 describe additional methods for entering data, entering variable information, computing new variables, recoding variables, and data transformation. In Chapter 5, you will enter data from an important database (*class_survey1.sav*) found in Appendix A.

This database will be used in many of the subsequent chapters. Chapter 7 describes and explains the Help Menu available in SPSS and how to find information on various statistical tests and procedures. Chapters 8 and 9 provide hands-on experience in creating and editing graphs and charts. Chapter 10 provides explicit directions for printing files, the output from statistical analysis, and graphs. Chapter 11 describes and explains basic descriptive statistics. Finally, Chapters 12 through 25 provide hands-on experience in employing the various statistical procedures and tests available in SPSS, including both parametric and nonparametric tests. Appendix A contains an essential database that is entered in Chapter 5 by the reader and then used and modified throughout the book. Appendix B provides the reader with a "one-stop" shopping spot for many of the important basic concepts of inferential statistical methods. Appendix C gives the answers and detailed explanations for the review exercises that are provided at the end of each chapter.

How to Use This Book △

As the reader will note in the first lesson in Chapter 1, we use a simple format to allow the reader to respond to requests. The reader will be moving the mouse around the computer screen and clicking and dragging items. The reader will also use the mouse to hover over various items in order to learn what these items do and how to make them respond by clicking on them. Things the reader should click on or select are in **boldface**. Other important terms in the book are in *italics*. Still other items, such as variable names, are enclosed in double quotes.

The reader will often be requested to enter information and data while working through the examples and exercises in this book. To help in this procedure, we often present figures that show SPSS windows and then show exactly, step-by-step, where to enter this information or data from the keyboard. And, at times, we use callouts in combination with screenshots to clearly show control points and where to click or unclick specific items.

In Summary △

The IBM SPSS Statistics program is an outstanding, powerful, and intuitive statistical package. A primary reason for our writing this book was to make the benefits of the SPSS program available not only to the novice but also to the more experienced user of statistics. We feel this second edition is appropriate for lower-division and upper-division courses in statistics and research methods. We also feel that it will benefit students at the master's and doctoral levels as an introduction to some of the more complex statistical methods and how they are handled by the SPSS statistical package.

ACKNOWLEDGMENTS

I first thank my students, who for many years followed my often hastily written instructions on how to get SPSS to do what it was supposed to do. Second, I thank my coauthor, who had the idea for the book and invited me to participate in writing the first edition. I also thank my teaching assistant Hilda Maricela Rodriguez for her careful and tireless review of all the SPSS steps and screenshots presented in the book.

—*James O. Aldrich*

I wish to thank my colleagues, Richard Goldman, Wendy Murawski, and Marcia Rea, in the Center for Teaching and Learning at California State University, Northridge, for planting the seed for this book in our minds and for their encouragement while this book was being written. In addition, I wish to thank Michael Spagna and Jerry Nader, Michael D. Eisner College of Education, for their ongoing support.

—*James B. Cunningham*

We wish to thank the professionals at SAGE Publications for their valuable contributions to the publication of this book. They were always there for us, from the initial drafts, throughout production, and finally to marketing. If Vicki Knight, Publisher, had not seen merit in our proposal, this work would not have been possible. Vicki always had words of encouragement as we sometimes struggled over difficult terrain. Yvonne McDuffee, Editorial Assistant for Research Methods, and Bennie Clark Allen, Production Editor, always kept us on track during the editing and production process. We also thank Gretchen Treadwell for her excellent proofreading. Janet Kiesel produced a perfect cover for the book. Many thanks to Nicole Elliott, Marketing Manager, and Jade Henderson, Marketing Associate, for their efforts in bringing our work to the attention of potential users. Special thanks to Shamila Swamy and her team from QuADS Prepress for attention to detail and excellent copyediting. We also wish to thank Wendy Allex for a superb job on indexing.

We also thank V. Monica Young (Author's Program) and Amy Bradley (External Submissions) at IBM Chicago for their timely assistance in programming and permissions requirements.

We, along with SAGE, would also like to acknowledge the contributions of the following reviewers:

- Ronald F. Dugan, The College of Saint Rose
- Mark G. Harmon, Portland State University
- Diane Ryan, Daemen College
- Richard Acton Rinaldo, Georgian College
- Sally Dear-Healey, SUNY Cortland
- Ashish Dwivedi, Hull University Business School, United Kingdom
- Victor E. Garcia, Texas A&M University–Corpus Christi
- Andrew Munn, University of Northampton
- Susan Serrano, Florida Southern College
- Geoffrey W. Sutton, Evangel University
- Tommy E. Turner, Jacksonville State University
- Herb Shon, California State University, San Bernardino

About the Authors

James O. Aldrich (Doctor of Public Administration, University of Laverne) is a retired lecturer on statistics and research methods at California State University, Northridge. He has served as the principal investigator and codirector of a National Cancer Institute research project. He held the appointment of Instructor in the Department of Pathology at the University of Southern California, School of Medicine. He has served in various committees for the Los Angeles chapter of the American Statistical Association and has also taught biostatistics, epidemiology, social statistics, and research methods courses for 20 years. The primary statistical software used for his coursework has been SPSS. SAGE recently published, in 2013, *Building SPSS Graphs to Understand Data*, coauthored with Hilda M. Rodriguez.

James B. Cunningham (PhD in Science Education, Syracuse University) is Professor Emeritus of Science and Computer Education and former chair of the Department of Secondary Education at California State University, Northridge, and of the Departments of Science and Mathematics in Washington State high schools. He is the author of *Teaching Metrics Simplified* and coauthor of *BASIC for Teachers*, *Authoring Educational Software*, *Hands-On Physics Activities With Real-Life Applications*, and *Hands-On Chemistry Activities With Real-Life Applications*. He used SPSS extensively during his tenure as director of the Credential Evaluation Unit in the College of Education. He is a past fellow in the Center for Teaching and Learning at California State University, Northridge.

FIRST ENCOUNTERS

1.1 INTRODUCTION AND OBJECTIVES △

Hi, and welcome to IBM SPSS Statistics. We assume you know little about variables, values, constants, statistics, and those other tedious things. But we do assume you know how to use a mouse to move around the computer screen and how to click an item, select an item, or drag (move) an item.

We have adopted an easy mouse-using and -typing convention for you to respond to our requests. For example, if you are requested to open an existing file from the SPSS *Menu*, you will see click **File**, select **Open**, and then click **Data**. In general, we will simply ask you to click an item, select (position the pointer over) an item, drag an item, or enter data from the keyboard. Note that in SPSS, the columns in the spreadsheets run vertically and the rows run horizontally, as in a typical spreadsheet such as Excel.

OBJECTIVES

After completing this chapter, you will be able to

Enter variables into the Variable View screen

Enter data into the Data View screen

Generate a table of statistics

Generate a graph summarizing your statistics

Save your data

△ 1.2 ENTERING, ANALYZING, AND GRAPHING DATA

We are going to walk you through your first encounter with SPSS and show you how to enter some data, analyze those data, and generate a graph. Just follow these steps:

If you see the IBM SPSS icon anywhere on the screen, simply click it; otherwise, locate your computer's program files, and open SPSS from there. Once the SPSS starts, a screen will appear, which can take different forms depending on the SPSS version you are using. There are some useful shortcuts in these SPSS opening windows, but for now click the **white "x" in the red box** in the upper right-hand corner to close the window. When the window closes, you will see the Data Editor spreadsheet on the screen. This screen can appear in two different ways depending on which tab is clicked at the bottom of the Data Editor screen. These two tabs, Data View and Variable View, are together called the SPSS Data Editor. When you wish to enter or view variable information, you click the Variable View tab, and when you wish to enter or view data, you simply click the Data View tab. Figures 1.1 through 1.4 provide pictures of these screens.

Let's get started with the bullet point part of this introduction to SPSS. We will insert various figures into the text when we wish to clarify certain actions required on your part.

- Start SPSS, close the opening window as discussed above.
- At the bottom of the Data Editor spreadsheet screen, there are two tabs; click **Variable View** (see Figures 1.1 and 1.2).

Figure 1.1   Upper-Left Portion of the Variable View Screen of the SPSS Data Editor

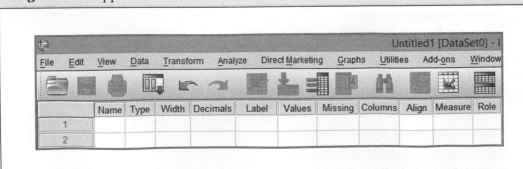

Figure 1.2 Lower Portion of the Variable View Screen of the SPSS Data Editor

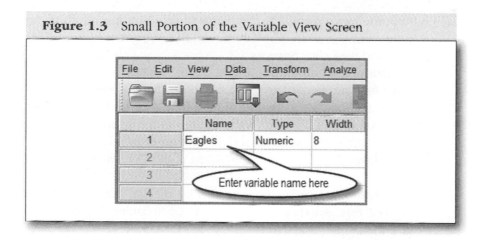

- At the top of the screen, type the word *Eagles* in the cell (this is the cell below *Name* and to the right of Row 1). The callout (balloon) shown in Figure 1.3 points to the cell in which you are to enter the variable name "Eagles." Cells are the little boxes at the intersection of *columns* and *rows*.

Figure 1.3 Small Portion of the Variable View Screen

- At the bottom of the screen, click **Data View** (note that the screen's appearance changes slightly).
- You will now enter the number of eagles observed on five consecutive days at the top of Holcomb Mountain. The callout in Figure 1.4 shows exactly where to type the number 3 (Row 1 and Column 1); for now, don't worry about the decimal points.
- Click in Row 2, and type *4*; click in Row 3, and type *2*; click in Row 4, and type *1*; and finally click in Row 5, and type *6*. Your screen should now look as shown in Figure 1.4. If you make a mistake in entering the numbers, just click the cell and reenter the correct number.

Figure 1.4   Small Portion of the Data View Screen

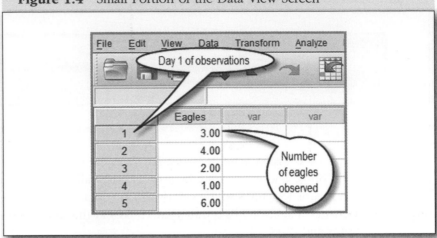

- After you have entered the five pieces of data, check carefully to see if the entries are correct. If they are, save your work as follows: Click **File**, and then click **Save As**.
- A window titled *Save Data As* will open, in which you will enter a name for your work (project). You could enter any name you wish, but for this exercise, enter the name *chapter1* in the *File Name* box. The *Look in* box (located in the middle of the window), showing where the file will be saved, should have an entry titled *Documents*. Click **Save**. Your data have been saved in the *Documents* section of your computer.
- An *Output* window opens; close this by clicking the **white "x" in the red box**. Another dialog box may open asking if you wish to save the output; click **No**.
- Let's continue with the exercise. On the SPSS *Menu* at the top of the screen, click **Analyze**, select **Descriptive Statistics**, and then click **Frequencies**. A window will appear titled *Frequencies*. Drag **Eagles** to the *Variable(s)* box, or click **Eagles** and then click the right arrow to place *Eagles* in the *Variable(s)* box (both methods work equally well).
- Click the **Statistics** button (the *Frequencies: Statistics* window opens). In the *Central Tendency* panel, click **Median** and **Sum**, then click **Continue**.
- Click **OK** (another screen opens, titled *Output IBM SPSS Statistics Viewer*, which shows the results of the analysis just requested). Look at Figure 1.5 for these results.

Figure 1.5 Frequency Statistics for 5-Day Eagle Observation

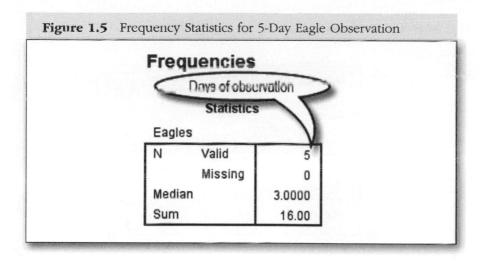

- On the Main Menu, click **Graphs**, select **Legacy Dialogs**, and then click **Bar**.
- The *Bar Charts* window opens; click **Simple**, and then click **Values of Individual Cases**. Click **Define**.
- The *Define Simple Bar: Values of Individual Cases* window opens. Click **Eagles** and drag it to the *Bars Represent* box, or click the right arrow to place *Eagles* in that box. Click **OK**. A simple bar graph will appear in the same Output IBM SPSS Statistics Viewer screen below the table, as shown in Figure 1.6.

After you have reviewed the graph, you will save the Output IBM SPSS Statistics Viewer screen, which contains the results of your analysis and the graph. Note that in the future we will often refer to this screen simply as the Output Viewer.

- In the screen, click **File**, and then click **Save As**.
- A window titled *Save Output As* will appear. In the *File name* box, type *chapter1*. Note that the file name is all lowercase and does not include any embedded spaces (blanks). The *Look in* box indicates the location where your file will be saved and should have an entry titled *Documents*. Click **Save**.
- After saving your work, your Output Viewer screen will remain. Click the **white "x" in the red box** found in the top right corner to make it go away.

Congratulations! You have just used SPSS (perhaps for the first time) to analyze some data and provide some statistical results and a graph.

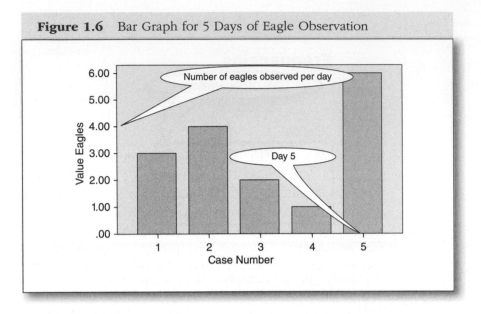

Figure 1.6   Bar Graph for 5 Days of Eagle Observation

Looking at the *Frequencies* table shown in Figure 1.5, we see that 16 eagles were observed over a period of 5 days with the median number per day of 3. The bar graph seen in Figure 1.6 provides the details regarding each day's observations. For example, we see that Day 5 yielded the most eagle sightings at 6, while the fewest were observed on Day 4, when only 1 was seen.

Admittedly, the statistical analysis and graph are not that exciting. But they do show you that SPSS is not difficult to use. Of course, you could have used a handheld calculator to do the same analysis in less than a minute. But suppose you had 50 different variables, such as height, weight, eye color, and so on, and thousands of cases for each of the variables! Using a calculator to analyze these data would be a monumental task. But SPSS can do it easily.

- If you wish to exit (quit using SPSS) at this time, click **File**, and then click **Exit**.

△ 1.3 SUMMARY

In this chapter, you learned how to enter variable names and data. You also learned how to generate a table of statistics and a graph summarizing those statistics. In the next chapter, you will learn to navigate in SPSS. You will be introduced to the Main Menu, the Toolbar editor, and the options available for these. Finally, you will be introduced to the various dialog boxes and windows in SPSS that allow you to enter information regarding your variables.

1.4 REVIEW EXERCISES △

1.1 You have classified the size of several fish that were caught in a "catch and release" fishing contest for children as small, medium, and large. The number of fish caught by the children are 32 small, 21 medium, and 11 large. *Note:* When inputting these data and information, you are *not* required to enter the names for the categories of the fish (small, medium, large). SPSS calls these categories *Labels* and *Label Values*. You will learn to input this information in a later chapter. Input the variable information and data, and build a frequency table and a bar graph. Name and save the database in the *Documents* section of your computer.

1.2 One day you are sitting in your professor's office getting help on regression analysis. His phone rings; he apologizes but says that he must take the call. As you wait for him to end his phone call, you scan his bookshelves and make mental notes of the titles. You arrive at the following: 15 books on introductory statistical analysis, 12 on advanced statistics, 3 on factor analysis, 8 on various regression topics, 13 on research methods, and 2 on mathematical statistics. You think to yourself, "Wow! This guy must have an exciting life!" As in the previous problem, don't concern yourself with the category labels for the textbooks. For now, just input the data and variable information, build a bar chart, generate a descriptive table, and name and save the database.

1.3 There was a quarter-mile drag race held at the abandoned airport last week. The makes of the winning cars were recorded by an interested fan. The results of her observations were as follows: Chevrolets won 23 races, Fords won 19 times, Toyota won 3, Hondas won 18, and KIAs won 8 races. As in the previous two problems, don't concern yourself with the categories' labels for the makes of the cars. Your task is to enter these data into SPSS, generate a bar graph and a frequency table, and then name and save the database.

Chapter 2

Navigating in SPSS

△ 2.1 Introduction and Objectives

As with any new software program you may use, it is important that you are able to move around the screen with the mouse and that you understand the meaning and purpose of the various items that appear on the screen. Consequently, we present a tour of the Variable View screen, the Data View screen, the Main Menu, and the Data Editor Toolbar. You will use these often as you complete the chapters in this book.

OBJECTIVES

After completing this chapter, you will be able to

Describe the Variable View screen and its purpose

Describe the Data View screen and its purpose

Select items from the Main Menu and the Data Editor Toolbar

Use the 11 items (*Name, Type, Width, Decimals, Label, Values, Missing, Columns, Align, Measure, and Role*) found in the Variable View screen to describe your variables

2.2 SPSS VARIABLE VIEW SCREEN △

Start SPSS, and click the Variable View tab at the bottom of the screen. Figure 2.1 shows a portion of the Variable View screen. We have entered the variable "height" in the first cell.

Figure 2.1 Upper Portion of the Variable View Screen Showing a Variable Named "height"

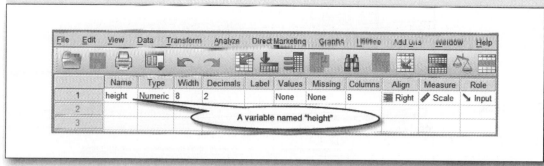

As you will recall from Chapter 1, you were briefly introduced to the Variable View screen when you entered the variable "Eagles." The *rows* represent variables, and the *columns* represent attributes (properties) and other information that you can enter for each variable. You must provide a name for each variable or SPSS will assign a default name, such as var1, var2, var3, and so on. It is in the Variable View screen that you enter all your variables and their properties. In Section 2.6, you are given all the details needed to properly enter the information on your variables.

Throughout this book, you will often be requested to enter information into a *cell*. Any cell you click is the active cell, displayed in color, indicating that it is ready to receive input from the keyboard. In Figure 2.2, you see an example showing a balloon pointing to the cell in which a variable named "Pre_treatment" has been entered.

2.3 SPSS DATA VIEW SCREEN △

A small portion of the Data View screen is shown in Figure 2.3.

Click the Data View tab if you are not already in that screen. It is in the Data View screen that you enter data for each variable. We have entered

Figure 2.2   Small Portion of the Variable View Screen Showing Two Named Variables

	Name			
	File Edit View Data Transform Analyze Direct Marketing			
1	Pre_treatment	Numeric	8	2
2	Post_treatment	Numeric	8	2
3				
4				

Variable View showing two named variables

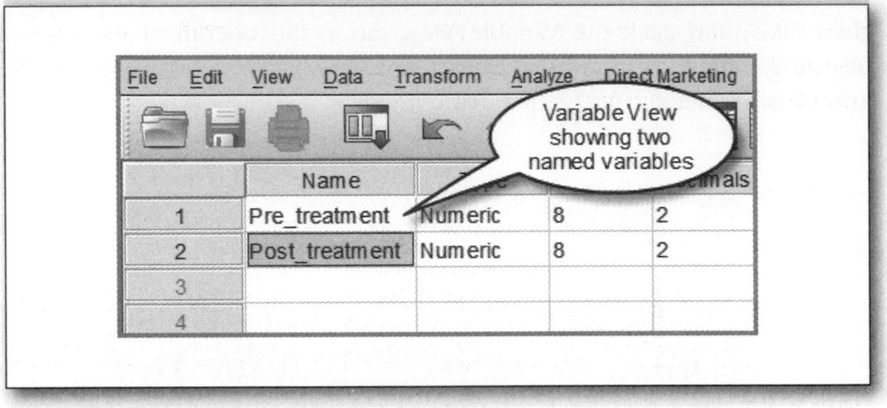

Figure 2.3   Small Portion of the Data View Screen With Data Entered

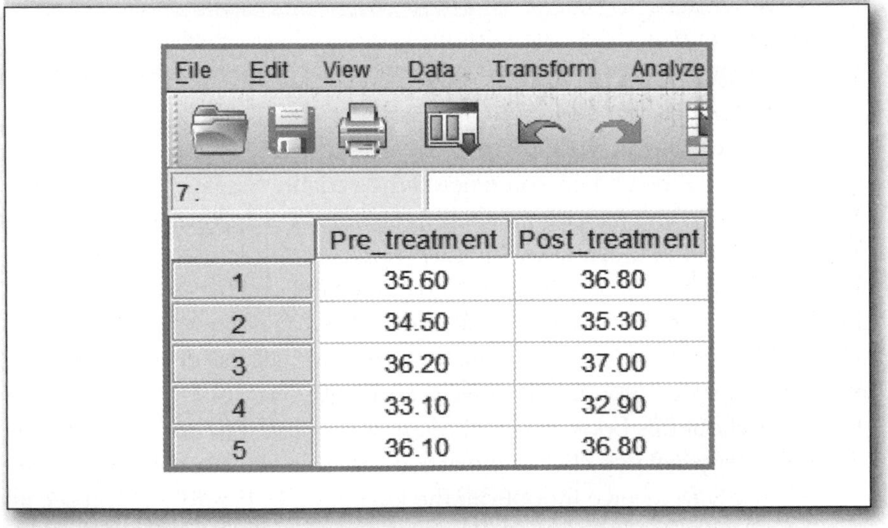

	Pre_treatment	Post_treatment
File Edit View Data Transform Analyze		
7 :		
1	35.60	36.80
2	34.50	35.30
3	36.20	37.00
4	33.10	32.90
5	36.10	36.80

five rows of data for two variables, "Pre_treatment" and "Post_treatment," as shown in Figure 2.3. The Data View screen is similar to the Variable View screen in that it shows *rows* and *columns*. However, in Data View, columns represent variables, and rows represent the cases, also called records, associated with each variable. A record may refer to a student, a teacher, a housewife, an automobile, a tree, or anything that can be measured or counted. Figure 2.3 shows records for five individuals and measurements on two variables called "Pre_treatment" and "Post_treatment."

2.4 SPSS MAIN MENU △

Let's take a look at the SPSS Main Menu, referred to hereafter as the *Menu*, as shown in Figure 2.4. This *Menu* is displayed at the very top of the Variable View and Data View screens.

Figure 2.4 Main Menu

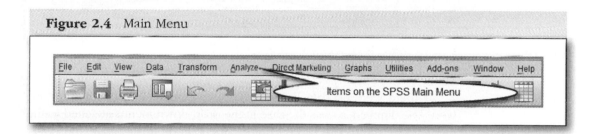

- Click **File**, and you will see a drop-down list of options you can choose.
- After clicking **File**, slide the mouse cursor over each of the items on the *Menu*—**Edit**, **View**, **Data**, **Transform**, and so on—until you have looked at each item on the *Menu*.

You may have noticed that some of the items on the drop-down menus were dimmed. This indicates that they could not be used at that particular time. There are various reasons for this, such as no open database, no statistical test underway, or perhaps no printing operation being done. As you progress through this book, you will see more of these icons undimmed and ready to use. And at this point, don't feel overwhelmed by the amount of information available on the *Menu* as you will only deal with a small portion in your work to become proficient in the use of SPSS. As you advance in using SPSS, you will be introduced to items on the *Menu* on an as-needed basis. The SPSS program is very intuitive, and after you have finished the first several chapters of this book, you will be breezing through the *Menu*.

2.5 DATA EDITOR TOOLBAR △

We next take a look at the Data Editor Toolbar, shown in Figure 2.5, which is a series of icons displayed horizontally across the page directly below the *Menu*. If you do not see this toolbar, do the following: On the *Menu*, click **View**, select **Toolbars**, and then click **Data Editor**.

Figure 2.5   Data Editor Toolbar

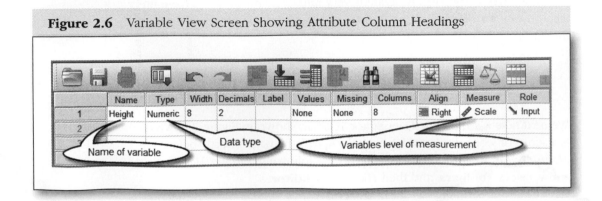

If there were no data in the Data View screen, some of these icons would be dimmed, as was the case in the drop-down menus attached to the *Menu*. Place the mouse pointer on the first icon on this toolbar, and hover over it. You will see *Open data document*, which is asking if you wish to open a document. Place the mouse pointer on the other icons, and hover over each so that you can see the purpose of these. Much of what you can do using the *Menu* can also be done using the Data Editor Toolbar. The toolbar simply makes your work easier by providing a simpler method. Older versions of SPSS may not include all these icons. But those we most frequently use are present in every version of SPSS. The Data Editor Toolbar is displayed in both the Variable View screen and the Data View screen unless you choose to hide this toolbar.

△   2.6 Variable View Screen: A Closer Look

We again show a portion of the Variable View screen in Figure 2.6.

Figure 2.6   Variable View Screen Showing Attribute Column Headings

Let's take a closer look and examine the options that are available in SPSS for describing and defining variables, such as the variable "Height." Think of a variable as a container that can hold values. To see how you can enter information regarding variables, do the following:

- Click **Variable View**.
- Click in Row1 below *Name*, and type the variable name "Height."
- Click the cell below *Type*. If you click in the left part of the cell, you will see a colored square (button).
- Click the button, and a window called *Variable Type* will open, as shown in Figure 2.7. (*Note:* It is more efficient to simply click the right-side portion of this and other similar cells as the dialog window then opens directly—there is no need to click a button).

In the *Variable Type* window, you can select certain settings to tell SPSS what type of numbers or information you wish to enter. In the absence of any additional information, SPSS has chosen *Numeric* as the type of data about to be entered.

- Click **OK** to close the window.
- Click the cell below *Width*. You can use the up–down arrows to set the width of a cell.

Figure 2.7 *Variable Type* Window

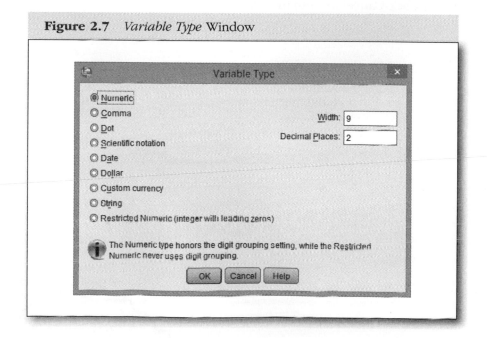

- Click the cell below *Decimals*. You can use the up–down arrows to change the number of decimal points in the values you have or will enter for that variable.
- Click the cell below *Label*. You can enter a longer identifying name for a variable. This can be important since this label will appear on much of the output, such as tables and graphs. If you choose not to enter a label, then SPSS reverts to the variable name, which can be sufficient in some cases.
- Click the cell below *Values*, and a window will open, as shown in Figure 2.8.

You can use this window to enter *labels* and *value labels* for variables (we mentioned these labels in the Review Exercises of Chapter 1). In the example shown in Figure 2.8, you see a *Value* of *1* and a *Label* of *tall*. This simply indicates that all people described as possessing the attribute of *tall-ness* will be entered under the variable "Height" as the number 1. We will describe and explain the *Value Labels* window in more detail in Chapter 5 when you enter a database.

- Click **Cancel** to close the window.
- Click the cell below *Missing*, and you will see a window, shown in Figure 2.9, in which you can enter information on missing values associated with the variables.
- Click **Cancel** to close the window.

Figure 2.8 *Value Labels* Window

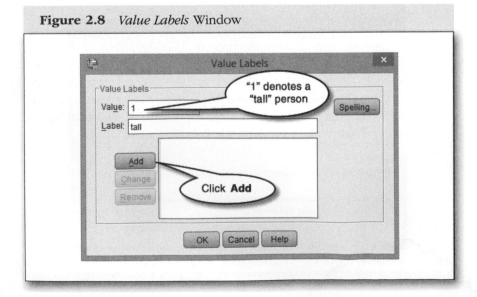

Figure 2.9　*Missing Values* Window

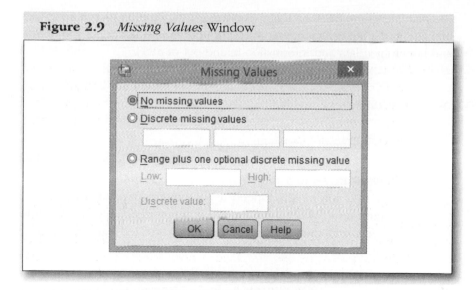

- Click the cell under *Columns*. You can use the up–down arrows to set the width of a column.
- Click the cell below *Align*. You can use the arrow to align information in a column. You may notice that we sometimes forgot to use this feature when presenting data entered into the SPSS Data View screen.
- Click the cell below *Measure*, which reads *Unknown*. You can use the arrow to indicate whether the level of measurement for a particular variable is *Scale*, *Ordinal*, or *Nominal*. We will have much to say about these three choices in Chapter 4.
- Click the cell below *Role*, which reads *Input*. A drop-down menu gives you a number of choices. These choices allow you to preselect how a variable is used in the analysis. The default choice for SPSS is *Input*. For the work in this book, we can leave this variable specification as *Input*. In the analytic procedures we use, the *Role* of our variables is specified when we set up the particular statistical analysis. If you want additional information, use the Help Menu, type in *role*, and select *Overview Variable Role Command*.

2.7 SUMMARY

In this chapter, you have learned to navigate the Variable View and Data View screens. You were introduced to various dialog windows and boxes used to enter information regarding variables, including *Name*, *Type*, *Width*, *Decimals*, *Label*, *Values*, *Missing*, *Columns*, *Align*, *Measure*, and *Role*. You investigated the Main Menu and the Data Editor Toolbar and the options available for each of these. In the next chapter, you will learn how to save your data and output and how to get data and information in and out of SPSS.

△ 2.8 REVIEW EXERCISES

2.1 You have designed a data-collecting instrument that has the following five variables measured at the *scale* level (*labels* are given in parentheses; *decimals* are set to 3 and *align* to *center*): (1) miles (speed in miles per hour), (2) kilometers (speed in kilometers per hour) (3) hours, (4) minutes, and (5) seconds. Input this information into the Variable View screen, and then enter four cases of fabricated data in the Data View screen.

2.2 You must set up the SPSS Data Editor to analyze the three variables listed below on 30,000 individuals. The variables are (1) age (label is *age in years*, no decimals, *center-aligned* and *scale* data), (2) education (label is *years beyond H.S.*, no decimals, *center-aligned* and *scale* data), and (3) family (label is *number of siblings*, no decimals, *center-aligned* and *scale* data). Make up and enter data for three cases—now you only have 29,997 more to enter!

2.3 You are the range safety officer at a long-distance firearms training facility. You have collected the ballistic information on four rifles—data are given below. You would like to set up a data file in SPSS to collect many hundreds of similar cases in the future. The variables are (1) caliber (2 decimals, *center-aligned* and *scale* data), (2) five hundred (2 decimals, label is *500-yard drop in feet*, *center-aligned* and *scale* data), (3) one thousand (2 decimals, label is *1,000-yard drop in feet*, *center-aligned* and *scale* data), and (4) weight (no decimals, label is *bullet weight in grains*, *center-aligned* and *scale* data). Set up the SPSS Variable View page for this officer. There is no need to enter data on this exercise— unless you're a military veteran!

CHAPTER 3

GETTING DATA
IN AND OUT OF SPSS

3.1 INTRODUCTION AND OBJECTIVES △

It is important that you save your data and output often in case your computer dies or the application you are using quits for no apparent reason. By "often" we do not mean every hour or so. We mean every 10 to 20 minutes. There may be occasions when you need to export your SPSS files to another application. Or you may need to import some files from other applications into SPSS. In addition, there are some handy SPSS sample files that were automatically included when SPSS was installed in your computer. We will be requesting that you use some of these sample files in many of the exercises presented in this book.

OBJECTIVES

After completing this chapter, you will be able to

Save and open your data and output files

Open and use sample files

Import files from other applications into SPSS

Export files from SPSS to other applications

Copy and paste data from SPSS to other applications

◭   3.2 Typing Data Using the Computer Keyboard

A simple method of entering data and other information into SPSS is to type them in using the computer keyboard. Whether you are a proficient typist or you use the "hunt and peck" or "peer and poke" method, your information must be entered into cells in the Variable View and Data View screens. We feel that the importance of typing in data, as opposed to opening existing databases, has been overlooked in many SPSS textbooks. There are elements of adventure and mystery when analyzing one's own data. There is the potential of making a thrilling new discovery by searching the unknown. In this chapter, you will learn to input data that you may have collected into the SPSS program.

◭   3.3 Saving Your SPSS Data Files

It is important that you understand the following convention of SPSS. When you save either a Data View screen or a Variable View screen, both screens are saved together whether you click **Save** or **Save As** on the Main Menu. If it is a new database or if you have made changes to an existing one, then you have the option to click **Save**. When clicking **Save**, the two screens are saved to the active location and with its current name. When using the **Save As** option, you are able to specify both a new name and a new location. To save these screens, you must do the following:

- Start SPSS, and open the database you created in Chapter 1 titled *chapter1.sav*
- Click **File** on the Main Menu, and then click **Save As**. A window titled *Save Data As* will open, as shown in Figure 3.1. A folder titled *Documents* will normally appear in the *Look in* box, indicating the SPSS default folder in which *chapter1* will be saved. If you wish to save the data in another location, perhaps a flash drive, then you can click on the button next to the *Look in* box (as shown in Figure 3.1), which will give you a drop-down menu of alternative locations. If you wish to change the file name or if you are saving new data, you would type the desired name in the *File name* box. Do *not* do this for this particular file.
- Click **Save**. Once you have clicked **Save**, a window will open, as shown in Figure 3.2. Since you have made no changes to the database, you may simply click **No**, **Yes**, or the **white "x" in the red box.** Other saving scenarios are self-explanatory.

Figure 3.1 *Save Data As* Window

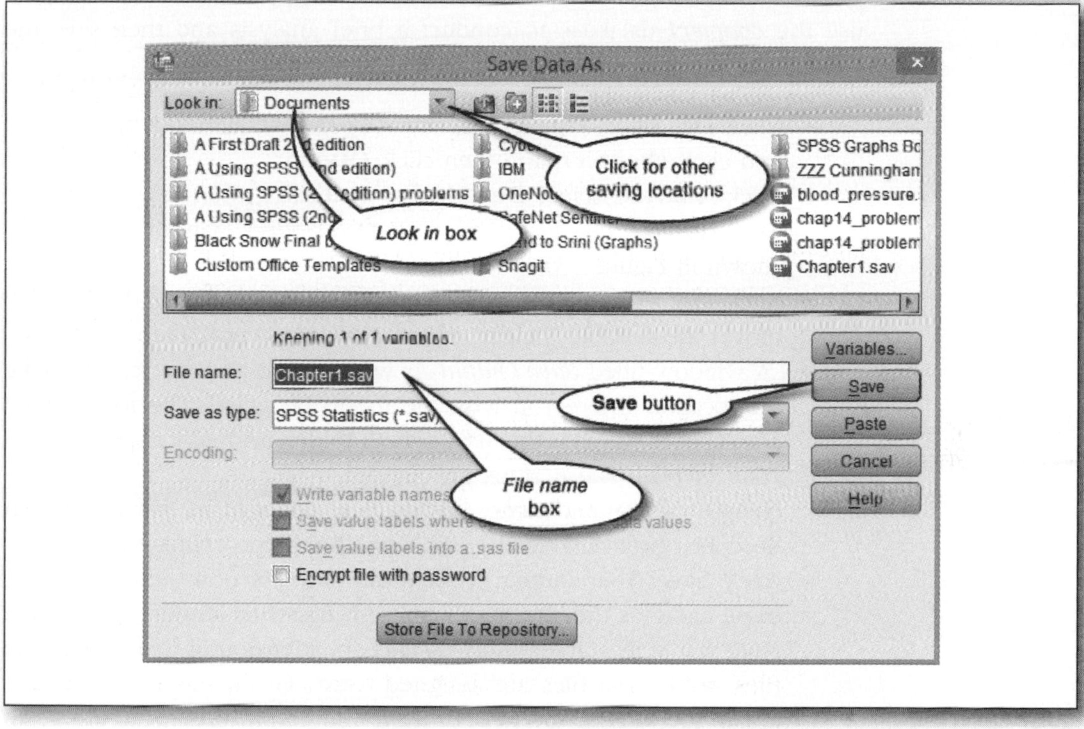

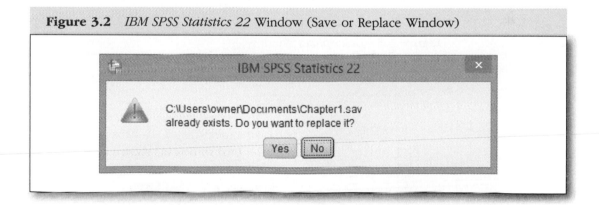

Figure 3.2 *IBM SPSS Statistics 22* Window (Save or Replace Window)

3.4 SAVING YOUR SPSS DATA FILES △

If you have requested an analysis or graph related to the data in the Data
View screen, that output is displayed on the Output IBM SPSS Statistics

Viewer screen. As mentioned in Chapter 1, for brevity we often refer to this screen as the Output Viewer. To demonstrate this process, we once again use the *chapter1* database to conduct a brief analysis and then save the Output Viewer screen that contains a statistical table.

- Click **File**, select **Open**, click **Data** (*Open Data* window opens), find and click **chapter1**, and then click **Open**.
- Click **Analyze**, select **Descriptive Statistics**, then select and click **Descriptives** (at this point, the *Descriptives* window opens, as shown in Figure 3.3).
- Click the arrow in the middle of the *Descriptives* window, which moves the variable "height" to the *Variable(s)* box, then click **OK**.
- A window titled *Save Output As* will appear, and you can indicate the location where you wish the file to be saved. The folder titled *Documents*, which is the SPSS default folder, normally appears in the *Look in* box. These steps are shown in Figure 3.4. You can also change the default name, in this case *Output3* to something that describes your data. In the *File name* box, type *chapter1*.
- Click **Save**. Your output, a statistical table, is now saved with the same name as the data file, making it easier to identify and locate. Note that SPSS automatically assigns an extension of ".spv" to output files, while data files are assigned ".sav". Given this fact, you may assign the same name to both *data* and *output* files without fear of them being overwritten.

Figure 3.3   *Descriptives* Window

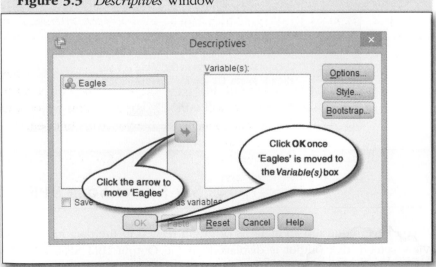

Figure 3.4   *Save Output As* Window

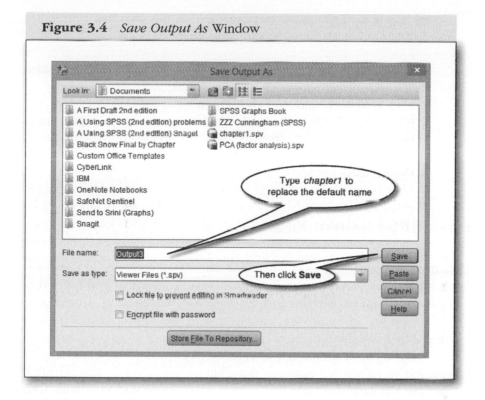

3.5 OPENING YOUR SAVED SPSS FILES △

Some additional practice on opening your existing SPSS files: You did this in the prior section with less specific instructions. To open a data file, from the Main Menu, do the following:

- Click **File**, select **Open**, and then click **Data**. A window titled *Open Data* will appear requesting that you locate the file you wish to open. The word *Documents* will normally appear in this window as the SPSS default folder that contains the file you wish to open.
- Click the file name you wish to open.
- Click **Open**. If this is not the location of your file, click the down arrow in the *Look in* box and scroll to locate your file.
- Click the file name, and it will appear in the *File name* box.
- Click **Open**.

If you are opening an output file, click **File**, select **Open**, and click **Output**. Then follow the steps for opening a data file.

For example, if you wish to open the data file *chapter1*, which you saved when reading the first chapter, do as follows:

- Click **File**, select **Open**, and then click **Data**. The folder titled *Documents* will appear in the *Look in* box.
- Click **chapter1**, and then click **Open**.
- If you wish to open the output file that you saved, click **File**, select **Open**, and then click **Output**.
- Click **chapter1.**
- Click **Open.**

△ 3.6 Opening SPSS Sample Files

In this book, we refer to a group of data that are to be entered in Data View as a *database*. Certain exercises in this book request that you enter data from the computer keyboard. These *databases* are listed in Appendix A. When you are in Data View, simply turn to the appropriate section of Appendix A and enter the data listed on that page.

To save you time and toil in entering larger sets of data that may be required to satisfy, for example, assumptions regarding certain statistical tests, we will have you open SPSS sample data files on your computer. With SPSS installed in your computer, you are given access to many useful SPSS sample files that were included in the installation. These are the SPSS files we will have you open for analysis. Follow the steps given below to access these SPSS files.

- Start SPSS.
- On the SPSS *Menu*, click **File**, select **Open**, and then click **Data** (the *Open Data* window opens).
- In the *Open Data* window, click the down arrow in the *Look in* box; then, scroll up or down to locate your computer's C drive, and click **C drive**.
- A window opens; double click **Program Files**.
- A window opens; scroll to and double click **IBM**.
- Double click **SPSS**.
- Double click **Statistics**.
- Double click **22**.
- A window opens in which you will see a long list of files that can now be opened.

Note: Opening these SPSS sample files can be tricky, especially on SPSS versions prior to 22. Also, the steps may vary slightly, but the general pattern is the same. Be sure to use the *double click* to proceed throughout the opening process. We also recommend that you save the most used sample files in your *Documents* section, which will then give you quicker access.

3.7 COPYING AND PASTING DATA TO OTHER APPLICATIONS △

A simple method to transfer data from SPSS to other applications such as Excel, and to transfer data from these applications to SPSS, is the familiar copy-and-paste procedure. We will describe the transfer of data from the file *chapter1.sav* to Excel:

- Start SPSS.
- Locate and open *chapter1.sav*.
- Next, open Excel (a spreadsheet will appear titled *Book 1*.)
- In SPSS, click on the first cell below "Eagles," then right click and drag down to select all five cells containing the Eagles data.
- Right click with your mouse pointer in one of the highlighted cells, and click **Copy**.
- Open Excel, right click in the first cell below *Column A*, then select and click **Paste**.
- The data from your SPSS file will be transferred to the five cells below *Column A*.

To copy data from Excel to SPSS, simply reverse the procedure.

3.8 IMPORTING FILES FROM OTHER APPLICATIONS △

Excel is a popular and readily available spreadsheet, so we will use it as an example of an application from which you may wish to import data files into SPSS.

- On the Main Menu, click **File**, select **Open**, and then click **Data**. A window titled *Open Data* will appear. In the *Files of type* box, use the arrow to scroll and then select the file type **Excel**. The file you wish to open should be in the *Documents* folder.
- Click **Excel file name**.

- Click **Open**. A window titled *Opening Excel Data Source* will appear.
- Click **OK**, and the file will be opened in the SPSS Data View screen.

Δ 3.9 EXPORTING SPSS FILES TO OTHER APPLICATIONS

An easy procedure to export SPSS files for use with other applications such as Excel is to use the *Save As* command from the Main Menu.

- With the file you wish to export open in SPSS, click **File** and then click **Save As**. A window titled *Save Data As* will open. Type a name for the file you wish to export in the *File name* box. Choose a destination for the file by clicking the arrow in the *Look in* box, which should be displaying the name *Documents*. If this is not the correct folder, click the arrow and scroll to the correct folder.
- Click the arrow in the *Save as type* box, and scroll to the name of the application that will be used to open the file you are saving. There may be different versions of the same application listed. You should choose the version that will most likely be used to open the file. For example, there are several selections for Excel. Not all users of Excel will be able to use the newest version listed, depending on the version of Excel they have installed in their computer. Consequently, it may be wise to choose an earlier version even if that version may lack some of the bells and whistles of the newer version.
- So click **Excel 97 through 2003**, and then click **Save**. Note that the file extension for Excel is ".xls." Your file has now been saved as an Excel file in SPSS.
- If you start Excel on your computer, you will be able to click **Open** and load this file.

Δ 3.10 SUMMARY

In this chapter, you learned how to save and open your data and output files. You learned how to open the sample files that were installed with SPSS. You also learned how to import files from other applications into SPSS and how to export files from SPSS to other applications. In the next chapter, we will describe levels of measurement, including *nominal, ordinal*, and *interval*, and the necessity of carefully describing your data using these levels of measurement.

3.11 REVIEW EXERCISES △

3.1 With this review exercise, you will open an SPSS sample file *workprog. sav*. Once it is opened, you will save it in your documents file, making it easy to access as you will need this database frequently as you progress through this book.

3.2 In this exercise, you must import an Excel file from your computer and show the appearance of the *Open Excel Data Source* window and the first six rows of the Variable View screen. There should be an Excel file used as a demonstration for the Excel data program within your system files. Its name is *demo [AB6401]*, and it will be opened as an SPSS spreadsheet; examine the file, and observe that you can analyze the data as in any other SPSS spreadsheet.

3.3 Open another one of SPSS's sample files called *customer_dbase.sav* (it has 132 variables and 5,000 cases), and save it in your documents file. It's another database that you will use several times throughout this book.

CHAPTER **4**

LEVELS OF MEASUREMENT

Thus far, we have led you through the basic procedures needed to enter variables and data and to navigate in IBM SPSS Statistics. Chapter 4, on levels of measurement, covers an essential bit of knowledge required to successfully use SPSS—specifying the correct level of measurement for each of your variables. SPSS provides the user with three choices when selecting these levels of measurement: (1) *nominal*, (2) *ordinal*, and (3) *scale*. The major purpose of the chapter is to assist you in selecting the appropriate level of measurement for each of your variables.

Each of the three levels provides the SPSS user with different amounts of analyzable information. Another important consideration is that different statistical procedures require that the data be collected at specific levels of measurement. The level of measurement is partially determined by the basic nature of the variable (more on this later); however, the analyst does have a certain degree of freedom when specifying one of the three levels.

By way of definition, we say that *level of measurement* is a phrase that describes how measurable information was obtained while observing variables. For example, measuring the length of a banana with a ruler yields *scale* data; ranking individual runners as finishing first, second, and third in a 100-yard dash provides *ordinal* data; and, finally, when the females in a room are counted, we have *nominal* data. The information presented in the following sections is concerned with the expansion of this basic definition of levels of measurement.

There are many considerations when assigning levels of measurement to variables. There are some hard empirical rules to follow, but there are also other considerations. Many of these considerations exceed the purpose of this book. Some are mentioned in this chapter, but others we leave to our readers to discover as they advance in the use of SPSS for statistical analysis. As an illustration of the challenges faced in assigning levels of measurement, we include a section in this chapter demonstrating SPSS's attempt to assist the user in this important task. However, you will see that there is *no* substitute for the user's judgment—the SPSS program can only "suggest" levels of measurement. This is one reason why we have spent so much ink on this topic.

OBJECTIVES

After completing this chapter, you will be able to

Describe the characteristics of *nominal*, *ordinal*, and *scale* levels of measurement

Distinguish between *nominal*, *ordinal*, and *scale* levels of measurement

Determine the level of measurement appropriate for a variable

Use SPSS features that *suggest* appropriate levels of measurement

4.2 VARIABLE VIEW SCREEN: MEASURE COLUMN   △

By this time, you should feel confident navigating the various menus and windows. You have used the Variable View screen when you named variables. You may have noticed that one column on the Variable View screen was called *Measure*. The function of the *Measure* column is the subject of this chapter. To help you understand the concept of levels of measurement, and the proper use of the *Measure* column in SPSS, you will need to open the data file you saved in Chapter 1 that you titled as *chapter1*.

- Start SPSS, and click **Cancel** to close the opening window.
- Click **File**, select **Open**, and then click **Data**.
- Click **chapter1.sav**, and then click **Open**.

- Click **Variable View** (at the bottom of the screen), and then inspect the *Measure* column, which shows the level of measurement. (Prior to clicking on the *Measure* column, it reads *Unknown*. After entering data, the most common SPSS default selection is *Scale*; however, in this case, the SPSS program made the correct decision and specified the *Nominal* level of measurement.)
- Click the cell below the *Measure* column. (You will see a drop-down menu displaying *Scale*, *Ordinal*, and *Nominal*, as shown in Figure 4.1.)

Figure 4.1   Variable View Screen: *Measure* Column Drop-Down Menu

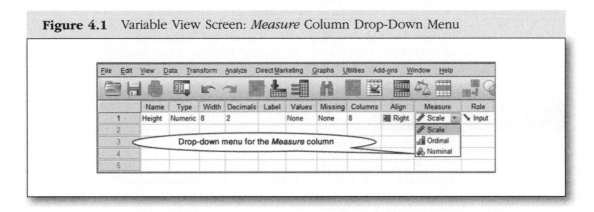

These three choices, as shown in Figure 4.1, are known as *levels of measurement*, which is the subject of this chapter.

In earlier versions (pre-22) of SPSS, the program would most frequently select *Scale* for numerical data entered in the Data View spreadsheet. SPSS did not always make the correct decision when it designated data as having been measured at the *scale* level. SPSS now has added a default called *Unknown*, which is present when you enter data. If you fail to designate a level of measurement, SPSS will assign levels depending on the format of the data and any attempted analysis you may do. The SPSS program could not then, and still cannot, consistently recognize the correct level of measurement that was used to measure your variables. Because of this, you, the person trying to make sense out of the data, must understand whether the variables were measured at the *nominal*, *ordinal*, or *scale* levels. Once you determine the correct level of measurement for your variables, you then specify this in the *Measure* column of the Variable View screen. We are going to spend time helping you understand how to do this for each of your variables. We will give you just enough information to ensure that SPSS will correctly understand your data.

SPSS recognizes three levels of measurement that should be considered as a hierarchy: *nominal* (lowest level), then *ordinal* (middle level), and finally *scale* (highest level). This hierarchy (lowest, middle, and highest) refers to the amount of analyzable information contained in the numbers (your data) resulting from your counting or measurement of your observations. Let's look at a few examples that demonstrate these levels of measurement, beginning with the lowest level, *nominal*.

4.3 VARIABLES MEASURED AT THE NOMINAL LEVEL △

When a variable is said to have been measured at the *nominal* level, the numerical values only represent labels for various categories of that variable. It should be noted that when a variable is measured at the *nominal* level, the categories are often referred to as attributes of that variable. It is a common practice to refer to variables as being "measured" at the *nominal* level, but in reality, we are only counting the occurrences of the variable's attributes. The numbers used to represent various categories (attributes) contain the least amount of analyzable information of the three levels of measurement. We can only count and calculate the percentage of the total number of observations that occupy the various categories. When trying to make sense of the attributes of variables measured at the *nominal* level, we may use data tables and bar graphs to display the numbers and/or percentages of the various attributes that were observed. Additional analytic methods, both *descriptive* and *inferential*, for the attributes of variables measured at the *nominal* level are described in future chapters.

Variables having separate categories that are not related in terms of quality or quantity are said to be measured at the *nominal* level. An example of a variable measured at the *nominal* level is gender. Gender has two possible categories: female, which could be labeled as Category 1, and male, labeled as Category 2. You should note that the categories for gender, female and male, do not differ, at least in any meaningful way, in quality or quantity. In other words, the value of 2 that was assigned to males has no numerical meaning; the value 2 may be twice the value of 1, but males are not twice the value of females. The numerical values only serve as *labels*.

It is important to note that numerical values such as the labels 1 = *female* and 2 = *male* can only be counted (not added, subtracted, multiplied, or divided). However, understand that SPSS will incorrectly calculate statistics requiring such mathematical operations. For this and other reasons, you must understand levels of measurement when using SPSS.

Some additional examples of variables having attributes that could be measured at the *nominal* level are religious affiliation, name of automobile owned, city of residence, university attended, high school attended, neighborhood of residence, and number of voluntary memberships in community organizations. It is important to note that the intention or purpose of the individual making the observations (collecting the data) is essential since the attributes of the *nominal* variables in the above list could also be measured at the *ordinal* level. As an example of this dual measurement potential, let's look at the variable "university attended," where the intention could ultimately determine the level of measurement. If the intention of the data collector is to infer that the quality of education depends on the university attended, then the numerical values representing the attributes (categories) take on the quality of ranked data, discussed in the next section. If this is not the purpose or intention of the data collection, then the attributes of the variable, "university attended," take on the quality of the most basic level—*nominal*.

Another example of this dual measurement potential would be the variable "name of automobile owned." During a job application process, a prospective employer might ask the make and year of the automobile you currently own. The purpose of this question may be to rank individuals on some quality—perhaps ambition or drive to succeed. However, the more common interpretation of such a variable would be a simple counting of individuals owning various types (categories) of automobiles. It is important to understand the intention and purpose of your data collection before assigning a level of measurement to the attributes of variables.

△ 4.4 VARIABLES MEASURED AT THE ORDINAL LEVEL

Data measured at the *ordinal* level are often called *ranked data* because the categories of such variables measure the amount (a quality or quantity) of whatever is being observed. Thus, the categories of *ordinal* data progress in some systematic fashion; usually, such variables are ranked from lowest to highest, but they might also be ranked from highest to lowest.

The *ordinal* level goes beyond the *nominal* in that it contains more analyzable information. The numerical values, which label the categories, are ordered or ranked in terms of the quality or quantity of the characteristic of interest they possess. Think of the situation where your intention is to measure the degree of military rank possessed by a group of U.S. Marines. When collecting data, you could specify the following: 1 = *private*, 2 = *corporal*, 3 = *sergeant*, and 4 = *general*. In this example, it can be seen clearly that as the numbers increase so does the degree of military rank. When using the *ordinal* level of measurement, it is important to note that the differences (distances) between ranks are undefined and/or meaningless. In this military rank example, the differences or distances between private and corporal are not the same as those between sergeant and general. The difference between a sergeant and a general is many times greater than the difference between a private and a corporal. If we input the numbers that were used to distinguish between these categories of military rank (1, 2, 3, and 4) into SPSS, we could do the same analysis that we did with the attributes of variables measured at the *nominal* level. However, in addition to tables, graphs, and the calculation of percentages of each attribute, we can directly calculate the median. The median will then accurately portray the middle point of the distribution, which in this example would be the rank at which 50% of the individuals counted fall above and 50% fall below.

Another example of *ordinal* data is socioeconomic status, which could be described in terms of lower, middle, and upper class. We could specify that 1 = *lower class* (income of $0 to $20,000), 2 = *middle* class ($20,001 to $100,000), and 3 = *upper class* (>$100,000). Socioeconomic status is now categorized as 1, 2, or 3. Individuals so classified according to their range of income are said to have been measured at the *ordinal* level.

One final example of a variable's attributes measured at the *ordinal* level would be a survey questionnaire seeking information about marital relationships. Suppose there was a question that sought to measure a person's satisfaction with his or her spouse. The variable might be titled "spousal satisfaction," and the attributes could be the following: 1 = *ready for divorce*, 2 = *thinking of divorce*, 3 = *satisfied*, 4 = *happy*, and 5 = *extremely happy*. The numerical data (1–5) could be entered into SPSS and analyzed using the mode and the median for basic descriptive information.

4.5 VARIABLES MEASURED AT THE SCALE LEVEL △

Data measured at the *scale* level of measurement, as defined by SPSS, actually consists of data measured at two separate levels known as the *interval* and *ratio* levels. SPSS does not require that you, the user, distinguish between data measured at the *interval* and *ratio* levels. For mathematicians, the difference between *interval* and *ratio* data is important, but for the types of applied research conducted when using SPSS, the differences are not important. Data measured at SPSS's highest level, *scale*, contain more analyzable information than the *nominal* and *ordinal* levels because the differences or distances between any two adjacent units of measurement are assumed to be equal. This quality of "equal distances" permits all the mathematical operations conducted on the *nominal* and *ordinal* levels plus many more. When using the *scale* level of measurement, you may now summarize your data by computing the mean and various measures of dispersion such as the standard deviation and the variance. The mean and a measure of dispersion provide a distinct picture of the collected data—making them more understandable.

It should be noted that we drop the references to categories and/or attributes when using the *scale* level of measurement. We now refer to the "units of measurement" that are used to determine the numerical values associated with our variables. An example is measuring the distance to the top of a mountain. The units of measurement for the variable "distance" could be given in feet, megalithic yards, meters, kilometers, or any other acceptable unit that might serve your purpose.

Examples of data that you could specify as scale data would be weight, height, speed, distance, temperature, scores on a test, and intelligence quotient (IQ). Let's look at height as a way of explaining the scale level of measurement. If you measure height in inches, you can state that it has equal intervals, meaning that the distance between 64 and 65 inches would be identical to the difference between 67 and 68 inches. Or for temperature, the difference between 72 and 73 degrees is the same as between 20 and 21 degrees. Height also has a true (absolute) zero point; thus, the mathematician would say that it is measured at the ratio level. For SPSS users, we only need to refer to it as data measured at the scale level.

Another example of a variable measured at the scale level is IQ. The difference between an IQ of 135 and 136 is considered the same as that between an IQ of 95 and 96. However, it should be understood that persons having an IQ of 160 cannot say that they are twice as intelligent as persons with an IQ of 80, because IQ is measured at the *interval*, not *ratio*, level. Remember that variables measured at the ratio level require a true (absolute) zero point. It would be difficult, if not impossible, to show that an individual has zero intelligence.

△ 4.6 Using SPSS to Suggest Variable Measurement Levels

We must be honest and tell our readers that we were hesitant to include a description of the SPSS feature that suggests *levels of measurement* for your variables. This hesitancy was partly based on the belief that some readers will skip over the information provided in previous sections and totally rely on SPSS to determine levels of measurement. We hope that you do not follow this approach. As we describe the steps necessary to use SPSS's program assistance in setting levels of measurement, please keep in mind the following SPSS admonition: "The suggested measurement level is based on empirical rules and is not a substitute for user judgment." For this demonstration, we use the SPSS sample file *workprog.sav*.

- Start SPSS, click **File**, select **Open**, then click **Data**.
- Locate *workprog.sav* in the SPSS sample files, and open it. (If you need help opening this sample file, return to Section 3.3 in Chapter 3.)
- Click **Data** on the Main Menu, then click **Define Variable Properties**. (The *Define Variable Properties* window opens, as shown in Figure 4.2.)

Figure 4.2 Define Variable Properties Prior to Moving Variables

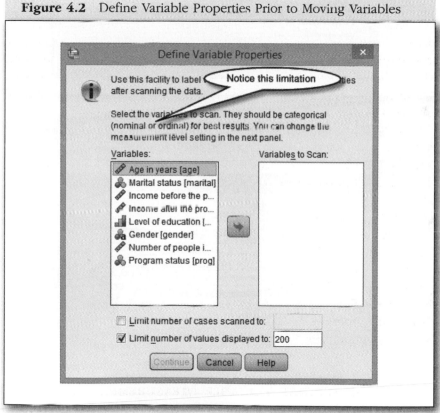

- Click *each variable* while holding down the computer's Ctrl key (to select all variables).
- Click the arrow to move the now highlighted variables to the *Variables to Scan* box, then click **Continue**. (Once **Continue** is clicked, a second *Define Variable Properties* window opens, as shown in Figure 4.3.)
- Click the **Suggest** button, as shown in Figure 4.3. (A third *Define Variable Properties* window opens, as shown in Figure 4.4.)

Figure 4.3   The Second *Define Variable Properties* Window (Current Variable Is "age")

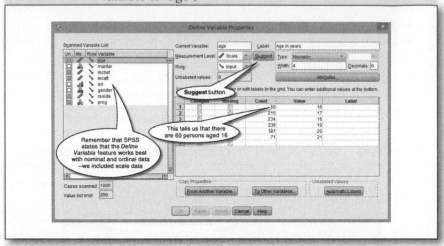

Figure 4.4   Third *Define Variable Properties* Window: Suggests Measurement Level for "age"

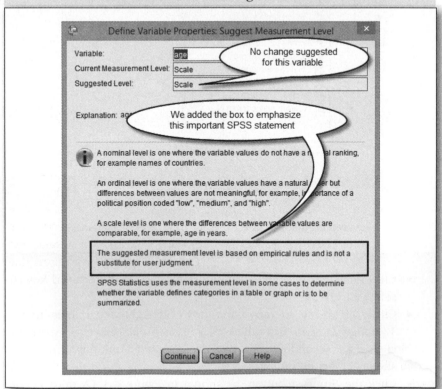

After you clicked the **Suggest** button in the second *Define Variable Properties* window, the final window opened, showing that SPSS agreed that the variable "age" was indeed a variable that should be measured at the *scale* level. Let's take this demonstration one step further and select a variable where SPSS's algorithm does not agree with the user's selected level. We do this, at least partly, to demonstrate the value of understanding levels of measurement.

- Click **Continue**, which returns you to the second *Define Variable Properties* window.
- Click the variable named "incbef" (which moves it to the *Current Variable* box).
- Click the **Suggest** button, and the *Define Variable Properties: Suggest Measurement Level* window opens once again, this time showing the variable as "incbef" (see Figure 4.5).

Figure 4.5   Upper Portion of the *Suggest* Window for the Variable "incbef"

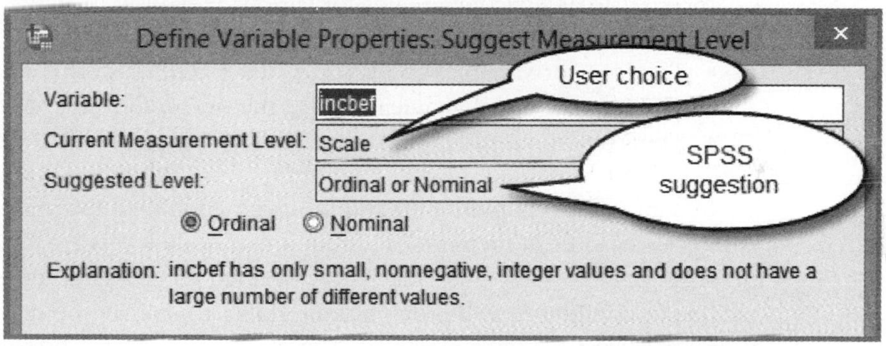

As you can see in Figure 4.5, the *Define Variable Properties* feature in SPSS made the recommendation to change the *level of measurement* from *scale* to *ordinal*. It did this to satisfy empirical rules that do not take into account things such as the user's intended analysis. You may be thinking that since this is an SPSS database, didn't SPSS assign the *scale* level to this variable and now it recommends a change to *ordinal?* Your thinking is correct, and it is another demonstration of the complexity of the process of assigning levels of measurement. Remember that the type of analysis is one important consideration when assigning levels of measurement to your variables. It should also be kept in mind that it is always possible to change the *scale* level of measurement to *ordinal* or *nominal*, but the reverse is not true.

Δ   4.7 Summary

In this chapter, you had the opportunity to learn how to distinguish between variables measured at the *nominal*, *ordinal*, and *scale* levels of measurement. It is very important for you to understand that the level of measurement you select for your variable determines the type of statistical analysis you may use in analyzing the data. Certain types of analysis are only appropriate for specific levels. An example would be that you never calculate the mean and standard deviation for data measured at the *nominal* level. The computer and SPSS would do it, but it is an incorrect statistical analysis. You also learned about the use of an SPSS feature (*Define Variable Properties*) that will "suggest" levels of measurement for your variables. The shortcomings of using this feature were illustrated.

Once the appropriate level of measurement has been inputted into SPSS, it is assigned an icon. These icons make the identification of levels of measurement for variables straightforward, thus facilitating the selection of the appropriate analytic techniques for various variables.

- The icon for variables measured at the *nominal* level is ⧉.
- The icon for variables measured at the *ordinal* level is ⧉.
- The icon for variables measured at the *scale* level is ⧉.

In subsequent chapters, you will be given additional examples of how variables measured at the nominal, ordinal, and scale levels are entered into the SPSS Variable View and Data View screens. As you gain more experience, your confidence will grow in your ability to specify the correct level of measurement for your variables. The following chapter presents some basic validation procedures for variables measured at the nominal, ordinal, and scale levels.

Δ   4.8 Review Exercises

4.1 The following is a list of variables that an investigator wishes to use to measure the health and survival of a particular species of earthworm: age, length, weight, moisture content, breed, environmental factors, and acid content of the soil. Your job is to assist this researcher in specifying the correct levels of measurement for these key variables and set up Variable View in SPSS.

4.2 A social researcher is interested in measuring the level of religiosity of a sample of senior citizens. Help her in establishing the levels of measurement for the following variables: pray (do you pray?), services (number of times you attend formal church services per year), money (donated to church), volunteer (hours per year of volunteer assistance), member (are you an official member of a church?), discuss (how many times each week do you discuss religious doctrine?), and times pray (how many times per week do you pray?). Input these data into the SPSS Variable View spreadsheet.

4.3 A political consultant wished to measure the level of politicization of the candidates for a job at the White House. He decided that the following variables would provide at least some of the evidence required to assess the extent of their interest in politics: vote (did you vote in the previous election?), letter (the most recent letter sent to a politician), meeting (the most recent political meeting attended), donate (how much money did you donate to a politician in the past year?), candidate (have you run for office?), and party (are you a member of a political party?). Your job is to input these variables and their labels into SPSS and specify their levels of measurement.

CHAPTER 5

ENTERING VARIABLES AND DATA AND VALIDATING DATA

△ 5.1 INTRODUCTION AND OBJECTIVES

An important component of learning how to use IBM SPSS Statistics is the correct entry of variables and data and then the validation (verification) of the entries. Entering and assigning attributes (properties) such as *Measure* and *Values* to a variable is accomplished in the Variable View screen. Entry of data is accomplished in the Data View screen. The Variable View screen and the Data View screen are known collectively as the *Data Editor*. Once variables and data are entered, SPSS can perform any number of statistical analyses and generate an extraordinary number of graphs of various types. However, if there are errors in either the data or the attributes assigned to variables, any analysis performed by SPSS will be flawed, and you may not even realize it. For example, if you assign *scale* in the *Measure* column to a nominal (numerical) variable, SPSS will calculate the mean, which is an incorrect analysis and therefore meaningless.

In this chapter, you will enter the database displayed in Appendix A (Tables A.1 and A.2), titled *class_survey1*. We will take you, step-by-step, through the process of correctly entering the variables, attributes, and data. If you do not assign attributes to each variable, SPSS will assign default values for some that may or may not be appropriate. For example, if you do not assign a level of measurement for a variable, SPSS will sometimes assign a level of measurement for that variable that may not be correct. The most recent SPSS versions will leave the level as *Unknown*.

OBJECTIVES

After completing this chapter, you will be able to

Enter variables and their attributes in the Variable View screen

Enter data associated with each variable in the Data View screen

Validate the accuracy of the data

5.2 ENTERING VARIABLES AND ASSIGNING ATTRIBUTES (PROPERTIES) △

You should now turn to Table A.1 found in Appendix A. This table displays all information needed to correctly enter the seven variables associated with the *class_survey1.sav* database. On the next page, you will find Table A.2, which displays all data associated with the variables. You will enter the variables and attributes (Table A.1) in the Variable View screen and then enter the associated data (Table A.2) for these variables in the Data View screen. Let's get started!

- Start SPSS, and close the opening window by clicking the **white "x" in the red box**.
- You should see the Variable View screen; if not, click the Variable View tab at the bottom.
- Click the cell below *Name*, and type *Class*.
- Click the right side of the cell below *Type*, and a window titled *Variable Type* will open. Click **Numeric**, and then click **OK**.
- Click the cell below *Width*, and then use the up–down arrows to select **8**.
- Click the cell below *Decimals*, and use the up–down arrows to select **0**.
- Click the cell below *Label*, and type *Morning or Afternoon Class*.
- Click the right side of the cell below *Values*, and a window will open titled *Value Labels*. Type *1* in the *Value* box, and type *Morning* in the *Label* box. Click **Add**. Then go back to the *Value* box, and type *2*, and in the *Label* box type *Afternoon*.
- Click **Add**. (Your window should now look like Figure 5.1.)
- Click **OK**.

Figure 5.1   Value Labels Window for Morning or Afternoon Class

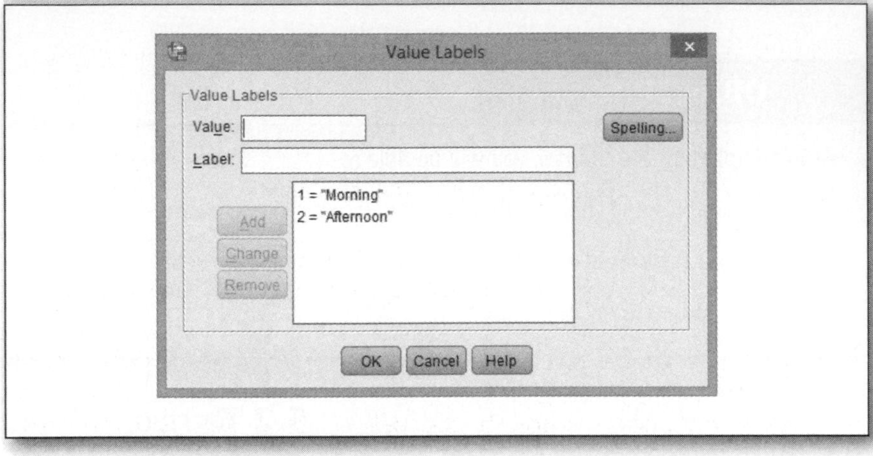

- Click the right side of the cell below *Missing*, and a window titled *Missing Values* will open. Click **No missing values**, and then click **OK**.
- Click the cell below *Columns*, and use the up–down arrows to select **8**.
- Click the right side of the cell below *Align*, and click the arrow; then click **Left**.
- Click the right side of the cell below *Measure*, and click the arrow; then click **Nominal**.

You have now entered all the attributes for the variable "Class."

- Click the second cell below *Name*, and type *exam1_pts*.
- Click the right side of the second cell below *Type*, and a window titled *Variable Type* will open. Click **Numeric**, and then click **OK**.
- Click the second cell below *Width*, and use the up–down arrows to select **8**.
- Click the second cell below *Decimals,* and use the up–down arrows to select **0**.
- Click the second cell below *Label*, and type *Points on Exam 1*.
- Skip the second cell below *Values*, and click the right side of the second cell below *Missing*. A window will open titled *Missing Values*. Click **No missing values**, then click **OK**.
- Click the second cell below *Columns*. Use the up–down arrows to select **8**.
- Click the right side of the second cell below *Align*, and click the arrow; then click **Left**.

- Click the right side of second cell below *Measure*, and click the arrow; then click **Scale**.

You have now entered all the attributes for the variable "exam1_pts."

- Click the third cell below *Name*, and type *exam2_pts*. For this variable, enter exactly the same information you entered for *exam1_pts*, except in the second cell below *Label*, where you will type *Points on Exam 2.*

You have now entered all the attributes for the variable "exam2_pts."

- Click the fourth cell below *Name*, and type *predict_grde*. All entries are the same as the previous variable, except *Label*, *Value*, and *Measure*. Click the fourth cell below *Label*, and type *Student's Predicted Final Grade*.
- Click the right side of the fourth cell below *Values*, and a window will open titled *Value Labels*. Type *1* in the *Value* box, and type *A* in the *Label* box. Click **Add**. Then go back to the *Value* box, and type *2*, and in the *Label* box type *B*. Click **Add**. Repeat for 3 = C, 4 = D, and 5 = F. Your window should now look like Figure 5.2.)
- Click **OK**.
- Click the right side of the fourth cell below *Measure*, and click the arrow; then click **Nominal**.

Figure 5.2   *Value Labels* Window for *Student's Predicted Final Grade*

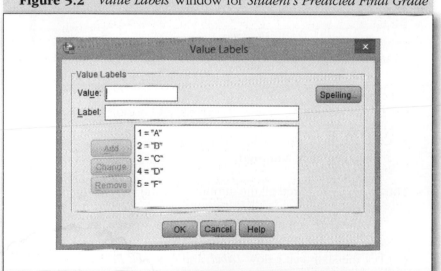

You have now entered the attributes for the variable "predict_grde."

- Click the fifth cell below *Name*, and type *Gender*. All entries are the same except *Label*, *Values*, and *Measure*.
- Click the fifth cell below *Label*, and type *Gender*.
- Click the right side of the fifth cell below *Values*, and a window titled *Value Labels* will open. Type *1* in the *Value* box, and type *Male* in the *Label* box. Click **Add.** Then go back to the *Value* box, and type *2*, and type *Female* in the *Label* box. Click **Add**. Your window should look like Figure 5.3.
- Click **OK**.

Figure 5.3   *Value Labels* Window for *Gender*

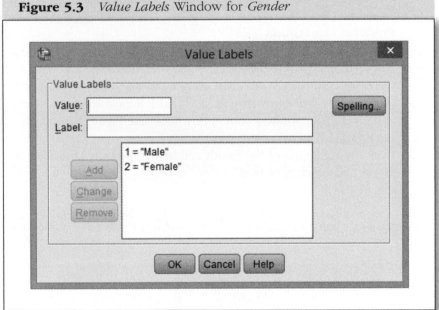

- Click the right side of the fifth cell below *Measure*, and click the arrow; then click **Nominal**.

You have now entered all the attributes for the variable "gender."

- Click the sixth cell below *Name*, and type *Anxiety*. All entries are the same as in the prior variable, except *Label*, *Values*, and *Measure*.
- Click the sixth cell below *Label*, and type *Self-Rated Anxiety Level*.

- Click the right side of the sixth cell below *Values*, and a window titled *Value Labels* will open. Type *1* in the *Value* box, and type *Much anxiety* in the *Label* box. Click **Add**. Then go back to the *Value* box, and type *2*, and in the *Label* box type *Some anxiety*. Click **Add**. Repeat for 3 = *Little anxiety* and 4 = *No anxiety*. Your window should now look like Figure 5.4.

Figure 5.4   *Value Labels* Window for *Self-Rated Anxiety Level*

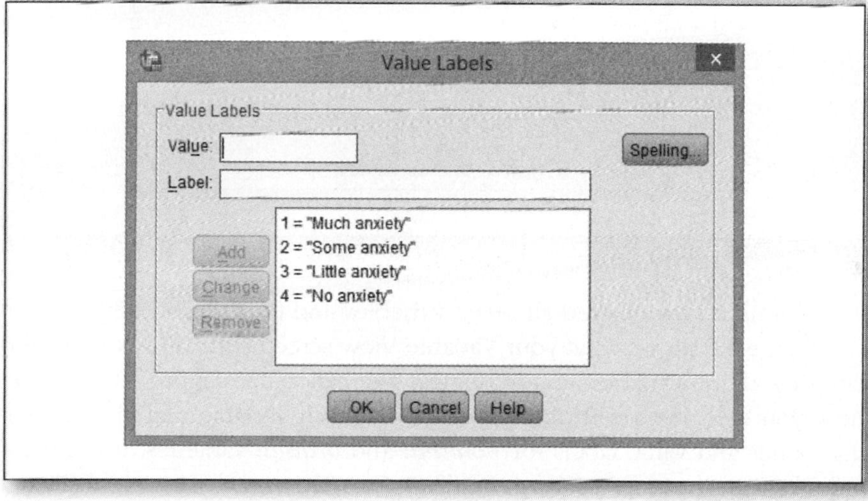

- Click **OK**.
- Click the right side of the sixth cell below *Measure*, and click the arrow; then click **Ordinal**.

You have now entered the attributes for the variable "anxiety."

- Click the seventh cell below *Name*, and type *rate_inst*. All entries are the same as in the prior variable except *Label*, *Values*, and *Measure*.
- Click the seventh cell below *Label*, and type *Instructor Rating*.
- Click the right side of the seventh **cell** below *Values*, and a window titled *Value Label* will open. Type *1* in the *Value* box, and type *Excellent* in the *Label* box. Click **Add**. Then go back and type *2* in the *Value* box, and type *Very good* in the *Label* box. Click **Add**. Repeat for 3 = *Average*, 4 = *Below average*, and 5 = *Poor*. Your window should now look like Figure 5.5.
- Click **OK**.
- Click the right side of the seventh cell below *Measure*, and click the arrow; then click **Ordinal**.

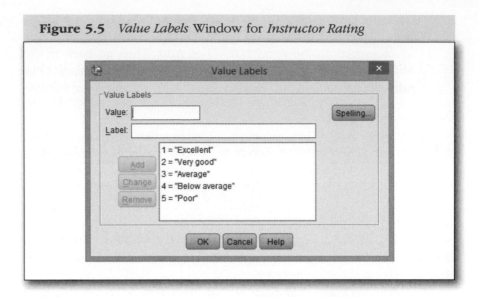

Figure 5.5 *Value Labels* Window for *Instructor Rating*

You have now entered all seven variables and their attributes.

Figure 5.6 shows what your Variable View screen should look like after entering all the variables and attributes. The callouts in Figure 5.6 indicate how you can check with the Appendix for each variable's label and also the values and value labels for *nominal* and *ordinal* variables. If the variables and attributes in your screen do not match, now is the time to make corrections.

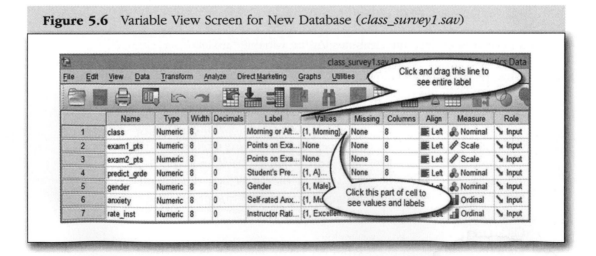

Figure 5.6 Variable View Screen for New Database (*class_survey1.sav*)

5.3 ENTERING DATA FOR EACH VARIABLE △

At this point, you will enter data from Table A.2 in Appendix A for each variable in what may seem a somewhat tedious process. Although tedious, it is extremely important that you accurately enter the data and verify your entries by visual inspection.

Click **Data View** at the bottom of the screen. The variables you have entered are displayed across the top of the screen. You will enter data, starting in the first cell (intersection of Row 1, Column 1) below *Class*, from left to right across columns. Use the Tab key or right arrow key to move from one cell to another. We strongly suggest that you place a ruler under the first row of data in the book to block out the other data. When you have entered the data for the last cell in a row, use the mouse button to click the second cell under the first variable, "Class," and enter data for that row. Repeat for all data.

An alternative approach is to enter all the data for the first variable moving down a column, then all data for the next variable, and so on. We prefer the first method because it involves less eye and head movement.

Before quitting SPSS, be certain to save your data using the file name *class_survey1*, because you will be using the variables and data in subsequent chapters.

5.4 VALIDATING DATA FOR DATABASES △

For smaller data sets with fewer than 50 cases and few variables, it is usually sufficient to validate the entries by visual inspection. But when there are hundreds or thousands of cases, and/or many variables, validation by inspection to determine if each case has been correctly entered can be an overwhelming, and often impossible, task. Fortunately, there are other methods such as using the *Validate Data* module. But this module is not included in the base SPSS installation and must be purchased and downloaded. On the Main Menu, click **Add-ons**, select **SPSS Data Validation**, and click **Validate Data**. Your computer browser will take you to a website that explains the process of validating data. There is a module titled *IBM SPSS Data Preparation* available for download that includes the *Validate Data* procedure, among others. The *Validate Data* procedure enables you to apply rules to perform data checks based on each variable's *level of measurement*. You can check for invalid cases and receive summaries regarding rule violations and the number of cases affected. In general, the validation process involves applying a set of rules to a data set and asking SPSS to determine if those rules have been violated. The *Validate Data* module is very useful, but unfortunately, it is not free.

We next describe some additional methods for data validation that do not require the use of the *Validate Data* module. In the following paragraphs, we will have you add fictitious and incorrect data to the database you just created and saved as *class_survey1*. We do this to demonstrate various basic validation procedures. The most direct method to check for data entry errors is to produce *frequency tables* for your variables that were measured at the *nominal* or *ordinal* levels. For those variables measured at the *scale* level, a *descriptive statistics table* is useful. This approach to validation also permits a quick inspection of the *variable labels* and the *values* assigned to each of the categories of your variables.

The hands-on experience will involve the use of the data from the class survey (named and saved as *class_survey1*). If the data file is not already open, open it now. Click **Data View**, and enter the erroneous data in Rows 38 and 40 (no data in Row 39) as shown in Figure 5.7. The data entered into these rows, and the empty row, will be used to illustrate the validation procedures for this new database consisting of 40 students. The data you just entered in Rows 38, 39 (missing data), and 40 represent some types of common errors made when entering data into a spreadsheet.

Figure 5.7 Class Survey Data View Screen: Erroneous Data Entered in Rows 38, 39, and 40

	class	exam1_pts	exam2_pts	predict_grde	gender	anxiety	rate_inst
37	2	41	72	4	2	1	1
38	3	105	102	5	3	5	5
39	.	.	.	.	.	.	.
40	3	101	107	1	3	5	0

Once the new data are entered, we begin the validation process by looking at the variables measured at the *nominal* and *ordinal* levels. *Caution:* Do not save this new database unless you assign it a new name—it's really not necessary to save it! We suggest that you just delete it.

Validation of Nominal and Ordinal Data

- Click **Analyze**, select **Descriptive Statistics**, and click **Frequencies**.
- Click **Morning or Afternoon Class**; then hold down the Ctrl key, and click **Student's Predicted Grade**, **Gender**, **Self-Rated Anxiety Level**, and **Instructor Rating** (these are the variables measured at the *nominal* and *ordinal* levels).
- Click the arrow to move the selected variables to the *Variable(s)* box.
- Click **OK**.

Once you have clicked **OK**, SPSS will produce six tables for your five variables. The first table is as shown in Figure 5.8. This table presents a summary of the five variables that you selected for this data check operation. There are five columns, each containing one variable and the number of values recorded for each.

Figure 5.8 Output for Class Survey Showing One Missing Value for Each Variable

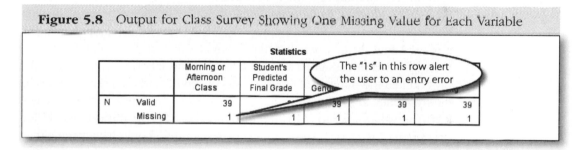

In the first row, you will see the labels of the variables you selected for the analysis. The next row is titled *N Valid*, which only indicates that 39 numerical values were entered. It should not be inferred that these values were somehow correct. The next row, titled *Missing*, informs you that there is one missing numerical value for each of the five variables. An inspection of the Data View screen reveals that the values for an entire case were omitted (missing data in Row 39).

The next step in our data validation procedure is to determine if there are additional data entry problems by looking at the frequency tables for each of our variables. The first frequency table in the SPSS output, titled *Morning or Afternoon Class*, is given in Figure 5.9.

First, look at the title of the table (*Morning or Afternoon Class*), which is the *label* that was entered in the Variable View screen. Make sure the spelling is correct and the title accurately describes the variable. Look at the rows; in the first column, you should see *Morning* and in the *Frequency*

Figure 5.9   Frequency Table: Output for Class Survey

Morning or Afternoon Class

		Frequency	Percent	Valid Percent	Cumulative Percent
Valid	Morning	18			
	Afternoon	19	47.5	48.7	94.9
	3	2	5.0	5.1	100.0
	Total	39	97.5	100.0	
Missing	System	1	2.5		
Total		40	100.0		

The "3" indicates a data entry error

column, *18*. The titles for the rows, such as *Morning*, are actually the value labels you attached to the categories of your variables. The error in data entry can be easily spotted because of the *3* that is printed, as a title, for a row of data. It indicates that the incorrect value of 3 was entered twice. Since you had not assigned any value label to this incorrect entry, SPSS simply counted the number of times "3" was observed and then reported this in the frequency table. For this variable, there were only two possible categories: The value label for *Morning Class* was *1*, whereas the value label for *Afternoon Class* was *2*. Thus, all the entered values should have been either 1 or 2. For small databases, it is then a simple matter to visually inspect the data file, locate the error, and correct the mistake. For larger databases, one should click the toolbar icon (called *Find*) shown in Figure 5.10. Note that to directly use the *Find* icon on the *Menu* you must be in the Variable View screen. An alternate method used to activate the *Find* function is to first click **Edit** on the Main Menu and then select and click the same icon found in the drop-down menu.

Figure 5.10   The *Find* Icon

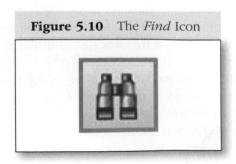

Clicking this icon opens the *Find and Replace - Variable View* window, as shown in Figure 5.11. Type *3*, the incorrect entry identified in the frequency table, in the *Find* box, and click **Find Next**. This procedure will then return you to the Data View screen, and the incorrect value will be highlighted. The frequency table identified multiple incorrect entries of 3, so you simply click **Find Next**, and SPSS will progress through the database stopping at each incorrect value. You could then inspect each case and make the necessary corrections.

Figure 5.11   *Find and Replace - Variable View* Window

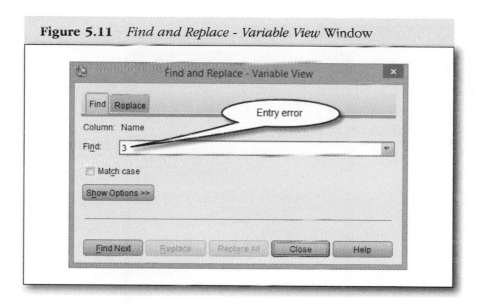

At this point, you should inspect the remaining four frequency tables in the Output Viewer and identify additional data entry errors for the remaining four variables.

Validation of Scale Data

To look for data entry errors for variables measured at the *scale* level, we will use the data entered for "exam1_pts" and "exam2_pts." To do this validation procedure, we check the data to see if there are any values that are less than or greater than some expected value. For *Exam 1* and *Exam 2*, we know that it was impossible to score less than 0 or more than 100 points; therefore, any values detected outside this range would be incorrect. The following procedure will check the entered data for any values outside the expected range.

- Click **Analyze**, select **Descriptive Statistics**, and click **Descriptives**.
- Click **Points on Exam One** and **Points on Exam Two** (variables measured at the scale level) while holding down the computer's Ctrl key.
- Click the arrow (moves variables to the *Variable(s)* box.
- Click the **Options** box; the *Descriptives: Options* window opens (see Figure 5.12).

Figure 5.12 *Descriptives: Options* Window

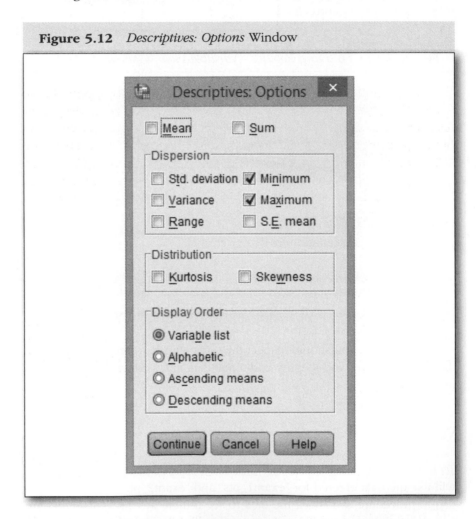

- Unclick all boxes that may be checked, then click **Minimum** and **Maximum** (the window should now look like Figure 5.12).
- Click **Continue**, and then click **OK**.

Figure 5.13 Output Viewer: Descriptive Statistics

Descriptive Statistics

	N	Minimum	Maximum
Points on Exam One	39	14	105
Points on Exam Two	39	23	107
Valid N (listwise)	39		

Once you click **OK**, a table titled *Descriptive Statistics* appears in the Output Viewer, as shown in Figure 5.13. Inspection of this table easily identifies any value that may be less than or greater than any expected value. In this example, we know that the total possible points on these exams are 100; thus, we can easily spot the data entry errors of 105 and 107. In the *N* column, you also see that there is a missing value for both variables because *N* = 39, and it should be 40. This informs us that somehow an entire case (data for one student) has been omitted.

5.5 SUMMARY △

In this chapter, you learned how to enter variables and their attributes. You also learned how to enter data for each variable. You entered an important database that you named *class_survey1.sav*, which will be used throughout many of the remaining chapters. Basic validation procedures for both *nominal* or *ordinal* (discrete) and *scale* (continuous) data were also presented. In the next chapter, you will learn how to work with variables and data that have already been entered into SPSS.

5.6 REVIEW EXERCISES △

5.1 The highway patrol officer wants to set up an SPSS file to record traffic violations. She wishes to record data at the *nominal*, *ordinal*, and *scale* levels of measurement. The first item of interest (the largest source of income for the highway patrol) is speeding. Input three variables that could record speed at each level of measurement. The next item of interest is vehicle violations—in the same database set up a variable at the correct level of measurement and with three categories if necessary. Impaired driving is another important violation. How would you measure and record information for this violation? Show all this in the same database.

5.2 A child psychologist is investigating the behavior of children in the play area of Balboa Park. Help him set up an SPSS file to measure the following variables on individual children: time the child was observed, pieces of play equipment used, other children in the play area, interaction with others, interaction with parent, and child's general demeanor. Input these variables into SPSS at measurement levels of your choosing but appropriate for the variable being analyzed.

5.3 The following sample data were collected by the owner of a private 640-acre forest reserve. He did a sample of 10 acres as a trial survey for the entire reserve. He needs to set up and test a computer file system using SPSS's Data Editor. The 10-acre sample was subdivided into 2.5 parcels, with each yielding the following data: hardwood trees, softwood trees, new-tree growth, stage of decay for fallen trees, soil moisture content, and crowded conditions. Your database will have four cases (4 × 2.5 = 10) and seven variables. Enter some fictitious data for the newly created variables on the four 2.5-acre plots.

CHAPTER 6

WORKING WITH DATA AND VARIABLES

6.1 INTRODUCTION AND OBJECTIVES △

Sometimes, it is desirable to compute new variables and/or recode variables from an existing database. When computing a new variable, you often add a number of variables together and calculate an average, which then becomes the new variable. Recoding may involve the changing of a variable measured at the *scale* level into a variable measured at the *nominal* or *ordinal* levels. In this chapter, you will recode a variable measured at the scale level into a nominal variable. You may wish to do such operations in order to make the data more understandable or perhaps for more specific *descriptive* or *inferential* analysis. In this chapter, you will use various SPSS commands to compute and recode data, which results in the creation of a new variable.

The SPSS user may sometimes find it necessary to insert a case or variable into an existing database. These procedures are also explained, which will provide many hands-on opportunities to use both the Data View and the Variable View screen.

OBJECTIVES

After completing this chapter, you will be able to

Compute a new variable by averaging two existing variables

Recode a variable measured at the scale level into a new nominal variable

Transform data using SPSS

Insert a missing case (row of data) into a data set

Insert a variable (column) into an existing data set

Cut and paste existing data within the Data View page

△ 6.2 COMPUTING A NEW VARIABLE

For this exercise, you will use the *class_survey1* database created in Chapter 5. You will combine the first and second exam scores and create a new variable. The new variable will be the average of the scores earned on Exams 1 and 2. *Important note:* Once you have created the new variable, you will give the database a new name and save it for use in future chapters.

- Start SPSS, and then click **Cancel** in the *SPSS Statistics* opening window.
- Click **File**, select **Open**, and then click **Data**.
- In the file list, locate and click **class_survey1.sav**.
- Click **Open**.
- Click **Transform** on the Main Menu, and then click **Compute Variable** (the *Compute Variable* window opens; see Figure 6.1).
- Type *avg_pts* in the *Target Variable* box (this is the name of your new variable as it will appear in the Data View screen).
- Click the *Type & Label* box, and the *Compute Variable: Type and Label* window opens and floats in front of the *Compute Variable* window (see Figure 6.2).
- In the *Label* box, type *Average Points on Exams* (this is your new variable's *label*; see Figure 6.2).
- Click **Continue**. (The window shown in Figure 6.2 disappears, and Figure 6.1 remains.)
- Click the parentheses key on the keypad (which will move the parentheses to the *Numeric Expression* box).

Figure 6.1 *Compute Variable* Window: Database Is *class_survey1.sav*

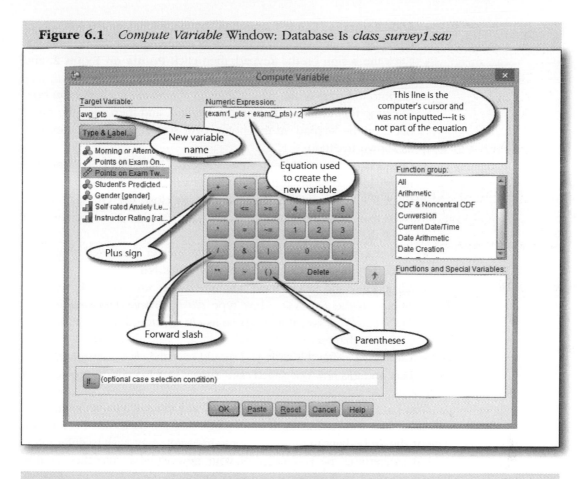

Figure 6.2 *Compute Variable: Type and Label* Window

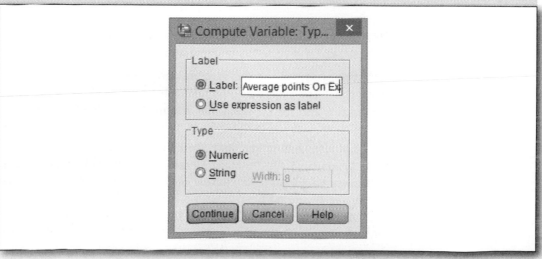

- Click **Points on Exam 1** (found in the variable list on the left side of the window), and then click the arrow to the right of the variable list.
- Click the **+** sign on the keypad, then click **Points on Exam 2**, and then click the arrow.
- Click to the right of the parentheses in the *Numeric Expression* box.
- Click the forward slash key on the keypad, and then click the number 2 on the keypad. At this point, your screen should look exactly as shown in Figure 6.1.
- Click **OK**, and the Output Viewer opens. (Note that only some file information appears in the Output Viewer.)
- Click the **white "x" in the red box** to close the Output Viewer. Note that both the Data View and the Variable View screen will show the newly added variable.
- A window opens when the Output Viewer is closed that asks whether you want to save the output. Click **No**.
- Click **File**, and select **Save As**. When the *Save Data As* window opens, in the *File Name* box type *class_survey2*. (*Important note:* Make sure you save this database as *class_survey2* as it will be used in future chapters.)
- Click **Save**. (After this click, you should have the *class_survey1.sav* database and this new one called *class_survey2.sav* in your documents file.)

You have now added a new variable ("avg_pts.sav") to your database and saved that database using another name that can be recalled at a future date. It should be kept in mind that sometimes you may wish to enter information concerning the properties of your new variable into the Variable View screen. No such action is required for this particular new variable.

△ 6.3 Recoding Scale Data Into a String Variable

For this exercise, you will use the variable ("avg_pts") that was created in the prior exercise. The objective is to take the average points (scale data) for the two exams and recode them into a letter grade (nominal data).

- If it's not already running, start SPSS, and open *class_survey2.sav*.
- Click **Transform** on the Main Menu, and then click **Recode into Different Variables** (the *Recode into Different Variables* window opens; see Figure 6.3).
- Click **Average Points On Exams**, and then click the arrow.
- Click the *Name* box, and type *Grade*.

Figure 6.3 *Recode into Different Variables* Window

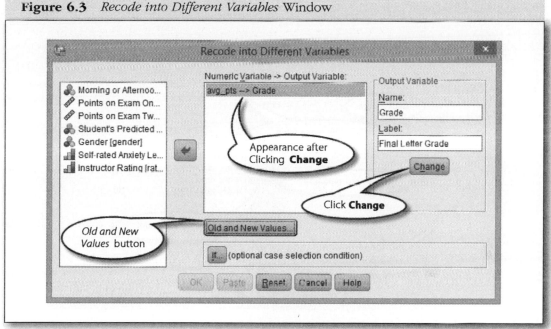

- Click the *Label* box, and type *Final Letter Grade*.
- Click **Change** (which moves the soon to be created new *output variable* to the center panel, which now looks like *avg_pts* --> *Grade*, and the *Recode into Different Variables* window now looks like Figure 6.3).
- Click **Old and New Values** (the *Recode into Different Variables: Old and New Values* window opens; see Figure 6.4).
- Click **Range**, and then type *0* in the upper box and *49.9* in the lower box.
- Check the *Output variables are strings* box (located on the lower right side of this window).
- Click the *Value* box, and type uppercase *F* in the box; then click **Add**.
- Click the *Range* box, and then type *50* in the upper box and *69.9* in the lower box.
- Click the *Value* box, and type uppercase *D* in the box; then click **Add**.
- Click the *Range* box, and then type *70* in the upper box and *79.9* in the lower box.
- Click the *Value* box, and type uppercase *C* in the box; then click **Add**.
- Click the *Range* box, and then type *80* in the upper box and *89.9* in the lower box.

Figure 6.4 *Recode into Different Variables: Old and New Values* Window

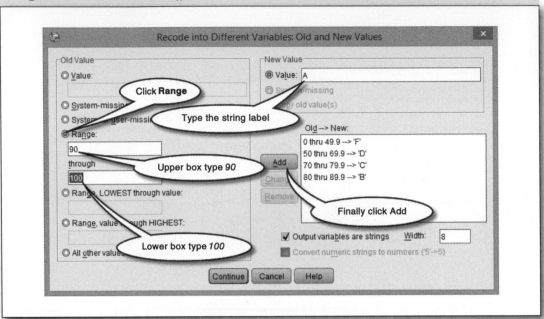

- Click the *Value* box, and type uppercase *B* in the box; then click **Add**.
- Click the *Range* box, and then type *90* in the upper box and *100* in the lower box.
- Click the *Value* box, and type uppercase *A* in the box (at this point, your window should look *exactly* like Figure 6.4 but without the callout balloons). Click **Add**.
- Click **Continue**, and then click **OK**.
- Output Viewer opens. (Note that only some file information appears in the Output Viewer.)
- Click the **white "x" in the red box** to close the Output Viewer.
- A window opens when the Output Viewer is closed that asks whether you want to save the output. Click **No**. (Once **No** is clicked the Variable View screen opens showing the new variable that you created.)
- Click **File**, and then click **Save As** (the *Save Data As* window opens).
- Click *class_survey2* in the data file, which moves *class_survey2* to the *File Name* box, and then click **Save**.

Once you click **Save**, a new database will overwrite the one you named in the prior exercise. This new database (*class_survey2*) now includes a scale variable called "avg_pts" and a *nominal* variable called "Grade," which was created using the recode function available in SPSS. You will use this database in future chapters, so be sure to save it as instructed.

6.4 DATA TRANSFORMATION △

Data transformation of values for a variable is obtained by applying mathematical modifications. If the variables are not normally distributed, you may apply nonparametric tests or you may change the scale of measurement using a transformation. There are a host of transformations available, and many of these simply reduce the relative distances between data points to improve normality. Three important transformations are the *square root transformation*, the *logarithmic transformation*, and the *inverse (reciprocal) transformation*. For example, if you apply a square root transformation to your data, the square root of each value is computed.

If you ever require a transformation of your data, SPSS can handle it easily. You will use the database with which you are familiar to perform a square root transformation of data values for the variable "exam1_pts." Actually, this variable does not require a transformation, but we wish to show you how to do it in case a situation arises with your actual data that may require a transformation.

- Open *class_survey2*.
- Click **Transform**, and then click **Compute variable**. A window titled *Compute Variable* will open.
- In the *Target Variable* box, type *sqrtexam1*.
- In the panel titled *Function Group* on the right, click **Arithmetic**.
- In the lower panel titled *Functions and Special Variables*, scroll down and click **Sqrt**.
- Click the upward-pointing arrow to place *SQRT* in the *Numeric Expression* box.
- In the left panel, click **Points on Exam One**, and then click the right arrow to place it in the parentheses following *SQRT* in the *Numeric Expression* box. Your screen should look like that in Figure 6.5.
- Click **OK**.

A new variable titled "sqrtexam1" will appear in the Data View screen with the square root values listed. At this point, you could use SPSS to perform any analysis on these data that may be required. You need not save this change to your database.

Figure 6.5   Compute Variable Screen With "sqrtexam1" as a New Variable

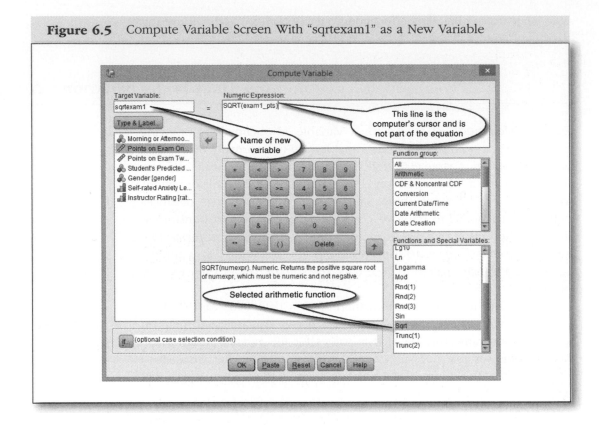

△   6.5 Split Cases for Independent Analysis

There are occasions when the data analyst finds it useful to be able to select specific cases for analysis, for instance, if you have 100 cases and you are only interested in including the first 50 cases. There are a number of ways to split cases into groups, such as using the copy-and-paste method to create a new database. Fortunately, there is a much better way to accomplish such case splitting. In this section, you will learn how to select blocks of cases for analysis.

A particularly useful application of case splitting will become apparent later in this book. As you will learn, there are different formats that must be followed when entering data for different types of analysis. For example, when doing many statistical tests, you must enter all the data in one column even though they may originate from three different groups. Often, you will have to examine the data as independent groups prior to performing the main analysis. We will show you how this is accomplished with the following bullet points:

- Start SPSS, click **File**, select **New**, and then click **Data**.
- Click the Variable View tab, and type *demo* for the name of your variable; set decimals to zero, and specify *scale* data.
- Click the Data View tab, and type *25, 12, 18, 34, 21, 36, 27, 23, 22, 34, 29,* and *30* in Rows 1 through 12. Consider that you now must analyze the first six cases only.
- Click **Data** on the Main Menu; then click **Select Cases** (next to the last icon), and the *Select Cases* window opens.
- Click the **Range** button, and in the box labeled *First Case*, type *1*; in the box titled *Last Case*, type *2*; then click **Continue**.
- Click **OK**, and the Output Viewer opens, showing the command and that it was executed. Return to the database, and view the result.

You will see diagonal lines through Cases 6 through 10, which are the ones you excluded from any analysis you may choose to do of the database. You are now free to analyze the first five cases in any manner necessary.

6.6 INSERTING NEW VARIABLES AND CASES INTO EXISTING DATABASES △

You may sometimes find it necessary to enter a variable (a new column) or case (a new row) into an existing database. Perhaps you just forgot a case or variable, or maybe you wish to update a database previously entered. SPSS provides two easy ways to accomplish this task. The first method is to click **Edit** on the Main Menu and then click **Insert Variable** or **Insert Cases** depending on your need. The second, and more efficient, method is to click on one of the icons found on the toolbar, which accomplishes the same task. Click the icon shown in Figure 6.6 when you need to insert a variable. Click the icon shown in Figure 6.7 if you must insert a case.

To obtain hands-on experience, open the database created in the prior section (*class_survey2.sav*), if it is not already running. Let's say you

Figure 6.6 The *Insert Variable* Icon

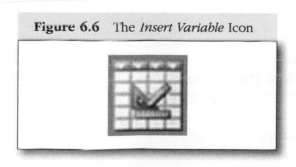

Figure 6.7 The *Insert Cases* Icon

forgot to enter a variable that should have followed the variable "anxiety." If you click in any column, the new variable is inserted to the left of that column.

- Start SPSS, and open *class_survey2.sav*.
- Click **Data View** to open the screen.
- Click anywhere in the column to the right of the "anxiety" variable.
- Click the *Insert Variable* icon (shown in Figure 6.6). A new column appears to the right of the "anxiety" variable column.
- You may now go to the Variable View screen and specify the variable properties.
- Next you may go to the Data View screen and enter data for your new variable.

To insert a case (a row of data), a procedure similar to that for entering a variable is followed. Using the same database as in the prior section, let's assume that you forgot to enter Case 10.

- In Data View, click **row 10**, then click the *Insert Cases* icon (Figure 6.7).
- A new row is created in Row 10 that is ready to receive data.

Important note: DO NOT save the changes made for the *Insert Variable* and *Insert Case* procedures as they will not be used in future chapters and will corrupt your database.

△ 6.7 Data View Page: Copy, Cut, and Paste Procedures

Sometimes it is convenient to use the copy-and-paste function when entering or manipulating data on the Data View page. This is especially useful when entering repetitive codes for your variables. The procedure used by SPSS is identical to that used in all common word-processing programs. We next demonstrate the copy-and-paste methods.

- Start SPSS.
- Click **File**, select **Open**, and then click **Data**.
- Click **class_survey1.sav**, and then click **Open** (click the Data View tab if not open).
- Click the cell in Row 1, Column 1, and drag the mouse cursor up to and including Case 18. (You now have a column of highlighted "1s"; recall that this is the code used for males.)
- Right click the mouse, and then click **Copy**.
- Move the mouse to Case 38, and right click the cell in Row 38, Column 1.
- Click **Paste**. (You will now see new *cases* added to the database; we will next use the cut function and return the database to its original version.)
- Click the cell in Row 38, Column 1, and drag the mouse up to and including Case 55.
- Right click the mouse, and then click **Cut** (do not save this altered database).

Important note: Once you have finished the above cut-and-paste exercises, make sure you DO NOT save the database. You will use the original *class_survey1* and *class_survey2* databases many times in future chapters, and any inadvertent changes you may make will have a negative impact on your future work.

6.8 SUMMARY △

In this chapter, you learned the basic techniques required to compute new variables from the values in an existing database. You first combined two variables that were measured at the *scale* level, found the average, and created a new variable. You then recoded this new variable into a letter grade to create a variable in *nominal* form. You also inserted new variables and new cases into an existing database.

In the next chapter, you will be given hands-on experience in using the various help features. Some of these help features are part of the SPSS program that was installed when you downloaded it onto your computer's hard drive. Other help features are based on links to online sites and as such must be used when your computer is online. All this will be explained in the following chapter.

△   6.9 Review Exercises

6.1 An urban planner was tasked with recording the walking speeds (in miles per hour) of people at a downtown government center. He recorded walking speeds of the same individuals over a period of 5 days. A sample of the data is provided below along with the variable information. There are three variables. You are to set up a database in SPSS and then use the *Compute Variable* feature to create a fourth variable called "avgspeed." There are five cases: Variable 1: *speed1* has the label of *1-walking speed* with speeds of 3.20, 3.18, 1.40, 3.26, and 2.57. Variable 2: *speed2* has the label of *2-walking speed* with speeds of 3.34, 3.61, 2.10, 3.12, and 2.82. Variable 3: *speed3* has the label of *3-walking speed* with speeds of 3.25, 3.24, 1.97, 3.41, and 2.98. Save the database, which has the new variable ("avgspeed") as speed—you may use it in future review exercises.

6.2 Use the data from the variable ("avgspeed") created in the previous exercise to recode the values into a *nominal* (string) variable. Your task is to use SPSS's *Recode into Different Variables* feature and form two categories for the average walking speeds of the individuals. The two categories are based on the average speed of 2.9 miles per hour for all walkers. All speeds above the mean are to be classified as *Fast*; those below the mean are classified as *Slow*. You will create a new nominal or string variable called "catspeed."

6.3 You have been given the following scale data (test scores): 100, 109, 114, 118, 125, 135, 135, 138, 139, and 140. You must set up a database in SPSS and then transform the data using the *Compute Variable* and *arithmetic* functions to calculate new variables giving the log and square root of the original test scores. You must end up with a database consisting of 10 cases and three variables named "test," "logtest," and "sqrttest."

CHAPTER 7

USING THE
SPSS HELP MENU

7.1 INTRODUCTION AND OBJECTIVES △

SPSS offers a rich and extensive variety of features on the Help Menu. You can obtain immediate assistance in understanding virtually any topic covered in SPSS. But the Help Menu is not the only method to obtain assistance on particular topics. For example, when you request SPSS to compute a statistic or generate a graph, various windows will open. These windows contain choices, one of which is a *Help* option that you can click to obtain assistance concerning the statistics or graph you wish to generate.

There are 10 options available in the Help Menu: *Topics, Tutorial, Case Studies, Statistics Coach, Command Syntax Reference, SPSS Developer Central, Algorithms, SPSS Home, Check for Updates,* and *Product Registration.* We will describe the first three options, and the purpose of the *Help* search box. In addition, we will describe the various types of assistance available in windows that open when you request SPSS to perform a statistical analysis or generate a graph.

OBJECTIVES

After completing this chapter, you will be able to

Describe the purpose of the following on the Help Menu: *Topics, Tutorial, Case Studies*

Describe the purpose of the *Help* search box

Use the *Topics* option to locate and gather information on a topic of interest

Use the *Tutorial* to select and run a tutorial concerning a major SPSS topic

Use the *Case Studies* option to study and analyze a topic of interest

Use the *Search* box to locate and study a topic of interest

Use the *Help* selection available in Windows that opens when selecting **Analyze** on the Main Menu

∆   7.2 HELP OPTIONS

Following is a brief description of the purpose of the first three selections in the Help Menu:

Topics: This provides access to the *Contents* and *Search* tabs, which you can use to find specific *Help* topics. When you select *Topics* in the Help Menu, your computer browser opens a window showing the topics available for your inspection. You may then choose from this list any topic that may be of interest. If you type a word or phrase in the *Search* box, SPSS will present a number of outcomes from your search terms. You can then click on any of these that might meet your information needs.

Tutorial: This offers illustrated, step-by-step instructions on how to use many of the basic features in SPSS. You don't have to view the whole tutorial from start to finish. You can choose the topics you want to view, skip around and view topics in any order, and use the index or table of contents to find specific topics. When you select *Tutorial* in the Help Menu, a window opens listing all the tutorial topics. *Note:* The tutorials also provide descriptions and explanations for the various SPSS sample files that were installed on your computer, as discussed in Chapter 3.

Case Studies: This consists of hands-on examples of how to create various types of statistical analyses and how to interpret the results. The sample data files used in the examples are also provided so that you can work through the examples to see exactly how the results were produced. You can choose the specific procedure(s) about which you wish to learn from the table of contents or search for relevant topics in the index.

7.3 USING HELP TOPICS △

Let's use *Topics* to get assistance concerning the proper method of entering variable properties, such as those you entered in the exercise described in Chapter 5.

- Click **Help** on the Main Menu, and then click **Topics**. A window opens that contains a tab titled *Help—IBM SPSS Statistics*, as displayed in Figure 7.1. Throughout this chapter, we will refer to this as the *Help* window.

Figure 7.1 Portion of the *Help—IBM SPSS Statistics* Window/Tab (Data Preparation Search)

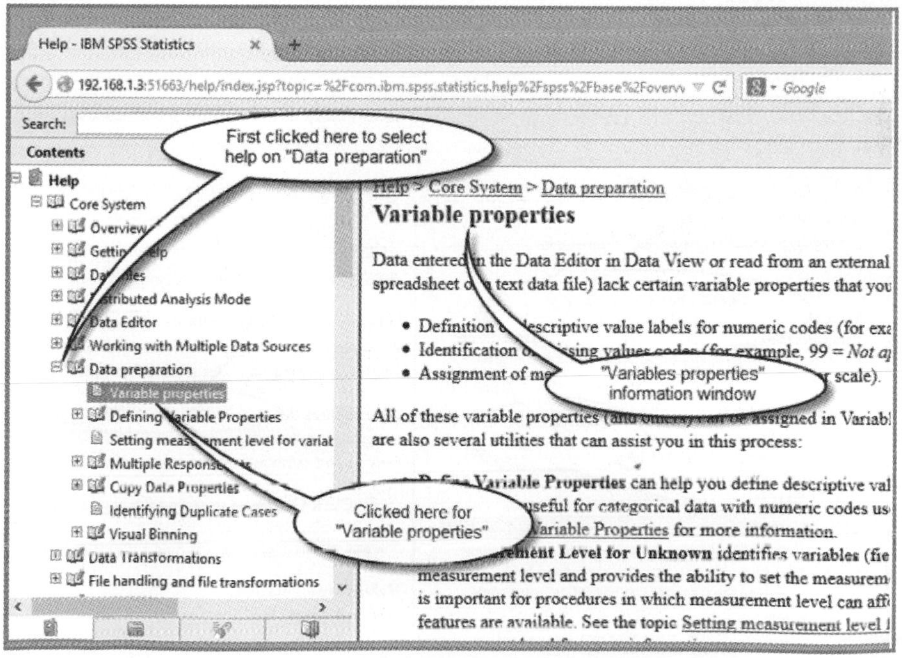

- Click the small box containing a + sign found to the left of *Data preparation*, and then click **Variable Properties**, found in the *Contents* section of the window. The portion of the window to the right will now display an explanation about several key factors concerning variables and SPSS. You will see information such as how to define value labels, identify missing values, and correctly assign levels of measurement. There are many additional topics covered that can be instantly obtained by simply clicking on that item in the *Contents* menu.

Let's assume that you wish to obtain some additional help on the assignment of levels of measurement to your data. You may recall that Chapter 4 was entirely devoted to this important topic. Let's see what SPSS has to offer that might be helpful in your future data manipulations.

- Click **Help** on the Main Menu, and then click **Topics**. The *Help* window will open.
- In the upper left-hand corner, locate the search box, and type in *levels of measurement*, then click **Go**. The result of clicking **Go** is shown in Figure 7.2.

Figure 7.2 Portion of the *Help—IBM SPSS Statistics* Window/Tab (Levels of Measurement)

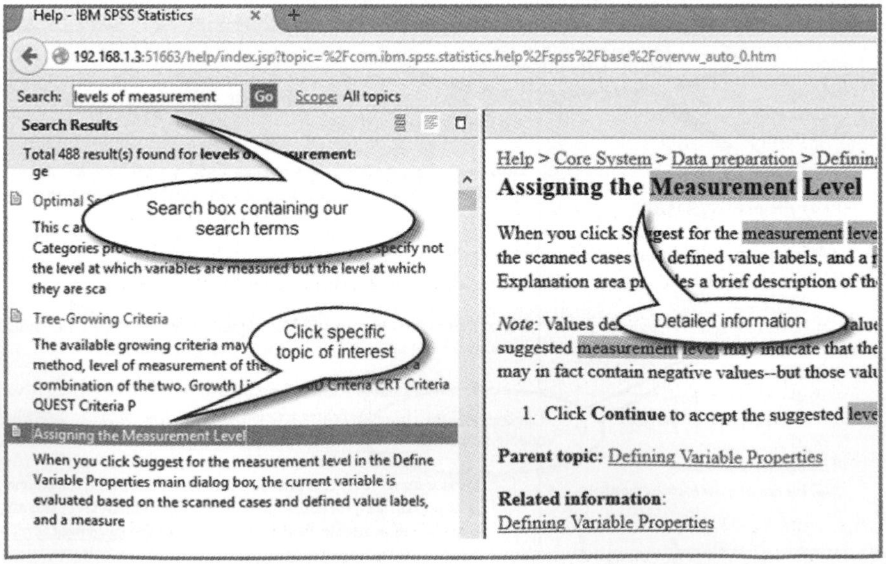

Looking at Figure 7.2 gives the impression that SPSS might just have some valuable information in its *Help* feature on this important topic. We recommend that you check it out for yourself.

Let's see what information we can find for the topic of the next two chapters, using SPSS to build charts, sometimes referred to as graphs. Actually, you will find that SPSS uses the terms *graphs* and *charts* to mean the same thing.

- Click **Help** on the Main Menu, and then click **Topics**. The *Help* window opens.
- In the *Search* box, type *chart building*, then click **Go**.
- Scroll to and click the second listing for *Building Charts*.

You will see a wealth of information that will be very helpful if you desire to make charts beyond our presentation in Chapters 8 and 9. This *Topics* section gives detailed information on how to use the SPSS *Chart Builder* feature to describe and explore data by building explanatory charts.

7.4 USING THE HELP TUTORIAL   △

Let's run a tutorial that will explain how to use *A Simple Cross Tabulation* in combination with *Counts versus Percentages*. Click **Help** on the Main Menu, and then click **Tutorial**. The *Help* window opens.

- In the *Contents* section of the *Help* window, **click** the box containing the + sign just to the left of *Tutorial*. A *Table of Contents* opens just below *Tutorial*.
- In this *Table of Contents*, click **Cross Tabulation**. Another *Table of Contents* opens; locate and click **A Simple Cross Tabulation**. The screen, called Crosstabs, now displays a concise explanation of the *Cross Tabulation* procedures. Many of these *Tutorials* also show the SPSS screenshots displaying the procedures required to accomplish specific tasks.
- At the bottom of the screen, click **Next**, which takes you to the second item in the *Table of Contents*; in this case, it is *Counts versus Percentages*.

Continue clicking **Next** at the bottom of each screen, and you will see all the items listed in the *Table of Contents* below *Cross Tabulation*. If you have questions regarding any topic in SPSS, it is likely you will find a tutorial to assist you in understanding. Some tutorials are rather lengthy and contain information that you may have to spend some time reviewing to achieve a working knowledge. Tutorials require your patience and fortitude, but most are worth the effort.

△ 7.5 Using Help Case Studies

Using the *Case Studies* feature provides you the hands-on opportunity to examine an actual database for a company, business, or other situation. The *Case Studies* feature generates statistics and graphs to help you learn how SPSS can best assist you when, for example, you have collected your own data for analysis. *Case Studies* use sample files that can be manipulated to answer questions. Essentially, one asks questions and then requests SPSS to provide answers, or at least output, that will help you answer these questions. The purpose and usefulness of using *Case Studies* are best illustrated with an example.

The following case study exercise uses frequencies to study *nominal*, *ordinal*, and *scale* data. Imagine that you manage a team that sells computer hardware to software development companies. In each company, your representatives have a primary contact. You have categorized these contacts by the department of the company in which they work (*Development, Computer Services, Finance, Other, Don't Know*). This information is collected in an SPSS sample file titled *contacts.sav*. Although you are not required to open this sample file for this demonstration, recall that the opening procedure is described in Chapter 3, Section 3.5. You will use frequencies to study the distribution of departments to see if it meshes with your goals. In addition, you will study the distribution of company ranks and the distribution of purchases. Along the way, you will construct graphs and charts to summarize results. The following bullet points show how to access this particular analysis utilizing the SPSS sample *contacts.sav*.

- Click **Help** on the Main Menu, and then click **Case Studies**. The *Help – IBM SPSS* window opens.
- Click the box containing the + sign to the left of *Case Studies*.
- Click **Statistics Base**, and then click **Summary Statistics Using Frequencies**.
- Click **Using Frequencies to Study Nominal Data**. This final click results in the display of extensive analysis of the data contained in the sample file. You may also click **Next** at the bottom of this screen to explore further analysis of the *contacts.sav* sample file.

7.6 GETTING HELP WHEN   △
USING *ANALYZE* ON THE MAIN MENU

A convenient and easily accessible method of obtaining help is available when you choose a topic from *Analyze* on the Main Menu. Make certain that there are data in the Data Editor (just about any database will work), and then do the following:

- Open any data file. The one used in this example is from the SPSS sample file titled *accidents.sav*.
- Click **Analyze** on the Main menu, and click **Missing Value Analysis**. A window titled *Missing Value Analysis* will open (see Figure 7.3).
- Click the *Help* box at the bottom of the window. Another window will open. You will see *Missing Value Analysis* highlighted in the left panel, and in the right panel, you will see information describing the topic of missing values and an example.

Figure 7.3 *Missing Value Analysis* Window Showing the **Help** Button

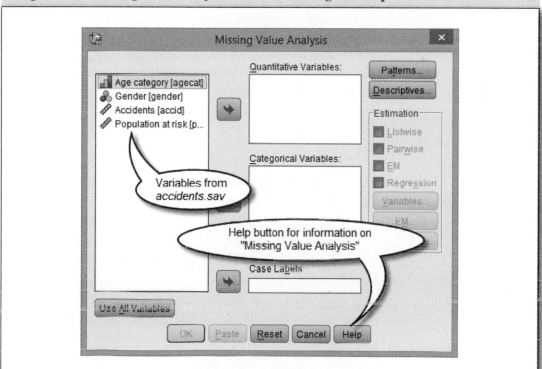

Here's another example:

- Open any data file, one from the SPSS sample file or one you created.
- Click **Analyze**, select **Descriptive Statistics**, and then click **Frequencies**. A window will open titled *Frequencies*.
- Click the *Help* box at the bottom of the window. Another window will open. You will see that *Frequencies* is highlighted in the left panel, and in the right panel, you will see a description of the purpose of the *Frequencies* procedure and an example.

Help is available for every option included in the *Analyze* menu.

Δ   7.7 Summary

In this chapter, you learned how to locate topics using the *Topics* selection on the Help Menu. You worked through a tutorial available when *Tutorial* is selected on the Help Menu, and you also investigated a case study by selecting *Case Studies*. In addition, you learned that help is available for the options listed when you click *Analyze* on the Main Menu. These methods of obtaining help will prove to be invaluable as you advance through and beyond this textbook.

The following two chapters present the SPSS steps used to create quality charts that clearly and succinctly describe and explain databases as well as the relationships between variables.

Δ   7.8 Review Exercises

7.1 How can you use the SPSS *Help* function to get information on transforming data in an attempt to get a more normal distribution of values?

7.2 You need help to find out how to summarize categorical (*nominal*) data for a PowerPoint presentation, and you want to use a graph made from data stored in SPSS. Hint: Use the SPSS Help on the Main Menu and the *Tutorial*.

7.3 You have a large database opened in SPSS, and now you must summarize and describe the data it contains. You decide to click **Analyze** on the Main Menu, then **Frequencies**; the *Frequencies* window opens, and you realize that you need help—what do you do next?

CHAPTER 8

CREATING GRAPHS FOR NOMINAL AND/OR ORDINAL DATA

8.1 INTRODUCTION AND OBJECTIVES △

IBM SPSS Statistics uses the terms *chart* and *graph* interchangeably, and we will do the same. SPSS offers different procedures to create graphs from a data set: *Chart Builder, Graphboard Template Chooser, Compare Subgroups, Regression Variable Plots,* and *Legacy Dialogs*. You can actually produce the same graphs when using the *Chart Builder, Graphboard Template Chooser,* or *Legacy Dialogs* method. The individual steps to produce the graph are, however, very different. The other two methods, *Compare Subgroups* and *Regression Variable Plots,* represent special cases and can be very useful for more specialized statistical analysis.

In this chapter, we prefer to demonstrate the *Chart Builder* because it is powerful and yet easy to master and use. The *Chart Builder* is the most recent graph-making feature in the SPSS program and enables the user to produce quite stunning graphs.

We will leave it as an exercise for you to investigate the use of *Graphboard Template Chooser* and/or *Legacy Dialogs* to produce graphs. Of the two methods, the *Legacy Dialogs* method is more straightforward, and by clicking on it you get a drop-down menu giving you 11 different graph choices.

To build the graphs presented in this and the following chapter, you will use the sample files that were included when SPSS was installed on your computer. If necessary, refer to Chapter 3 for directions in locating and opening these sample files. In the following introductory section, you will use the *class_survey1.sav* database to help familiarize yourself with basic dialog windows of the *Chart Builder*.

You have heard the adage that a picture is worth a thousand words. Well, a properly made graph depicting a systematic summary and analysis of data is also worth a thousand words while at the same time making sense of thousands of numbers.

OBJECTIVES

After completing this chapter, you will be able to

Describe the steps required to create a graph using the *Chart Builder*

Use *Chart Builder* to create a 3-D (three-dimensional) pie graph for nominal data

Use *Chart Builder* to create a population pyramid graph for nominal and ordinal data

Interpret large databases through the use of graphs

∆   8.2 A Brief Introduction to the Chart Builder

- This introduction to the *Chart Builder* in SPSS will apply to this and the following chapter. Start SPSS, Click **File**, select **Open**, and click **Data**.
- Locate and click the file titled *class_survey1.sav*. Click **Graphs** on the Main Menu, then click **Chart Builder**. Depending on the program setting, a small window may or may not open titled *Chart Builder*. Read its function as you may wish to disable this window; then click **OK**.
- Two windows open, one next to the other. The one on the left is titled *Chart Builder* (it is a different window from the one having the same name in the prior step), and the other is called *Element Properties*. You can activate one or the other by simply clicking it (try clicking back and forth). Leave the *Chart Builder* window active; this is the one shown in Figure 8.1 and is the main chart-building window.

- First make sure that **Gallery** is clicked (found in the middle of the window, as shown in Figure 8.1).
- Click **Bar** (found in the *Choose From* panel in the lower left portion of the *Chart Builder* window).
- Double click the first icon (**Simple Bar**) found on the right. (This moves a simple bar to the *Chart preview* window located above.) Click and drag the first variable (**Morning or Afternoon Class**) to the box titled *X-Axis?* (found in the *Chart preview* window).
- Click **OK**. Your graph of the *nominal* variable appears in the Output Viewer. It is not necessary that you save this output.

Figure 8.1 *Chart Builder* Window: Key Features and Control Points

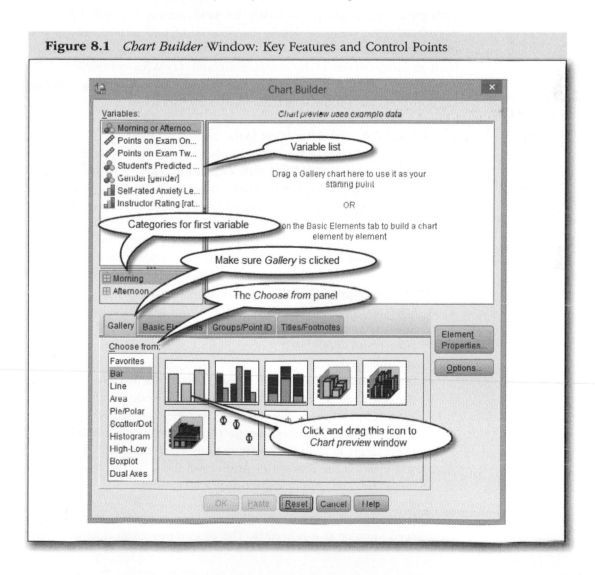

Of course, there are many more features that allow for the enhancement and embellishment of the basic graph you built here. You will be introduced to several of these features as you proceed through this and the following chapter. For now, just look at the graph, and answer the question as to which class (*morning or afternoon*) was more popular?

∆ 8.3 Using The Chart Builder to Build a Simple 3-D Pie Graph

Let's create a 3-D pie graph using the SPSS sample file called *poll_cs_sample.sav*. This database details the number of poll respondents in five different geographic regions of a county. We will answer two questions from an interpretation of the completed graph: (1) What percentage of the 9,449 survey respondents live in the eastern portion of the county? (2) Which region has the largest number of respondents?

- Start SPSS, click **File**, select **Data**, and click **Open**.
- Locate and click (if needed, go to Section 3.5 in Chapter 3 for help) the SPSS sample file titled **poll_cs_sample.sav**.
- Click **Graphs**, and select **Chart Builder**. A window titled *Chart Builder* will open. At the bottom of the window, make certain that **Gallery** is selected.
- Note that you can click the **Reset** button at any time if you make a mistake or if the settings from a previous graph-building operation remain. (This feature is found at the bottom of the *Chart Builder* window.)
- Click **Pie/Polar** (found in the *Choose from* panel).
- Click and drag the **Pie Graph** icon to the large white panel at the top-right portion of the *Chart Builder* window. You will see a simulated graph appear in this panel, with boxes representing the *x*-axis and the *y*-axis.
- Click and drag the variable **County** (from the *Variables* panel) to the box titled *Slice by?* (this is the *x-axis* for your pie chart).
- In the *Element Properties* window, click the black arrow to the right of *Count*, then click **Percentage (?)**, and finally click **Apply**. *Note:* Don't forget to click the **Apply** button, which activates any changes you make in the *Element Properties* window.
- Click **OK**. The Output Viewer displays Figure 8.2.

This basic graph is quite difficult to interpret, especially since it is not reproduced in color as it is in SPSS. The procedures given next will show

Figure 8.2 Basic Graph (No Embellishments) of Poll Respondents by County Regions

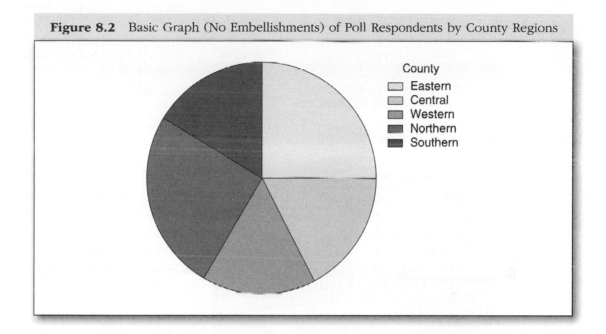

how to embellish this basic graph, which makes it much easier to interpret and will provide us with the information needed to answer our questions about this "County" variable.

Before making improvements to the basic graph by using the *Chart Editor*, we offer a couple of explanations concerning what you will accomplish and why. You will change the pie graph from a flat 1-D picture to a 3-D representation. This is done, in part, because we feel it simply improves its appearance. However, another benefit of the 3-D pie graph is that it increases the reader's ability to visualize proportions by adding the appearance of volume to the graph. You will also separate a slice from the others to emphasize a particular category that may be important to the study. This procedure is known as "exploding" a slice. You will also change the slice colors to unique patterns since most books and scientific articles are not published in color. Changing the slice patterns will also make it easier to answer the questions presented at the beginning of this section.

Follow the procedures given below to make several changes in the basic pie graph shown in Figure 8.2.

- Double click the graph as it appears in the Output Viewer (the *Chart Editor* opens, as shown in Figure 8.3).
- Click the small box to the left of *Eastern*, and a faint yellow border appears around the slice that represents the *Eastern* region (*Eastern*

Figure 8.3   *Chart Editor* Window Showing the Key Control Points

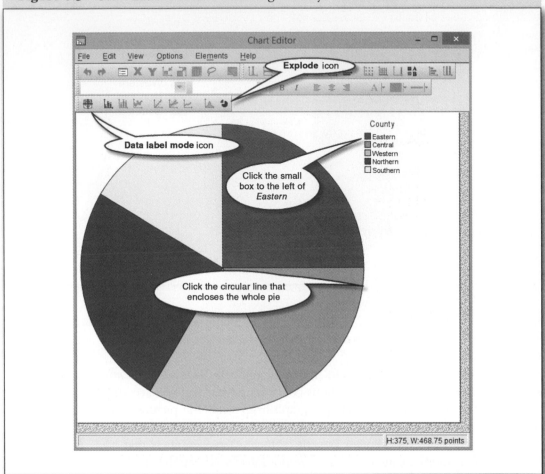

is found in the upper-right corner of the graph, and this area is referred to as the *Legend* area of your graph).

- Click the **Explode Slice** icon, which is the last icon in the *Chart Editor* (if the icon is not present, click on the *Elements* menu choice in the top of the *Chart Editor*, then click **Explode Slice**).
- Click the **Data Label Mode** icon (the one that resembles a target), then click on each slice of the pie in the graph to insert the percentages.
- Click the **Data Label Mode** icon to turn this function off.
- Click on the circular line that encloses the whole pie (you select the whole pie). The *Properties* window should open, as shown in Figure 8.4.

Figure 8.4 *Properties* Window Appearance Before Clicking the Depth & Angle Tab

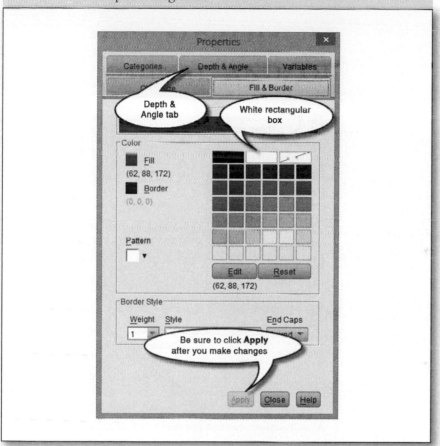

- In the *Properties* window, click on the Depth & Angle tab (once you click the Depth & Angle tab, the window takes on another appearance, as shown in Figure 8.5).
- In the *Effect* panel, click the small circle to the left of *3-D*.
- In the *Angle* panel, click and drag the slider to adjust the angle of the pie to a negative 35 (you can select your own preference for the pie angle by selecting a different number).
- Click **Apply**, as shown in Figure 8.5.
- In the *Properties* window, click the Fill & Border tab.
- Click the small box to the left of the *County* region of *Eastern* (a faint yellow frame appears around the box and in the corresponding pie slice).

Figure 8.5   *Properties* Window Appearance After Clicking the Depth & Angle Tab

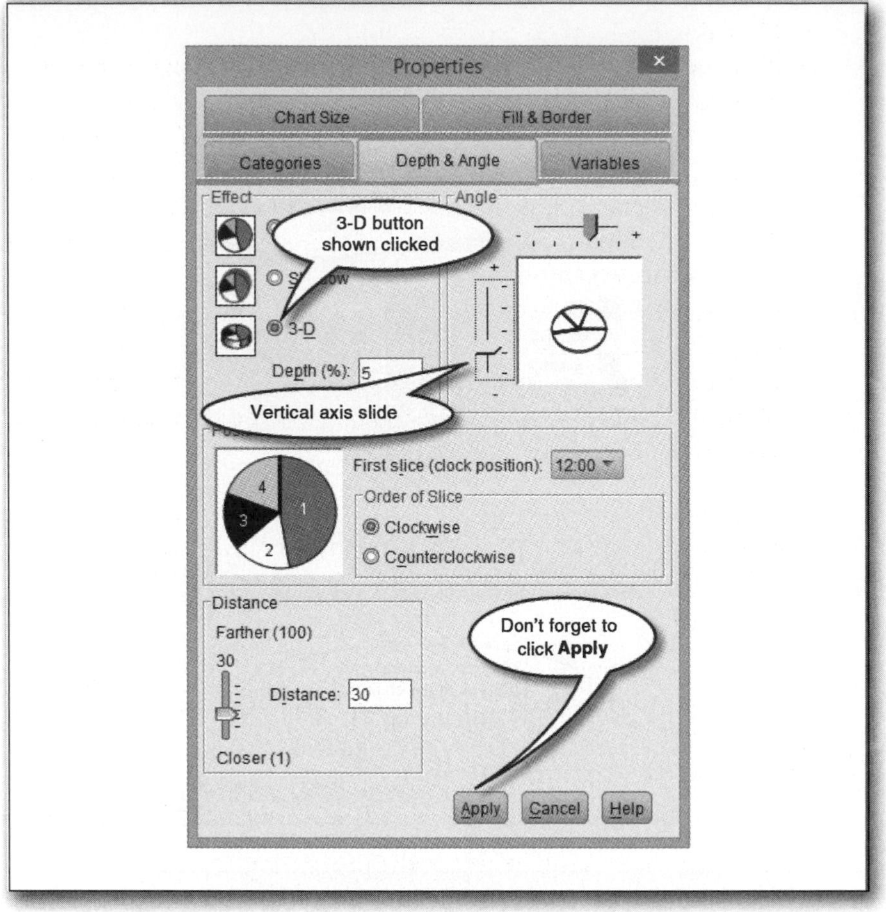

- In the *color* section, click the white rectangular box in the top row of available colors.
- Click the black arrow beneath *pattern*, then click the second pattern in the first row.
- Click **Apply**.
- Click the small box to the left of the *County* region of *Central*.
- Click the white rectangular box in the *color* panel, click the black arrow beneath *pattern*, click the third pattern in the first row, and click **Apply**.
- Click the small box to the left of *Western*.
- Click the white rectangular box, click the black arrow beneath *pattern*, click the first pattern in the second row, and click **Apply**.

- Click the small box to the left of *Northern*.
- Click the white rectangular box, click the black arrow beneath *pattern*, click the second pattern in the second row, and click **Apply**.
- Click the small box to the left of *Southern*.
- Click the white rectangular box, click the black arrow beneath *pattern*, click the second pattern in the fourth row, and click **Apply**.
- Click directly above the small box of *Eastern* (the mouse pointer should be lower than the word *County*) (a yellow rectangular frame with little squares on it surrounds the entire legend).
- Position the mouse pointer in any of the little squares until you get a *four-sided pointer*.
- Click and drag the rectangular frame closer to the graph to improve the graph's appearance.
- Click the **white "x" in the red box** located in the top-right corner of the *Chart Editor*. The finished graph now appears in the Output Viewer exactly as shown in Figure 8.6.

Figure 8.6   Final Pie Graph for Percentage of People by County Regions

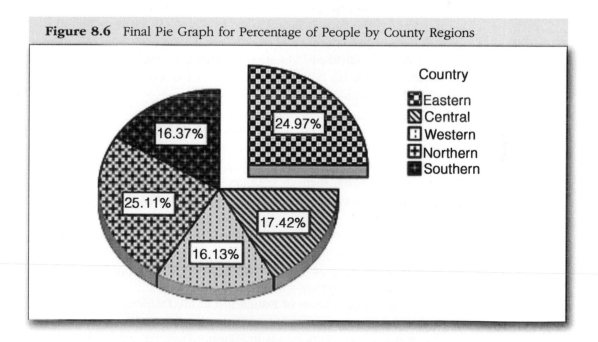

You just created a pie graph containing the information needed to answer our questions about the different regions of the County. Question 1: What percentage of the 9,449 survey respondents live in the *Eastern* portion of the county? Answer: The slice representing the *Eastern* region of the County shows that 24.97% of the respondents live in this part of the County. Question 2:

Which region has the largest percentage of respondents? Answer: The *Northern* region has the highest percentage of respondents (25.11).

△ 8.4 Building a Population Pyramid

In this section, you will learn to build a graph type known as a *population pyramid*. This type of graph can be used for both continuous (*scale*) and discrete (*nominal* and *ordinal*) data. In this particular graph-building exercise, you make a graph depicting the relationship between two discrete variables. To build this population pyramid, you use the SPSS database called *workprog.sav*. In this section, you build a graph showing the number of individuals in three different categories of education: (1) some college, (2) high school degree, and (3) did not complete high school. These three categories are then split (divided) into persons who are married and those who are single. The entire distribution consists of 1,000 individuals.

Before building the population pyramid, let's propose a couple of questions regarding the data contained in the *workprog.sav* database. Question 1: Which category of education shows the most difference between married and unmarried individuals? Question 2: What does the graph tell us about a possible relationship between marital status and level of education? Now, let's build the graph and answer the questions.

- Start SPSS, click **File**, select **Data**, and click **Open**.
- Locate and click the file titled *workprog.sav* (if needed, see Chapter 3, Section 3.5 for help).
- Click **Graphs**, then click **Chart Builder**. A window titled *Chart Builder* will open. At the bottom of the window, make certain **Gallery** is selected.
- Click **Histogram**, then click and drag **population pyramid** (the fourth icon) to the *Chart preview* panel.
- Click and drag the **Marital Status** variable to the *Split Variable* box in the *Chart preview* window.
- Click and drag **Level of Education** to the *Distribution Variable* box in the *Chart preview* window (following this step, the *Chart Builder* window should appear, as in Figure 8.7).
- Click **OK** (Output Viewer opens).
- Double click the graph (*Chart Editor* opens).
- Click any number on the horizontal axis (all numbers are then highlighted).

Figure 8.7   Population Pyramid *Chart Builder* Window

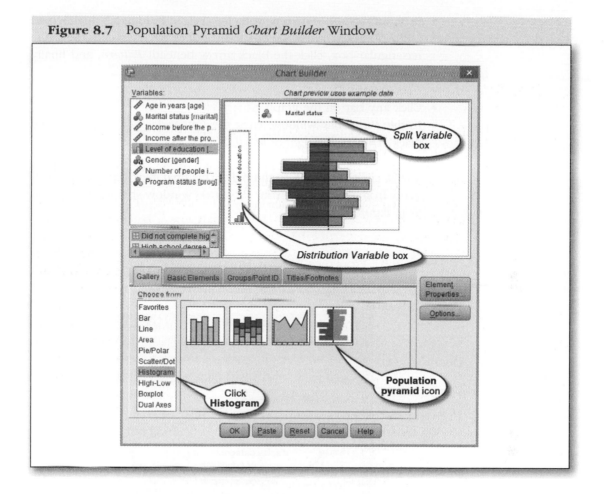

- In the *Properties* window, click the *Scale* tab, and in the *Range* panel, change *Minimum* from *50* to *0*.
- Click **Apply**.
- Click any of the rectangular bars in the *unmarried* category (a frame appears around all the bars in the graph). Click the same bar a second time (only the *unmarried* category remains framed).
- In the *Properties* window, click the Fill & Border tab, and in the *color* panel, click the box to the left of *Fill*, then click the white rectangular box.
- Click **Apply**.
- Click any bar in the *Married* category.

- In the *Properties* window with the Fill & Border tab highlighted (already clicked), click the box to the left of *Fill*, then click the white rectangular box, click the black arrow beneath *Pattern*, and finally click the third *pattern* in the last row.
- Click **Apply**.
- Click any bar on the graph.
- In the *Properties* window, click the Categories tab, then click the black arrow to the right of *Marital status*, click **Level of education**, and in the *Categories* panel, click **post-undergraduate degree**; finally, click the **red X** to the right. This final click removes this category from the analysis. Your *Properties* window should now appear as in Figure 8.8.

Figure 8.8   Properties Window Shown With the Categories Tab Activated (Clicked)

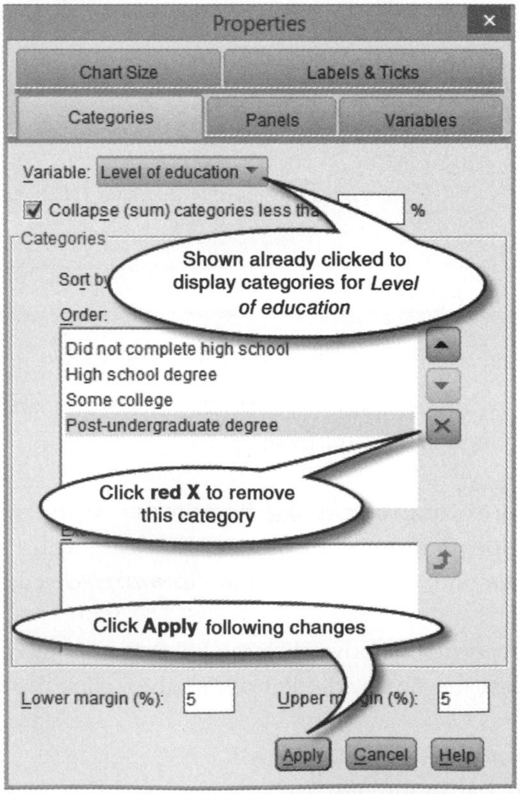

- Click **Apply**.
- Click any bar, then click the **Show Data Labels** icon (the second icon in the last row of icons in the *Chart Editor*).
- Click the **white "x" in the red box** in the upper right-hand corner of the *Chart Editor*. (Output Viewer opens, shown in Figure 8.9.)
- Click the graph in the Output Viewer, and a frame appears. You can then adjust the size of your graph by clicking and dragging the frame lines. Your population pyramid graph is now completed, as shown in Figure 8.9.

Figure 8.9   Population Pyramid for Age, Marital Status, and Gender

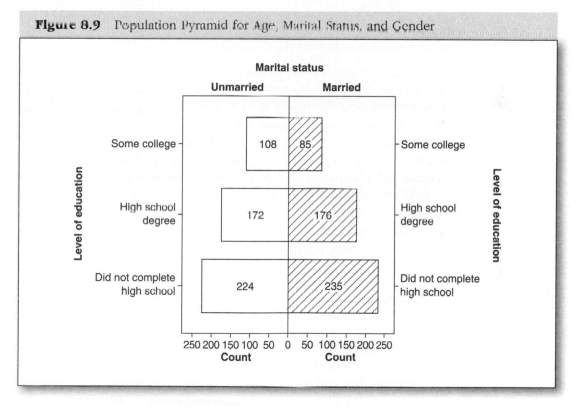

You have now built a graph that both explores and describes the relationship between two *discrete (categorical)* variables. Shall we see if we can answer the two questions just by looking at the graph shown in Figure 8.9? Question 1: Which category of education shows the largest difference between married and unmarried individuals? Answer: The greatest difference between married and unmarried is in the educational category of "Some college." There were 108 unmarried and 85 married in the "Some college" category. Question 2: What does the graph tell us about a possible relationship between marital status and level of education? Answer: Looking at the graph in Figure 8.9 and thinking in a proportional manner, we see that there is little difference between levels of education and marital status. We could even say that we strongly suspect that there is no relationship between marital status and level of education for this population of 1,000 individuals. This should, of course, be confirmed with a chi-square statistical test (we did this and found no significant relationship). At this point, we could say that looking at marital status by itself would not help in predicting level of education.

△ 8.5 Summary

In this chapter, you learned to create graphs when you have *nominal* or *ordinal* data. You used the SPSS feature known as the *Chart Builder*. You were introduced to the use of the *Chart Builder* by creating a simple bar graph. You next acquired more advanced graph-building techniques by transforming the attributes of a single *nominal* variable into a 3-D *pie graph*. The graph you built was suitable for publication in professional journals. You also built a *population pyramid* graph that split a *categorical* variable measured at the *ordinal* level (educational level) into the *nominal* categories of married and unmarried. In the next chapter, you will expand your graph-making capabilities by building various graphs used to explore and summarize *scale (continuous)* data from another SPSS sample file.

△ 8.6 Review Exercises

8.1 You are given the task of building a pie chart that summarizes the five age categories of 582 individuals recorded in an SPSS sample file titled *satisf.sav*. The variable is named "agecat" and labeled *Age category*. Build a 3-D pie graph that displays the names of the categories, the numbers of observations, and the percentages of the total for each slice. Also, answer the following questions: (a) What percentage of people are 18 to 24 years old? (b) How many individuals are 50 to 64 years old? (c) What is the largest age category? (d) What is the quantity, and its respective percentage, for the individuals who are 25 to 34 years old? (e) What is the quantity, and its respective percentage, for the smallest category?

8.2 You must summarize the data for 8,083 individuals. Specifically, you are tasked with building a 3-D pie graph showing the number and percentage of people in each of four regions. Open *dmdata3.sav*, select the discrete variable named "Region," and build a pie graph. Embellish the graph to display the names of the categories, the numbers of observations, and the percentages of the total for the slices to answer the following questions: (a) What is the largest region, and what percentage of the total respondents does it represent? (b) Which region is the smallest, and how many individuals does it have? (c) What is the number of customers, and its respective percentage, for the *North*? (d) What is the number of customers, and its respective percentage, for the *West*? (e) Rank the regions from the smallest to the biggest.

8.3 You must visually display the relationship between two categorical variables using the population pyramid graphing method. The variables you use are found in the SPSS sample file called *workprog.sav*. The variable "prog" (program status) is the split variable, and "ed" (level of education) is the distribution variable. Build the population pyramid to compare these discrete variables split into two groups of program status, 0 and 1. It has been determined that the categories of 0 and 1 are approximately equal; therefore, these distributions can be directly compared. Also, look at the finished graph, and answer the following questions: (a) By looking at the graph, does it appear that program status and level of education are related? (b) Which of the six categories contains the most observations? (c) Which category has the least number of observations?

CHAPTER 9

GRAPHS FOR
CONTINUOUS DATA

△ **9.1 INTRODUCTION AND OBJECTIVES**
_____

In the previous chapter, you used the IBM SPSS Statistics *Chart Builder* to create graphs to summarize and explore data measured at the *nominal* and *ordinal* levels. In this chapter, you will once again use the *Chart Builder* to build graphs, but for data measured at the *scale* level. We have decided to stick with *Chart Builder* since we believe that this approach is easier and more intuitive than either the *Graphboard Template Chooser* or the *Legacy Dialogs* choices. It is definitely to your advantage to become familiar with its use in the summarization and analysis of data.

In this chapter, you will use the SPSS sample file called *workprog. sav*. You will build a *histogram* first, then build a *boxplot* using the same variable ("Income before the program in dollars per hour"). By building the two different graphs, you will gain insight as to which method best answers different questions about the data. This chapter also introduces the method of data graphing known as *paneling*. With this approach, you can take a single continuous variable (*scale* data) such as "Age" and then categorize it using discrete (*nominal* and/or *ordinal*) variables. For our *paneling* graph–building demonstration, we will once again use the *workprog.sav* sample file. We will categorize "age" by the *nominal* variables of "Marital status" and "Gender."

OBJECTIVES

After completing this chapter, you will be able to

Create a histogram to visually display the distribution of continuous (scale) data

Create a boxplot to visually display the distribution of a continuous variable

Compare the histogram and boxplot in answering certain questions about data

Create a paneled graph for continuous data categorized (divided) by two nominal variables

Use additional features in the *Chart Builder Properties* and *Element Properties* windows

Adjust the scale, range, and increment values on the graph's *x*-axis and *y*-axis

9.2 CREATING A HISTOGRAM △

The *workprog.sav* database consists of 1,000 individuals (the cases) who were measured on eight variables. To build the histogram and then answer some basic questions, you will use the variable labeled "Income before the program." The SPSS name for this variable is "incbef." The following questions deal with the distribution of hourly wages recorded for this variable. Question 1: Which range of dollars per hour contains the fewest individuals? Question 2: Which range of dollars per hour contains the most individuals?

Let's build the histogram and answer these basic questions.

- Start SPSS, and click **File**, **Open**, and **Data**.
- Go to the SPSS sample files found in the C drive, then locate and click **workprog.sav** (if needed, see Section 3.5 in Chapter 3 for help in locating and opening this file).
- Click **Graphs**, then click **Chart Builder** to open the *Chart Builder* window.
- Click **Histogram**, then click and drag the **Simple Histogram** icon to the *Chart Preview* window.
- Click and drag **Income before the program** to the *x-axis* box.
- In the *Element Properties* window, click **Display normal curve**.
- Click **Apply**.
- Click **OK** (the basic graph now appears in the Output Viewer).
- Double click the graph to open the *Chart Editor*.

- Click the *x-axis label* **Income before the program** (a faint frame appears around this label), then click it a second time, and at the end, add the phrase *dollars per hour*.
- Click any number on the *x*-axis (*Properties* window opens).
- In the *Properties* window, click the Scale tab; then in the *Range* panel, set *Minimum* at 5, *Maximum* at 15, *Major increment* at 1, and *Origin* at 5.
- Click **Apply**. (Figure 9.1 shows the *Properties* window following the changes, and **Apply** has been clicked.)
- Click any number on the *y*-axis, click the Number Format tab in the *Properties* window, then change *Decimal Places* to read 0 (zero).
- Click **Apply**.

Figure 9.1   *Properties* Window Showing Scale Tab and *Range* Panel Changes

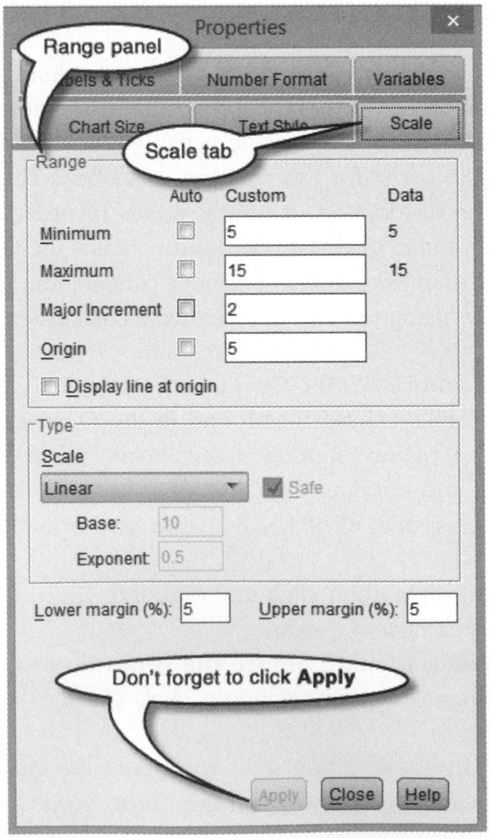

- Click the Scale tab, and in the *Range* panel, change *Major increment* to 50.
- Click **Apply**.
- Click the Show Grid Lines tab (found in the first row of icons in the *Chart Editor* as the fifth icon in from the right side). The *Properties* window opens; look to see that *Major ticks only* is checked.
- Click the Lines tab, and in the *Lines* panel, click the black arrow below *Weight*, and change *Weight* to 0.5; then click the black arrow by *Style*, and click the first dotted line.
- Click **Apply**.
- Click any of the bars on the graph (a faint frame appears around all the bars in the histogram).
- In the *Properties* window, click the Fill & Border tab.
- In the *Color* panel, click the white rectangular box.
- Click the black arrow beneath *Pattern*, then click the first pattern in the third row.
- Click **Apply**.
- Click the **white "x" in the red box** in the right-hand corner of the *Chart Editor*; this click closes the Chart Editor, and the graph is moved to the Output Viewer.
- Click the graph (a frame appears around the graph); then click and grab the lower right corner of the frame (marked by a small black square), hover the mouse pointer until you see a double-headed arrow, then move it diagonally up and to the left to reach the approximate size of the graph in Figure 9.2.

Figure 9.2 Histogram Showing the Distribution of Income Before the Program

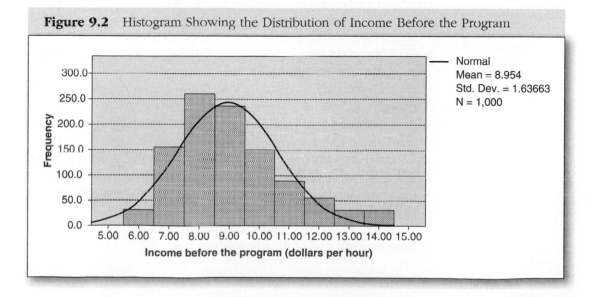

Now, let's see if we can answer our questions posed at the beginning of this section just by looking at the graph shown in Figure 9.2. Question 1: Which range of dollars per hour contains the fewest individuals? The range containing the fewest number of individuals is $13.50 to $14.50 per hour. Question 2: Which range of dollars per hour contains the most individuals? Most individuals in this population earned between $7.50 and $8.50 per hour. Of course, there are many other questions that might be proposed and answered; we simply wish to demonstrate the usefulness of such a graphing method.

△ 9.3 CREATING A BOXPLOT

The boxplot graph you are about to build uses an approach that differs from the histogram when attempting to decipher a mass of continuous data. The boxplot places an emphasis on the variability of the data as measured by various ranges. In this graph-building exercise, you create a boxplot using the same data used for the histogram. Doing this will allow us to explore the differences between these two graph types.

We will use the graph you build to answer two questions about the variability of the data. Question 1: What are the lower and upper limits, in dollars per hour, for the middle 50% (interquartile values) of this population? And what is the value of the interquartile range? Question 2: What are the lower and upper limits, in dollars per hour, for the first 25% (first quartile) of the population?

Let's build the boxplot and see if we might answer these basic questions.

- Start SPSS, click **File**, click **Open**, then click **Data**.
- Go to the SPSS sample files in the C drive, locate and click **work-prog.sav** (see Section 3.5 for help if needed).
- Click **Graphs**, then click **Chart Builder** to open the *Chart Builder* window.
- Click **Boxplot**, then click and drag the **1-D Boxplot** icon (the third icon) to the *Chart Preview* panel.
- Click and drag **Income before the program** to the *x-axis* box. (Note that for this type of graph, SPSS places the *x*-axis on the vertical line.)
- Click **OK** (the basic graph now appears in the Output Viewer).
- Double click the graph to open the *Chart Editor*.
- Click the **Transpose chart coordinate system** icon (second in from the right end of the top row of icons).

- Click the **Show Grid Lines** icon (fifth icon in from the right end of the top row of icons).
- In the *Properties* window, click the small circle to the left of **Major ticks only** (it should already be selected; if not, then click it, and also click **Apply**).
- In the *Properties* window, click the Lines tab, and in the *Lines* panel, click the black arrow beneath *Weight*, and click **0.25**; then click the black arrow beneath *Style*, and click the first dotted line.
- Click **Apply**.
- Click any whisker of the boxplot (a faint line appears around both whiskers).
- If not open now, click the **Properties window** icon (third from the left).
- In the *Properties* window, click the Lines tab if not already highlighted.
- In the *Lines* panel, click the black arrow beneath *Weight*; click **2**, then click **Apply.**
- Click any number on the *x*-axis.
- In the *Properties* window, click the Scale tab, then change *Major Increment* to 1.
- Click **Apply**.
- Click **interquartile range** (the *Properties* window opens with the Fill & Border tab highlighted).
- In the *Color* panel, click the white rectangular box.
- Click the black arrow beneath *Pattern*, then click the first pattern in the second row.
- Click **Apply**.
- Click the *x*-axis label (a faint frame appears around the *label*), then click the label a second time, and edit it to read *Income before the program (dollars per hour)*.
- Click the **white "x" in the red box** in the upper right-hand corner of the *Chart Editor* (the graph is moved to the Output Viewer).
- Click the graph (a frame appears around the graph), and then click and grab the lower right corner of the frame (marked by a small black square); hover the mouse pointer until you see a double-headed arrow, and move it diagonally up and left to reach the approximate size of the graph in Figure 9.3.

The first thing to note is that the case numbers for the outliers are easily identified. It is a simple matter to go to the database and examine these cases more closely. Next, let's look at the graph and answer

Figure 9.3   Boxplot for Income Before the Program (Dollars Per Hour)

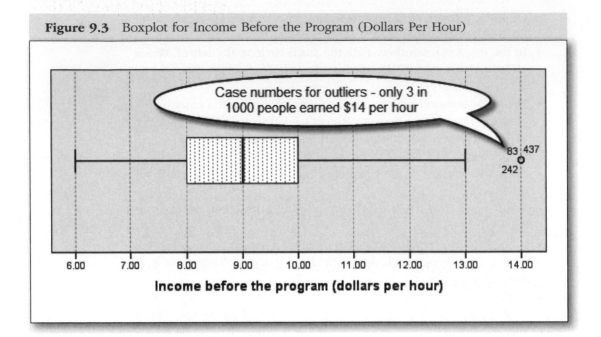

our questions from the beginning of this section. Question 1: What are the lower and upper limits, in dollars per hour, for the middle 50% (interquartile values) of this population? And what is the value of the interquartile range? Answer: The graph informs us that the middle 50% in this population earned between $8 and $10 per hour. The value of the interquartile range is 10 − 8 = $2 per hour. Question 2: What are the lower and upper limits, in dollars per hour, for the first 25% (first quartile) of the population? Answer: The lowest wage is $6 per hour; the highest for this first quartile is $8 per hour. This means that 25% of the population and the lowest paid individuals earn somewhere between $6 and $8 per hour.

We can also compare the usefulness and differences of the histogram and boxplot graphs by answering a few more questions. Question 1: What are the measures of central tendency used with the boxplot and the histogram? Answer: The boxplot uses the median, while the histogram uses the mean as a measure of central tendency. Question 2: What are the measures of variability for the boxplot and the histogram? Answer: The boxplot uses ranges to measure variability, while the histogram uses standard deviations. Question 3: What advantage of using the boxplot can you identify by looking at these two graphs? Answer: The boxplot identifies any outliers and extremes, which the histogram does not. Knowing your outliers and extremes can prove very useful in guiding further investigation of the database.

9.4 CREATING A PANEL GRAPH △

There are many different types of *paneled* graphs, each of which can be used for unique purposes. The basic differences are based on whether you panel on rows *or* columns or rows *and* columns. It is not the purpose of this book to explain the differences and the reasons for choosing one type over the other, so we have arbitrarily selected a graph-building example where you will build a graph based on rows *and* columns.

For this graph-building exercise, you will use the same sample file as above, *workprog.sav*. You will graph the *scale* (continuous) data called "Age," which will then be subdivided into categories by the *nominal* (discrete) variables of "Marital status" and "Gender."

Here are a few questions regarding ages as categorized by marital status and gender. Question 1: How many married males 15.5 to 16.5 years of age were enrolled in the work program? Question 2: Which age category has the most married females, and how many are there in that category? Question 3: How many unmarried females are 18.5 to 21.5 years of age?

You next build the paneled graph that will answer these (and many other) questions (see Figure 9.4).

- Start SPSS, click **File**, click **Open**, and then click **Data**.
- Go to the SPSS sample files in the C drive, then locate and click **workprog.sav** (see Section 3.5 for help if needed).
- Click **Graphs**, then click **Chart Builder**.
- Click **Histogram** in the *Choose from* list.
- Click and drag the **Simple Histogram** icon (the first icon) to the *Chart preview* panel.
- In the middle of the *Chart Builder* window, click the **Groups/Point ID** button, then click the box to the left of *Rows panel variable* and the box to the left of *Columns panel variable*.
- Click and drag **Age in Years** to the *x*-axis.
- Click and drag **Gender** to the panel on the right (this is the row panel).
- Click and drag **Marital Status** to the top panel (this is the column panel).

Figure 9.4   *Chart Builder* Window for Row and Column Paneling

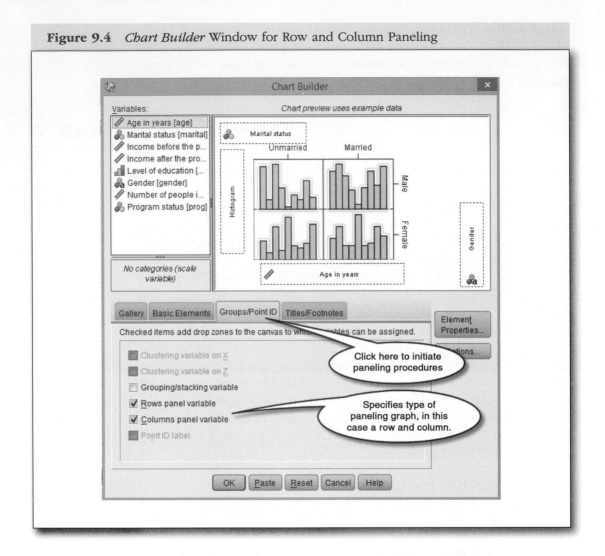

- Click **OK** (the *Chart Builder* closes, and the Output Viewer opens with the basic graph).
- If the Output Viewer is not showing, then click **Output** (found on the task bar at the bottom of the computer screen).
- Double click the basic graph to open the *Chart Editor* window.
- Click any histogram bar (a faint frame surrounds all bars).
- Click the Fill & Border tab in the *Properties* window, and in the *Color* panel, click the box to the left of *Fill*, then click the white rectangular box to the right, the black arrow beneath *Pattern*, and the third pattern in the first row (diagonal lines).

- Click **Apply**.
- Click any number on the *y*-axis.
- In the *Properties* window, click the Scale tab, and in the *Range* panel, change *Major Increment* to 10, then click **Apply**.
- Click **Number Format**, and change *Decimal Places* to 0 (zero).
- Click **Apply**.
- Click the **Show Grid Lines** icon (in the first row of icons, click the fifth icon in from the right side).
- In the *Properties* window with the Grid Lines tab highlighted, click the small circle to the left of *Both major and minor ticks*, then click **Apply**.
- Click any of the grid lines (all grid lines become highlighted).
- In the *Properties* window with the Lines tab highlighted and in the *Lines* panel, click the black arrow beneath *Weight*, then click **0.25**.
- Under *Style*, click the black arrow and the first dotted line, then click **Apply**.
- Click the very center of the graph between the panels (a frame should appear around all panels), then click the Panels tab in the *Properties* window.
- In the *Level Options* panel, click the black arrow to the right of *Level*, and click **Marital status**, then change *Panel spacing (%)* to 3, and finally click **Apply**.
- Click the black arrow to the right of *Level*, click **Gender** (from the drop-down menu), change *Panel spacing (%)* to 3, and then click **Apply**.
- Click any number on the *x*-axis.
- In the *Properties* window, click the Scale tab, and in the *Range* panel, change *Major Increment* to 1, then click **Apply**.
- Click the **white "x" in the red box** in the upper-right corner to close the *Chart Editor* and to move the edited graph to the Output Viewer.
- Click the graph, and adjust its size to approximate the graph in Figure 9.5.

Let's answer our questions now that we have built the paneled graph. Question 1: How many married males 15.5 to 16.5 years of age were enrolled in the work program? Answer: By looking at the *x*-axis and *y*-axis of the lower right panel, we determine that there are 15 married males between the ages of 15.5 and 16.5 years. Question 2: Which age category has the most married females, and how many are there in that category? Answer: The upper-right panel (married females), which indicates the highest bar, presents 66 females 18.5 to 19.5 years of age. Question 3:

Figure 9.5 Paneled Histogram for Age Categorized by Marital Status and Gender

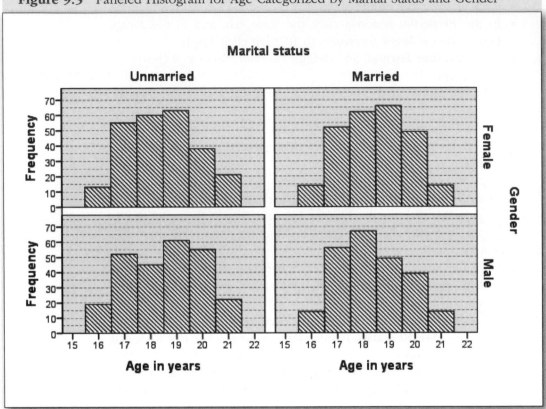

How many unmarried females are between 18.5 and 21.5 years of age? Answer: By looking at the upper-left panel for this age-group, we add 63 + 38 + 21 = 122 for the number of unmarried females.

△ 9.5 SUMMARY

In this chapter, you learned many new graph-making techniques where *axis scales*, *ranges*, and *start and finish points* were manipulated. This required the use of many features in the *Element properties* and *Properties* windows. You built graphs for continuous data (SPSS scale data), such as the histogram and boxplot, and also compared the information provided by these two types of graphs. You also built a graph that displayed the distribution of a *continuous* variable ("Age") and then categorized age by the *nominal* variables of "Marital status" and "Gender." Such graphs are known as *paneled* graphs.

9.6 REVIEW EXERCISES △

9.1 For this exercise, you will build a simple histogram using the SPSS sample file known as *autoaccidents.sav*. Open the file, select the variable named "accident" and labeled *Number of accidents in the past 5 years*, and build the simple histogram. Use the graph you build to answer the following questions: (a) Are the data skewed to the right, skewed to the left, or normal? (b) What is the average number of accidents for this group? (c) How many of this group reported one accident? (d) How many people had four accidents? (e) Of the 500 people in this group, how many had no accidents? (f) How many people had three accidents?

9.2 Use the same database as in the previous exercise (*autoaccidents.sav*), and select the variable named "Age" having the label *Age of insured*. Build the boxplot, and answer the following questions: (a) What are the minimum and maximum ages, excluding outliers and extremes? (b) What is the age of the outlier? (c) What are the limits of the interquartile range? (d) What is the interquartile range for this age distribution? (e) What is the median value? (f) Does the graph depict normally distributed data or perhaps a negative or positive skew?

9.3 For this exercise, you will open the SPSS sample file called *workprog.sav* and select *two* discrete variables and *one* continuous variable to build a histogram paneled on both columns and rows. For the *x*-axis use the variable labeled as *Age in years* and named "Age." The discrete variable named "Marital" and labeled as *Marital status* will be the *column-paneled* variable. The *row-paneled* variable is named "Ed" and labeled as *Level of education*. Look at the finished graph, and answer the following questions: (a) Which of the row panels (*Level of Education*) contains the most people? (b) Which of the six groups had the most participants in the age interval 17.5 to 18.5 years, and how many are in that group? (c) Which group had a distribution of ages that most closely resembled a normal distribution? (d) What is the shape of the distribution for unmarried individuals with a high school degree?

CHAPTER 10

PRINTING DATA VIEW, VARIABLE VIEW, AND OUTPUT VIEWER SCREENS

△ 10.1 INTRODUCTION AND OBJECTIVES

Printing raw data and the information that describes your variables (input) and the results of data analysis (output) is a major task made easy (most of the time) by SPSS. This chapter continues the theme and purpose of the prior chapters, namely, how can we make our observations (the data) more understandable and therefore useful? You will expand on this theme by printing the results of your analysis, which can then be used in statistical reports or perhaps for a more careful and relaxed examination while having a cup of coffee with a friend.

There are three major sections in this chapter: (1) How do you print data from the Data View screen? Printing a copy of the data you entered can be very helpful when checking the accuracy of your work. You are also shown how to select and print a portion of your data. (2) How do you print variable information? Having a printed version of the information used to describe your variables can be very useful as you proceed with the task of data analysis. A printed summary of variable information also serves as a *codebook* by providing a list of all the variables and their codes for your categorical variables. Another advantage is that the *codebook* makes it easier to share your work with a colleague for comment and/or discussion. (3) How do you print tables, graphs, and other analysis from the Output Viewer screen? These are the critical portions of your analysis that are intended to provide a succinct summary of all the work. Printing and embellishing selected portions of your output provides a convenient communication method with colleagues and the world.

OBJECTIVES

After completing this chapter, you will be able to

Print all entered data as displayed in the Data View screen

Print descriptive information for variables from the Variable View screen

Print output such as tables and graphs from the Output Viewer screen

Use the mouse to select and print data

10.2 PRINTING DATA FROM THE VARIABLE VIEW SCREEN △

Printing data that have been entered into the Data View screen in a format easy to interpret is not always a simple task. In general, when your database has more than six variables and more than 36 cases, SPSS spreads the printing over multiple pages. You just happen to have a database containing seven variables and 37 cases that you entered in Chapter 5. It is known as *class_survey1.sav* and will be perfect to illustrate what we mean when we say that the printing is easy, just one click, but the collation and interpretation of a data-printing job can be challenging.

- Start SPSS, and click **Cancel** in the opening window.
- Click **File**, select **Open**, and click **Data**.
- In the file list, locate and click **class_survey1.sav** (the database opens).
- Click the Data View tab if it has not already been clicked.
- Click **File**, and then click **Print Preview** (the Data Editor screen opens; see Figure 10.1).
- Click **Next Page** (shown in Figure 10.1), which opens another page that displays the seventh variable ("rate_inst"), creating another entire page as shown in Figure 10.2.

Figure 10.1   Print Preview for the Data View Screen: First of Four Pages

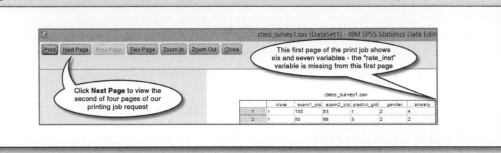

Figure 10.2   Print Preview for the Data View Screen: Second of Four Pages

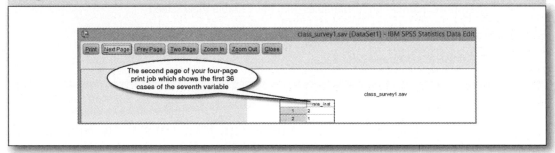

Figure 10.3   Print Preview for the Data View Screen: Third of Four Pages

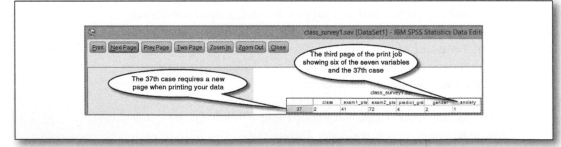

- Clicking **Next Page** brings up the third page—the 37th case, as shown in Figure 10.3.
- Clicking **Next Page** again brings up the final page of your printing job (see Figure 10.4).

Figure 10.4 Print Preview for the Data View Screen: Fourth of Four Pages

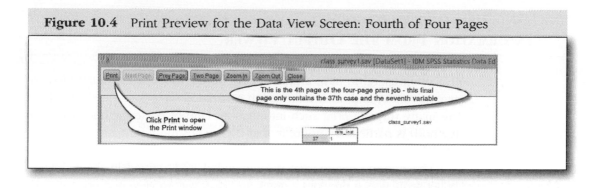

After following this procedure, one can surmise the complexity of a print job having hundreds of cases and dozens of variables. We would like to be able to describe an easy way to assemble the four pages (Figures 10.1 through 10.4) into one understandable format. Unfortunately, we cannot. If you must have a hard copy of your data, we can only advise you to get out your scissors and cellophane tape. The scissors-and-tape procedure takes time, but keep in mind that you rarely need a hard copy of your data.

To complete the printing job of the four pages as depicted in Figures 10.1 through 10.4, do the following:

- Click **Print** as shown in Figure 10.4, and the *Print* window opens (see Figure 10.8).
- Click **OK** (your printer is activated, and your printing job is completed).

Printing a Selected Portion of Your Data

There are occasions when you may wish to print some subset of your database. To print the desired portion, click the mouse, and drag the pointer over the desired cases and variables that you wish to print. These areas will be automatically highlighted. Then, follow the normal printing sequence.

- Click **File**, and then click **Print**.
- Click **Selection**, and then click **OK**.

The click-and-drag process allows you to save ink and paper if you only require a hard copy of a portion of your data.

△ 10.3 PRINTING VARIABLE INFORMATION FROM THE OUTPUT VIEWER

It is often useful to have a printed copy of the *labels, value labels, level of measurement*, and other information associated with your variables. There are many ways to print such information. One of these methods (the *File* method) is particularly useful in that it provides the information in a printable format.

The printing method we recommended is described in the following bullet sequence. It provides a quick and easy way to view and print all variable information from any SPSS file. If you have an open file, you can easily obtain variable information by selecting *working file*. Let's proceed on the assumption that the *class_survey1.sav* database is open—if not, open the file.

- Click **File**, and select **Display Data File Information**.
- You now have a side menu choice of *Working File* or *External File*.
- Click **Working File**.
- If the Output Viewer does not automatically open, click the **SPSS** icon at the bottom of your screen, and click the *Output* window at the bottom of your screen.

The Output Viewer opens with two tables: (1) Variable Information (Figure 10.5) and (2) Variable Values (Figure 10.6). The *Working File* side menu item was selected since this was the file that was currently open and active. You would select *External File* if you desired information on any other file.

Figure 10.5 Variable Information for the Class Survey Database

Variable Information

Variable	Position	Label	Measurement Level	Role	Column Width	Alignment	Print Format	Write Format
class	1	Morning or Afternoon Class	Nominal	Input	8	Left	F8	F8
exam1_pts	2	Points on Exam One	Scale	Input	8	Left	F8	F8
exam2_pts	3	Points on Exam Two	Scale	Input	8	Left	F8	F8
predict_grde	4	Student's Predicted Final Grade	Nominal	Input	8	Left	F8	F8
gender	5	Gender	Nominal	Input	8	Left	F8	F8
anxiety	6	Self-rated Anxiety Level	Ordinal	Input	8	Left	F8	F8
rate_inst	7	Instructor Rating	Ordinal	Input	8	Left	F8	F8

Variables in the working file

Figure 10.6 Variable Values for the Class Survey Database

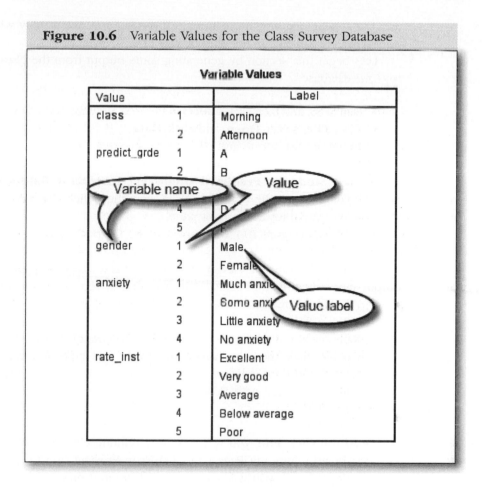

To complete the printing operation, do the following:

- Click **File**, and then click **Print**.
- Click **OK**.

The two tables shown in Figures 10.5 and 10.6 provide all the information about the specifications of your variables.

10.4 PRINTING TABLES FROM THE OUTPUT VIEWER △

When printing from the Output Viewer, there are several ways to enhance the appearance of your print job. Of course, if the printing is only for your use, the appearance may not matter. In this case, you may simply wish to print whatever appears in the Output Viewer.

Let's begin this section by generating some output from the *class_survey1.sav* database.

- Start SPSS, and then click **Cancel** in the *SPSS Statistics* opening window.
- Click **File**, select **Open**, and click **Data**.
- In the file list, locate and click **class_survey1.sav**.
- Click **Analyze**, select **Descriptive Statistics**, and click **Frequencies**.
- Click **Student's Predicted Grade** and **Instructor Rating** while holding down the computer's Ctrl key, then click the arrow (this moves variables to the right panel).
- Click **OK** (Figure 10.7 will appear in the Output Viewer).

Return to the Output Viewer, and follow these steps to produce an output more suitable for inclusion in a statistical report. In the Output Viewer, do the following:

- Right click and delete all items prior to *Frequency Table*.
- Double click the title **Frequency Table** and type *Statistics Class Survey*, and then delete the words *Frequency Table* (i.e., *Frequency Table* is replaced with *Statistics Class Survey*).
- Click **File**.

Once you click **File**, you will see three printing choices provided on the drop-down menu: (1) *Page Setup*, (2) *Print Preview*, and (3) *Print*. In the majority of cases, you will simply select *Print* and proceed from there. However, we now briefly discuss the other two choices.

The first option, as mentioned above, is *Page Setup*, which provides the opportunity to specify paper size, portrait or landscape orientation, and margin size. The second option, *Print Preview*, simply shows what the output will look like once it is printed.

Note: If you have any graphs in the Output Viewer, you will see a fourth printing option, *Page Attributes*. Clicking on **Page Attributes** opens the *Page Attributes* window. You can then add custom headers and footers to your output. By clicking the *Options* tab in this window, you may also specify the number of graphs printed on each page. This printing option can be very useful when printing a large number of graphs. For this exercise, we select the *Print* option.

- Click **Print** (*Print* window opens as shown in Figure 10.8).
- Click **OK**.

Figure 10.7 Output as It Appears in Output Viewer

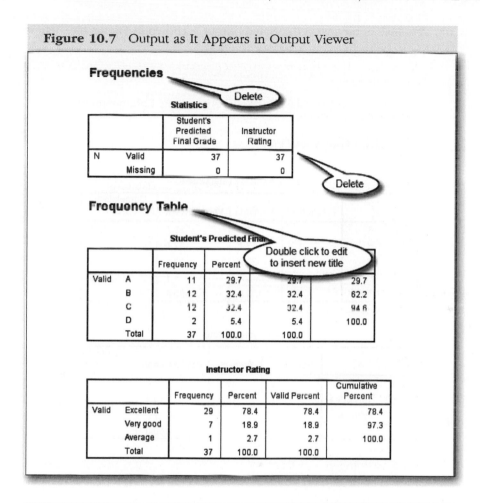

Frequencies

Delete

Statistics

		Student's Predicted Final Grade	Instructor Rating
N	Valid	37	37
	Missing	0	0

Delete

Frequency Table

Student's Predicted Fina

Double click to edit to insert new title

		Frequency	Percent		
Valid	A	11	29.7	29.7	29.7
	B	12	32.4	32.4	62.2
	C	12	32.4	32.4	94.6
	D	2	5.4	5.4	100.0
	Total	37	100.0	100.0	

Instructor Rating

		Frequency	Percent	Valid Percent	Cumulative Percent
Valid	Excellent	29	78.4	78.4	78.4
	Very good	7	18.9	18.9	97.3
	Average	1	2.7	2.7	100.0
	Total	37	100.0	100.0	

Figure 10.8 *Print* Window

Print

Printer: Win32 Printer : Canon MG3200 series

Print Range
● All
○ Selection

Number of copies: 1

☐ Collate

OK   Cancel

The final, and enhanced, printed output is shown in Figure 10.9, and it is ready to be included in a statistical report.

Figure 10.9   Printed Class Survey Output After Enhancement

Statistics Class Survey

Student's Predicted Final Grade

		Frequency	Percent	Valid Percent	Cumulative Percent
Valid	A	11	29.7	29.7	29.7
	B	12	32.4	32.4	62.2
	C	12	32.4	32.4	94.6
	D	2	5.4	5.4	100.0
	Total	37	100.0	100.0	

Instructor Rating

		Frequency	Percent	Valid Percent	Cumulative Percent
Valid	Excellent	29	78.4	78.4	78.4
	Very good	7	18.9	18.9	97.3
	Average	1	2.7	2.7	100.0
	Total	37	100.0	100.0	

△   10.5 Summary

In this chapter, you learned basic SPSS printing procedures. You are now able to print data from the Data View screen, variable information from the Variable View screen, and various types of SPSS outputs given in the Output Viewer. The following chapter will provide some basic information on descriptive statistics and how they are used by SPSS.

△   10.6 Review Exercises

10.1   Open the SPSS sample file called *workprog.sav*, and print the data for the first six variables of the first 10 cases.

10.2 You have a meeting with your research group, and you wish to discuss some of the variables you have selected for a project. The database is stored in an SPSS sample file called *workprog.sav*. What is the quickest and easiest way to print out all the information about your study variables?

10.3 You must analyze several variables of the *workprog.sav* database (SPSS sample file) and then prepare a clean handout for your colleagues. You need basic descriptive statistics on all your scale data and a frequency table for level of education for the categorical variables. Generate the output, give it the title "Work Program Study," and print the "cleaned" output.

CHAPTER **11**

BASIC DESCRIPTIVE STATISTICS

Δ 11.1 INTRODUCTION AND OBJECTIVES

This chapter presents the basic statistical tools available in SPSS that are used to make data more understandable. The better we understand our data, the more useful they become in assisting us in discovering patterns and making informed decisions.

We begin with a definition of descriptive statistical analysis. *Descriptive* statistical analysis is any statistical or mathematical procedure that reduces or summarizes numerical and/or categorical data into a form that is more easily understood.

Descriptive statistics is that branch of statistics that you use when the data have been collected and you now wish to describe them. Contrast this with the other major branch of statistics, known as inferential statistical analysis. *Inferential* statistics uses sample data to generate approximations of *unknown* values in the population. In other words, it goes beyond descriptive statistical analysis. This does not mean that inferential analysis is somehow superior but only that it serves a different purpose. In fact, many inferential techniques require that you first conduct a descriptive analysis of the sample data obtained from a population—more on this aspect of descriptive analysis in Section 11.4.

There are many descriptive statistical techniques designed to accomplish data reduction and summarization. For instance, you may find it useful to generate frequency tables, graphs, and scatterplots. Graphs were covered in Chapters 8 and 9; therefore, tables are emphasized in this chapter. When conducting descriptive statistical analysis, to make the data more understandable, calculation of the *mode, median, mean, standard deviation, variance, range, skewness,* or *kurtosis* can be very helpful. This chapter gives you the skills needed to use the power of SPSS to describe any distribution of data (actually a bunch of numbers). Another important descriptive statistic is the correlation coefficient, which describes the strength of a relationship between two variables (bivariate analysis). This type of descriptive analysis includes scatterplots, which were mentioned above. Bivariate analysis is covered in Chapter 19, whereas this chapter covers those methods used to describe single variables, one at a time (univariate analysis). In Chapter 22, you are also given an introduction to a more advanced exploratory/descriptive method known as *factor analysis*. Factor analysis is a process where you attempt to reduce the number of variables that measure some outcome of interest.

OBJECTIVES

After completing this chapter, you will be able to

Describe and interpret measures of central tendency

Generate and interpret frequency tables

Describe and interpret measures of dispersion (variability)

Generate and interpret descriptive statistical tables

Describe and summarize variables that are measured at the nominal, ordinal, and scale levels

Determine if a variable's values approximate the normal distribution

Use SPSS to build a normal P-P plot

Conduct the one-sample Kolmogorov-Smirnov test for normality

11.2 MEASURES OF CENTRAL TENDENCY △

Measures of central tendency are used when we wish to describe one variable at a time. An example of such data description would be the class survey that we have used in prior chapters. Think in terms of the average score on the first exam, the number of females and males, or perhaps the level of anxiety. Each of these individual variables could be summarized using a measure of central tendency. The goal, of course, is to better understand the survey respondents—to tell us what these students are like. The following section discusses three measures of central tendency: the *mode*, the *median*, and the *mean*.

The Mode

The *mode* is the value of the most frequent observation. The mode can be used to help understand data measured at any level. To obtain the mode, you first count or tally the number of observations in each distinct category of a nominal or ordinal variable or the most frequently observed score for scale data. The mode is the category of the variable that contains the most observations. Let's say you have a variable named "fruit" and the categories are 1 = *mangoes*, 2 = *apples*, and 3 = *oranges*. Furthermore, you have a basket filled with these fruits; you count the fruits and find that there are 23 mangoes, 15 apples, and 19 oranges in the basket. The mode is "1," which is the *value* that you assigned to mangoes.

SPSS can quickly find the mode of any variable by producing a frequency table. In addition to the mode, SPSS's frequency table method also provides a count and then calculates the percentages found in each of the variable's categories. Using this approach, SPSS can reduce many thousands of cases into a format that can be better understood in milliseconds. In the following demonstration, we use the *class_survey2* database that was created in Chapters 5 and 6. Start SPSS, and click **Cancel** in the *SPSS Statistics* opening window.

- Click **File**, select **Open**, and click **Data**.
- In the file list, locate and click **class_survey2.sav**.
- Click **Analyze**, select **Descriptive Statistics**, and then click **Frequencies**.
- Click **Student's Predicted Grade** and **Gender** while holding down the computer's Ctrl key.
- Click the arrow to move the selected variables to the *Variable(s):* panel. Once the arrow is clicked, the window should show the *Frequencies* window, as in Figure 11.1.
- Click **OK**. Following this click, the Output Viewer opens, which displays a table titled *Student's Predicted Final Grade* (see Figure 11.2) and another titled *Gender* (see Figure 11.3).

Figure 11.1 *Frequencies* Window for Analysis of Two Nominal Variables

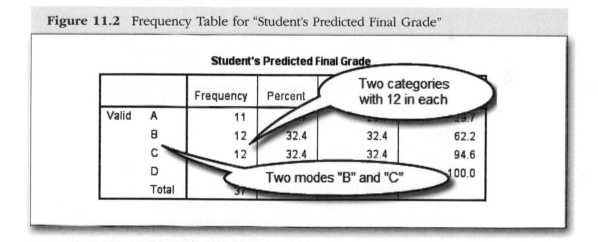

Figure 11.2 Frequency Table for "Student's Predicted Final Grade"

In our quest to better understand the data, let's begin by looking at the frequency table shown in Figure 11.2. There are four categories, and we find that the same number of students predicted grades of "B" and "C." The identification of the mode informs us that there are two modes—we have a *bimodal* distribution for the variable. How does the frequency table add to

Figure 11.3   Frequency Table for "Gender"

		Frequency	Percent	Percentage of females	
Valid	Male	11	29.7		29.7
	Female	26	70.3	70.3	100.0
	Total	37	100.0	100.0	

our understanding of the data for this single variable? One thing we can say is that there were 24 students (32.4% + 32.4% = 64.8%) who believed that they would earn a grade of "B" or "C." There was not a single student who predicted a grade of "F," which was one of the survey options. Furthermore, we may say that the students appeared to be rather optimistic in that almost 30% predicted a grade of "A."

The next variable, "Gender" (see Figure 11.3), indicates that 26 of the 37 students (70.3%) were female—the mode. At this point in our analysis, we could say that we had a class of mostly female students who were rather optimistic about receiving an excellent grade in the statistics class. Note that further analysis would be required to determine whether gender was related to predicted grade. Remember that our *descriptive* analysis, up to this point, only describes one variable at a time.

Next, we examine the median as a measure of central tendency.

The Median

The *median* is the value that occupies the middle point of a distribution. It is the point that divides the distribution in half. Half the values fall at or below the median, and half fall at or above it. We next demonstrate the use of SPSS to determine the value of the median.

Assuming that *class_survey2* is still open, use the sequence given below to obtain the median for selected variables from this database.

- Click **Analyze**, select **Descriptive Statistics**, and then click **Frequencies**. The *Frequencies* window opens. Note that the two variables from the prior exercise may remain in the *Variable(s)* panel. If this is the case, then click the **Reset** button at the bottom of the window to remove these two variables.

- Click **Points on Exam One** and **Points on Exam Two** while holding down the computer's Cul key.
- Click the arrow to move the selected variables to the *Variable(s):* panel.
- In the *Frequencies* window, unclick **Display frequency tables**, then click **Statistics** (see Figure 11.4).

Figure 11.4 *Frequencies* Window for Analysis of Two Scale Variables

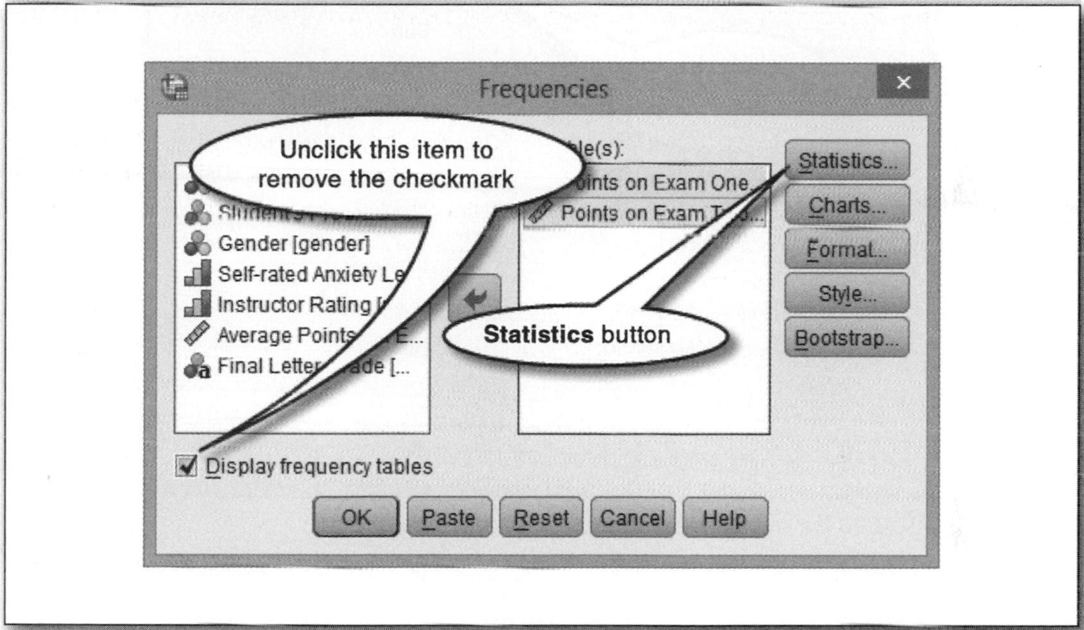

- Once **Statistics** is clicked, the *Frequencies: Statistics* window opens. Click **Median** (everything else should be unclicked, as shown in Figure 11.5).
- Click **Continue**.
- Click **OK**. (Figure 11.6 shows the Output Viewer.)

The table reports that the median value for "Points on Exam One" is 67, whereas the median for "Points on Exam Two" is 81. With this information, we can say that half the students scored less than or equal to 67 points on the first exam; on the second exam, the median was 81 for an improvement of 14 points.

Figure 11.5   The *Frequencies: Statistics* Window

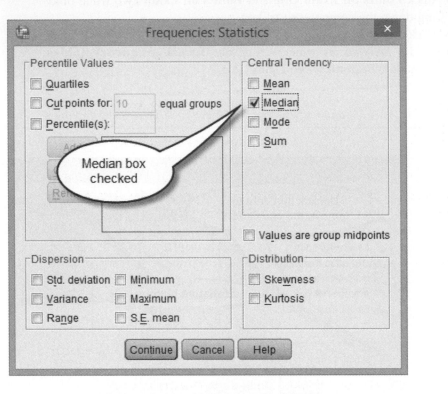

Figure 11.6   Statistics Table for the Median

Statistics

		Points on Exam One	Points on Exam Two
N	Valid	37	37
	Missing	0	0
Median		67.00	81.00

The Mean

Another common name for the mean is the *average*. The simplest way to find the mean of a distribution is to add all the values and divide this sum by the number of values. You will find that sometimes the mean is used with categorical variables, such as *ordinal* data. However, many statisticians consider this practice to be mathematically unsound. As was discussed in Chapter 4, on levels of measurement, it is difficult, or impossible, to demonstrate equal intervals (technically required to calculate the mean) with ordinal data.

The current section considers numerical data (SPSS's *scale* data), the only legitimate use of this descriptive statistic. The *mean*, as another measure of *central tendency*, has as its purpose the making of a distribution of values more understandable. Once again, we assume that the *class_survey2* database is open before following the procedure below.

- Click **Analyze**, select **Descriptive Statistics**, and then click **Frequencies**.
- Click **Points on Exam One**, **Points on Exam Two**, and **Average Points On Exams** while holding down the computer's Ctrl key.
- Click the arrow to move the selected variables to the *Variable(s):* panel. All three scale variables should now be in the *Variable(s)* box, ready for analysis.
- Click **Statistics** (the *Frequencies: Statistics* window opens) in this new window; click **Mean**, **Median**, and **Mode**, and then click **Continue**.
- Click **OK**. (Figure 11.7 shows the Output Viewer.)

Figure 11.7   Statistics Table for the Mean, Median, and Mode

Statistics

		Points on Exam One	Points on Exam Two	Average Points On Exams
N	Valid	37	37	37
	Missing	0	0	0
Mean		65.95	78.11	72.0270
Median		67.00	81.00	73.0000
Mode		50	93[a]	57.00[a]

a. Multiple modes exist. The smallest value is shown

As shown in Figure 11.7, the means for these three exams are 65.95, 78.11, and 72.03. If the purpose of descriptive statistics is to make data more understandable, then how do these means accomplish this? For one thing, when looking at the raw scores for the 37 students, it is impossible to compare the performance on the two tests in any meaningful way. Using the mean, we can easily see that there was a 12-point improvement on the second test.

▵   11.3 Measures of Dispersion

There are three basic measures of dispersion discussed next: *range, standard deviation*, and *variance*. The range can be calculated on *ordinal* or *scale* data, whereas the *standard deviation* and *variance* are calculated only with variables measured at the scale level. Perhaps the easiest way to think of these measures of dispersion is that they tell us how *spread out* the values are. With scale data, we are usually concerned with how the data are dispersed around the *mean*. One of the most important concepts in descriptive statistics is that to accurately describe scale data, you must calculate both a measure of dispersion (standard deviation is the best) and a suitable measure of central tendency—most often the mean. If this is not done, then an incomplete and potentially misleading view of the values may result.

We now use SPSS to do the hard work and calculate the measures of dispersion. We also calculate several other descriptive statistics that describe the shape of the distribution. The descriptive statistics (other than the dispersion measures) will also be explained. It is necessary to use a slightly different command sequence than that used for the measures of central tendency—so pay careful attention. Make sure that the *class_survey2* database is open, and then follow the procedure given next.

- Click **Analyze**, select **Descriptive Statistics**, and then click **Descriptives** (note that the final click for this sequence is *Descriptive* and not *Frequencies* as in prior exercises).
- Click **Points on Exam One**, **Points on Exam Two**, and **Average Points On Exams** while holding down the computer's Ctrl key.
- Click the arrow to move the selected variables to the *Variable(s):* panel.
- Click **Options**, and the *Descriptives: Options* window opens (see Figure 11.8).
- Click **Mean, Std. deviation, Variance, Range, Minimum, Maximum, S.E. mean, Kurtosis,** and **Skewness**. (Figure 11.8 shows how your screen should look prior to the next bullet.)
- Click **Continue**.
- Click **OK** (the Output Viewer produces the analysis shown in Figure 11.9).

Figure 11.8 The *Descriptives: Options* Window

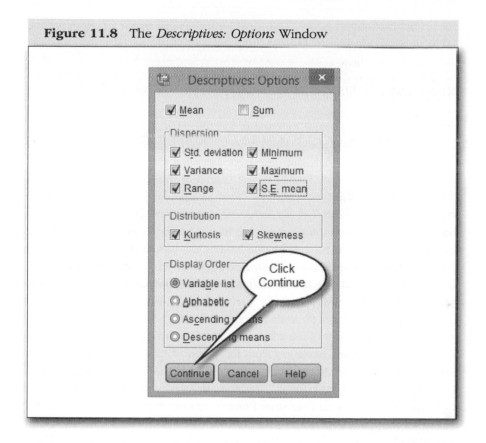

The Output Viewer opens with the statistics that were requested in Figure 11.8 and are now shown in Figure 11.9. The table provides statistics for the three variables measured at the scale level.

Let's describe the values contained in this output and how they qualify as descriptive statistics. We must always keep in mind that the intention of the information presented in Figure 11.9 is to help us better understand

Figure 11.9 The *Descriptive Statistics* Table for Three Variables Measured at the Scale Level

Descriptive Statistics

	N	Range	Minimum	Maximum	Mean		Std. Deviation	Variance	Skewness		Kurtosis	
	Statistic	Statistic	Statistic	Statistic	Statistic	Std. Error	Statistic	Statistic	Statistic	Std. Error	Statistic	Std. Error
Points on Exam One	37	86	14	100	65.95	4.266	25.951	673.441	-.280	.388	-1.124	.759
Points on Exam Two	37	77	23	100	78.11	2.721	16.551	273.932	-1.021	.388	1.936	.759
Average Points On Exams	37	76.00	24.00	100.00	72.0270	3.08385	18.75831	351.874	-.451	.388	-.291	.759
Valid N (listwise)	37											

these distributions of scores. We next examine the major portions of the table presented in Figure 11.9.

Range, Mean, Standard Deviation, and Variance

We have 37 values (scores) for each of the three variables (two tests and the average for those two tests). The range for each is given as 86, 77, and 76, respectively. The range is the difference between the maximum and minimum values and as such measures the spread of the test scores.

The mean of the test scores for "Exam One" is 65.95 with a *Std. Error* (standard error) of 4.266. The *Std. Error* indicates how well the sample mean of 65.95 estimates the unknown population mean. Additional details regarding the definition of the *Std. Error* can be obtained in any introductory statistics book.

In addition to the *Range* (as described above), we have two more measures of variability for our test scores: The first is the *Variance*, and the second is the *Std. Deviation* (standard deviation). The variance is simply the average of the squared deviations from the mean. For "Exam One," the variance is 673.441, whereas the square root of the variance is 25.951—the standard deviation. This represents the average deviation from the mean of 65.95 when all individual scores are taken into account. The standard deviation is most useful when comparing multiple distributions. Therefore, let's compare our standard deviations for "Points on Exam One" and "Points on Exam Two"—25.951 and 16.551.

If all the grades were the same on the first test, the standard deviation would be 0. As the variability of test scores increases, so does the size of the standard deviation. The standard deviation of 25.951 indicates that there was some deviation from the mean. The range of 86 also provides evidence that the scores were spread out (from 14 to 100) over the possible values. Comparing "Exam One" with the distribution of scores for "Points on Exam Two," we see that the test resulted in less variability, having a standard deviation of 16.551 and a range of 77 (23–100). The data clearly indicate that the scores on the second exam were not as spread out as those on the first exam. Whether the differences in the standard deviations are *statistically significant* is another question and would have to be addressed via a chi-square hypothesis testing.

The Shape of the Distribution (Skewness)

Before leaving the *descriptive statistics* table given in Figure 11.9, we briefly examine *skewness* and *kurtosis*. *Skewness* is a way to describe the shape of the distribution in relationship to the normal curve. The normal curve represents a symmetrical distribution of values. It is a curve that results in exactly the same proportion of area under the curve on both sides of the mean. The *mean, median,* and *mode* are equal in the normal distribution; see Figure 11.10(b). Data that are approximately normally distributed, as seen in Figure 11.10(b), make it possible to estimate proportions of values by using the normal curve as a mathematical model for the data.

When raw data deviate from the normal distribution, we have a *skewed* distribution. The distribution could have a *negative skew,* as shown in Figure 11.10(a). A negative skew means that the majority of the values tend to be at the high end of the *x*-axis, which results in the median being greater than the mean and a more representative measure of central tendency than the mean. If we look at the skewness number for our "Exam One" data in Figure 11.9, we find a minus 0.280 with a *Std. Error* of 0.388. The minus sign indicates that the distribution of test scores has a negative skew, similar to that depicted in Figure 11.10(a).

The skewness numbers for the three test score variables given in the table shown in Figure 11.9 all indicate minus skews (−0.280, −1.021, and −0.451, respectively). You might wish to confirm the fact that the mean is less than the median by looking at Figure 11.7. A positive skew, as shown in Figure 11.10(c), would indicate that the test scores were at the lower end of the *x*-axis, with a mean greater than the median. This was not the case for the variables in the class survey.

Figure 11.10 Common Distribution Shapes Comparing Normal and Skewness

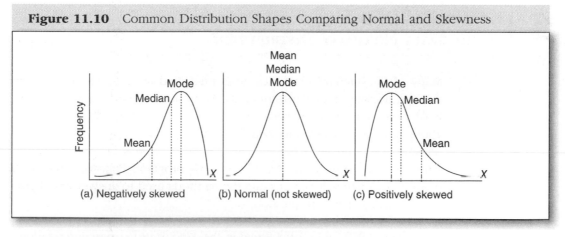

The Shape of the Distribution (Kurtosis)

Another descriptive statistic used to describe the shape of the distribution is *kurtosis.* As with the skewness, the kurtosis value can be negative, zero, or positive. Figure 11.11(a) shows what a distribution of values would look like if it had *negative kurtosis*. Basically, it indicates that there is an abundance of cases in the tails—giving it the flat top appearance. Its formal name is *platykurtic*. Figure 11.11(b) shows the normal curve, and Figure 11.11 (c) shows a *positive kurtosis*. A minimum of cases in the tails results in a *positive kurtosis*, which is known as a *leptokurtic* distribution.

Figure 11.11   Three Common Distribution Shapes for Kurtosis

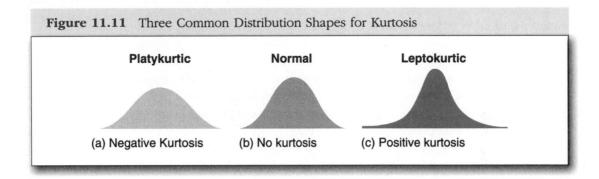

For now, it is sufficient that the reader understands that these descriptive statistics tell us how far a distribution of values deviates from the well-proportioned normal curve.

△ 11.4 The Big Question: Are the Data Normally Distributed?

Many of the inferential statistical tests presented in later chapters require that the data be normally distributed. Therefore, the question of normality is of major importance when making inferences (informed speculations) about unknown population values. There are many ways to answer the question of normality; some involve the use of *kurtosis* and *skewness*, whereas some use the power of the SPSS program. Following the SPSS procedure below will demonstrate the methods used to answer the important normality statistical question.

- Start SPSS, and click **Cancel** in the *SPSS Statistics* opening window.
- Click **File**, select **Open**, and click **Data**.
- In the file list, locate and click **class_survey2.sav**.
- Click **Analyze**, select **Descriptive Statistics**, and then click **P-P Plots**.

- Click **Average Points On Exams**, and then click the arrow to move this variable to the *Variables:* panel.
- Click **OK** (when the Output Viewer opens, direct your attention to the *P-P Plot* as shown in Figure 11.12).

Figure 11.12   Normal P-P Plot for "Average Points on Exam"

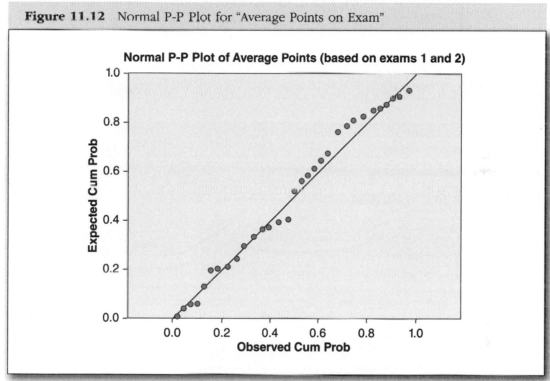

Visual inspection of the *P-P Plot* tells us that there is an excellent possibility that the data may be normally distributed. The closer the points are to the diagonal line, the greater the probability of normality. However, rather than relying on subjective judgment regarding the interpretation of the P-P Plot for normality, we next perform a *nonparametric* test for normality.

- Click **Analyze**, select **Nonparametric**, and click **One-Sample** (the *One-Sample Nonparametric Tests* window opens and is shown in Figure 11.13).
- Click the **Fields** tab.
- Click **Use custom field assignments**.

- Click **Average Points**, and then click the arrow (which moves this variable to the *Test Fields* panel; see Figure 11.13 for the window's appearance following this click).
- Click **Run** (the Output Viewer displays Figure 11.13).

The Output Viewer opens, showing the results of the one-sample Kolmogorov-Smirnov test, as indicated in Figure 11.14. The purpose of the Kolmogorov-Smirnov test is to determine if the distribution of values

Figure 11.13   *One-Sample Nonparametric Tests* Window for "Average Points on Exam"

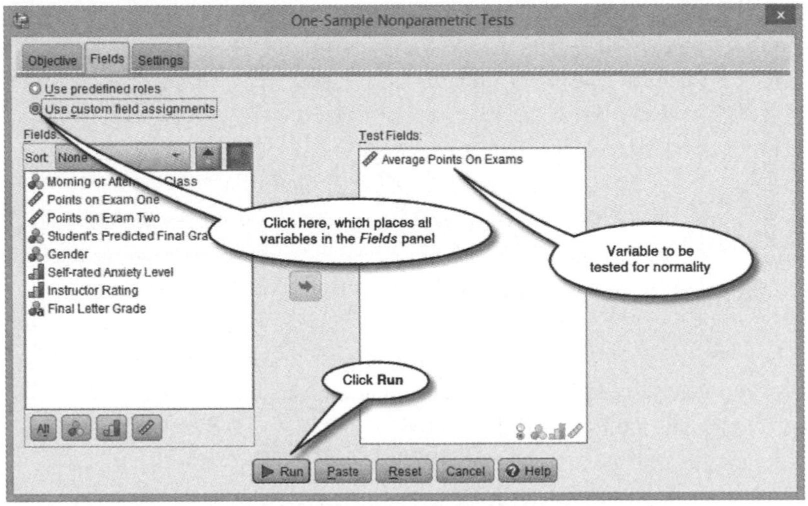

Figure 11.14   One-Sample Test for Normality for "Average Points on Exams"

Hypothesis Test Summary

	Null Hypothesis	Test	Sig.	Decision
1	The distribution of Average Points On Exams is normal with mean 72.027 and standard deviation 18.76.	One-Sample Kolmogorov-Smirnov Test	.200[1,2]	Retain the null hypothesis.

Asymptotic significances are displayed. The significance level is .05.

approximates the normal curve. We now have additional evidence that the data are normally distributed, as the null hypothesis of normality is *not* rejected.

Questions about a distribution's shape are very important since many of the tests presented in later chapters require that the data approximate the shape of the normal curve.

11.5 DESCRIPTIVE STATISTICS FOR THE CLASS SURVEY   △

This chapter opened by indicating that we would show you how SPSS can help make data more understandable. We went on to show how such knowledge would make it possible to discover patterns in the data and assist one in making informed decisions. The data used in the demonstration resulted from a class survey of intermediate statistics students. Let's summarize the SPSS findings that were generated and see if the findings make the data more understandable. What can we say about these students as a result of our descriptive statistical analysis of the data?

We had a rather optimistic group of students, as 62% anticipated either an "A" or a "B" in this traditionally challenging class. There were also an equal number of students anticipating a grade of "B" or "C" (32% for both these grades). It was also discovered that 70% of the students were female. We might speculate that the students may have become more serious on taking the second test as the median score increased by 14 points—from 67 to 81. However, it might also be that the second test was less challenging. Both of these ideas, which account for the test score differences, are pure conjecture. Such conjecture is permissible at this stage of the analysis, as offering plausible explanations for any observed pattern can be a part of the descriptive process. The mean score also followed the same pattern as the median in that there was a 12-point increase from the first to the second exam. The fact that the mean and median values were so close gave us a hint that the data's distribution followed the normal curve.

Measures of dispersion also supported the idea of more serious attitudes toward the material (or an easier second exam) as the variability of test scores declined for the second exam, from 26 to 17, a change of 9 points. It was also shown that the distributions were similar to a normal curve and perhaps subject to further standard statistical tests investigating these differences. These initial observations of normality were confirmed by using the *P-P Plots* function of SPSS. The nonparametric one-sample test (Kolmogorov-Smirnov test) provided additional evidence that these data approximated the shape of the normal curve.

△ 11.6 Summary

In this chapter, several basic descriptive statistical procedures were described. These procedures were introduced with the idea of summarizing variables so that the analyst could make sense of the data and derive useful information. Descriptive statistical techniques appropriate for variables measured at the nominal, ordinal, and scale (interval and ratio) levels were described. Testing procedures to determine the shape of any distribution, such as whether the data approximate the normal curve, were also described. The following chapter introduces the concept of hypothesis testing and its relationship to inferential statistical analysis.

△ 11.7 Review Exercises

11.1 Open the SPSS sample database called *bankloan.sav*, and calculate the following statistics for all variables measured at the *scale* level: *mean, std. deviation, variance, range, minimum, maximum, S.E., mean, kurtosis,* and *skewness.* Print the SPSS descriptive statistics output produced in answer to your request.

11.2 Open the SPSS sample database called *bankloan.sav*, and build frequency tables for all variables measured at the *nominal* or *ordinal* level.

11.3 Open the SPSS sample database called *bankloan.sav*, and determine if the variables "age" and "household income in thousands" are normally distributed.

CHAPTER 12

ONE-SAMPLE *t* TEST AND A BINOMIAL TEST OF EQUALITY

12.1 INTRODUCTION AND OBJECTIVES △

This is the first time we directly address the *inferential* statistical technique known as *hypothesis* testing. You may recall that in the previous chapter we presented the results of a hypothesis test for normality with a very brief explanation. This chapter will give many more details, and you will learn about the use of specific hypothesis-testing procedures when using the *one-sample t test* and the *binomial test of equality*. We caution the reader that the concepts of hypothesis testing and their relationship to SPSS are especially important. Understanding the concepts and procedures in this chapter will be a tremendous help in clarifying the information presented in subsequent chapters.

Prior to presenting the purposes of the *one-sample* and *binomial* tests, we will digress slightly to discuss hypothesis testing and its relationship to inferential statistics. The result of the hypothesis testing procedure is that we discover evidence (or do not) that supports an assertion about population value. Such evidence is derived from data obtained from the samples of that population. The reader should realize that our intention in this textbook is not to impart a broad or total understanding of the research process,

However, we do intend to give you the knowledge required to select the correct hypothesis test, apply that test, and then answer specific research questions. These questions will often deal with assertions about population values based on the sample data.

In the following pages, we describe the purpose of the *one-sample t test* and the *binomial test of equality*. The purpose of the one-sample *t* test is to test whether a sample mean is significantly different from some hypothesized value in a population. The purpose of the binomial test of equality is to determine whether 50% of the values fall above and 50% below the hypothesized population value.

OBJECTIVES

After completing this chapter, you will be able to

Describe the data requirements for using the one-sample *t* test

Write a research question and a null hypothesis for the one-sample *t* test

Conduct and interpret the one-sample *t* test using SPSS

Describe when it is appropriate to use the binomial test of equality

Conduct and interpret the binomial test as an alternative to the one-sample *t* test

△   12.2 Research Scenario and Test Selection

A review of several different studies conducted in Southern California resulted in the belief that jackrabbits successfully crossing a busy highway travel at an average speed of 8.3 miles per hour (mph). A conservation scientist in Northern California had the idea that the jackrabbits of Northern California travel at a significantly different speed. Armed with a radar gun, the scientist positioned himself by the side of Interstate 80, somewhere north of Lake Tahoe. His task was to record the speed of jackrabbits as they crossed the busy highway. Twenty jackrabbits, selected at random, were clocked, and their average speed was determined to be 8.7 mph. What is the appropriate statistical test to answer this conservation scientists' question?

The reasoning leading to the selection of the correct test that will answer the scientist's question follows. The speeds of the Southern and Northern California jackrabbits were measured at the *scale* level in miles per hour.

Since scale data are a requirement of all t tests, it is therefore possible to use one of the t tests. It is clear that the conservation officer has only *one random sample*—how about the *one-sample t* test? The mean rabbit speed of this sample (8.7 mph) could then be compared with the hypothesized speed of 8.3 mph, which was based on the speed of the Southern California rabbits.

Also, the various t tests require that the original populations be approximately normally distributed and that their variability is approximately the same. In the current situation, the assumption of normality and equal variances is based on prior research. The normality requirement is lenient as the t test is quite insensitive to nonnormal data. Statisticians describe this quality by saying that the test is *robust*. Given that the data were measured at the scale level and that we can justify the assumptions of *normality* and *equal variances* for both populations, the one-sample t test is selected.

12.3 RESEARCH QUESTION AND NULL HYPOTHESIS △

The research question is the researcher's idea or, in other words, the reason for doing the research. In the world of scientific inquiry, the research question is referred to as the *alternative hypothesis* and designated as H_A. As mentioned above, the researcher's idea is that the average speed of Northern California jackrabbits is different from the average speed of Southern California rabbits of 8.3 mph. The alternative hypothesis is written in a statistical format as follows: H_A: $\mu \neq 8.3$. This simply states that the mean speed of the Northern California rabbits does not equal 8.3 mph. We are attempting to develop statistical evidence that will support the alternative hypothesis.

The null hypothesis always states the opposite of the researcher's idea (H_A). Therefore, the null hypothesis, designated as H_0, states that the average speed of Northern rabbits is 8.3 mph. The null hypothesis written in a statistical format is H_0: $\mu = 8.3$, which indicates that the mean speed of the Northern California rabbits equals the mean speed of the Southern California rabbits. In most cases, and indeed in this instance, the researcher wishes to reject the null hypothesis (H_0), which would then provide evidence in support of the alternative hypothesis (H_A).

12.4 DATA INPUT, ANALYSIS, AND INTERPRETATION OF OUTPUT △

Let's begin by setting up a new SPSS database that records the speeds of 20 jackrabbits crossing a busy section of Interstate 80 in Northern California. The speeds (in mph) are given in Figure 12.1.

Figure 12.1   Speeds, in Miles per Hour, of 20 Northern California Jackrabbits

9.53	7.50	6.21	8.95	10.53	6.30	5.20	12.51	6.35	10.23
9.56	6.57	11.78	10.56	7.24	6.19	10.86	7.25	8.34	12.78

- Start SPSS, and click **Cancel** in the SPSS opening window.
- Click **File**, select **New**, and click **Data**. (clicking **New** will bypass the *Open Data* window step and therefore require fewer clicks to get started on your project).
- Click **Variable View**.
- Click the cell in Row 1 and Column 1, and type *rabbits*.
- Click the cell in the *Label* column, and type *Jackrabbit speeds* (select 2 decimals and *scale* as the level of measurement).
- Click **Data View**, and in the *rabbits* column type the rabbit speeds (shown in Figure 12.1) of 9.53, 7.50, and so forth through 12.78 (you should have 20 rows of data once you complete this data entry).
- Click **File**, click **Save As**, and then type *rabbits* in the *File name* box.
- Click **Save**.

Now that you have entered the data and saved the file, let's answer the researcher's question using the power of SPSS to do the required calculations. Assuming that the *rabbits.sav* database is open, do the following.

- Click **Analyze**, select **Compare Means**, and then click **One-Sample T-Test** (the *One-Sample T Test* window opens).
- Click **Jackrabbit Speeds**, then click the arrow (moves the variable to the *Variable(s)* box).
- Click the *Test Value:* box, and type *8.3*. (Following these operations, the *One-Sample T Test* window should appear, as seen in Figure 12.2.).
- Click **OK** (the Output Viewer opens; see Figures 12.3 and 12.4).

The most important aspect of this rabbit research scenario is the answer to the research question: The researcher's idea (and the basis for the research question) is that the average speed of Northern California jackrabbits is different from the average speed of Southern California jackrabbits of 8.3 mph. One value in Figure 12.4 answers this important question: It is the value of .435 found in the column titled *Sig.* (*2-tailed*). *Sig* is an abbreviation

Figure 12.2 The *One-Sample T Test* Window

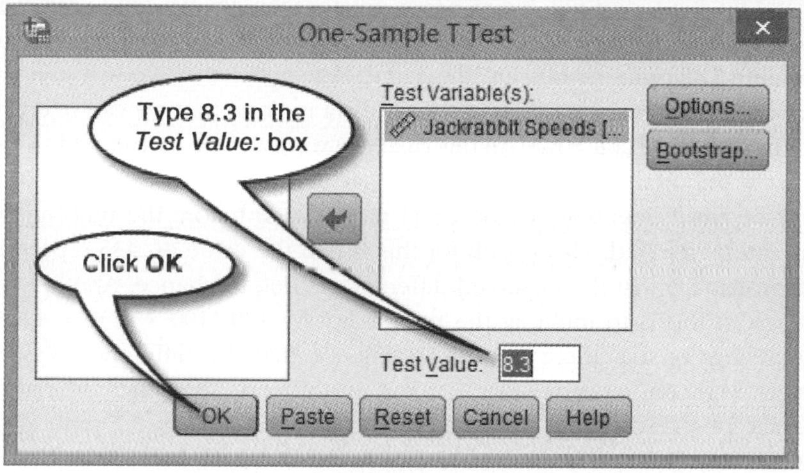

Figure 12.3 One-Sample Statistics for Northern California Rabbits

One-Sample Statistics

	N	Mean	Std. Deviation	Std. Error Mean
Jackrabbit Speeds	20	8.7130	2.31356	.51733

Figure 12.4 One-Sample *t* Test Results

One-Sample Test

Test Value = 8.3

	t	df	Sig. (2-tailed)	Mean Difference	95% Confidence Interval of the Difference	
					Lower	Upper
Jackrabbit Speeds	.798	19	.435	.41300	-.6698	1.4958

This value tells us that the null hypothesis *cannot* be rejected

Note this value

for *significance*. Let's explain how the value of .435 answers the research question. The value of .435 directly refers to the H_0 (null hypothesis) and gives us the information required to decide whether the null hypothesis should be rejected or not. Since .435 is greater than .05, we *fail* to reject the null hypothesis, which stated that the average speeds of the Northern and Southern California rabbits are the same. Remember that we can generate support for our research idea only if the data indicate that we can reject the null hypothesis (H_0: $\mu = 8.3$). In this case, we were unable to accomplish this.

It is important that you recognize that the number in the *Sig.* column must be small (less than .01 or .05 in most cases) before the null hypothesis can be rejected. The reason for this is that the value of .435 represents the probability that the observed difference is due to chance. Another way to look at this is to multiply the probability by 100 ($100 \times .435 = 43.5\%$) and recognize that there is a 43.5% chance that the difference was the result of random movement of the data. In this case, we conclude that the average observed speed for Northern California rabbits of 8.7 mph is not significantly different from the average speed of 8.3 mph recorded for the Southern California rabbits. Our conclusion is that the differences are attributable to random movements in the data. In other words, we say that the null hypothesis remains in force—the average speed of Northern California rabbits is 8.3 mph (H_0: $\mu = 8.3$).

As a way to further explain the concept of finding small values in the *Sig. (2-tailed)* column, let's assume that .003 was displayed in the *Sig. (2-tailed)* column of Figure 12.4. The value of .003 could only have occurred if our sample data resulted in an average speed for Northern California rabbits that was much different from 8.3 mph. In this case, the value of .003 (a small number) would indicate that there was only a .003 probability (0.3% chance) that the observed difference was attributable to chance. Therefore, we could reject the null hypothesis and would have evidence in support of the alternative hypothesis that the speeds were significantly different.

To summarize, we say that small numbers (less than .01 or .05) in the *Sig. (2-tailed)* column indicate that the detected differences are due to something other than chance (perhaps rabbit characteristics in this exercise), whereas large numbers indicate that the differences are due to chance.

A Word About Confidence Intervals

In looking at Figure 12.4, you will also see a column labeled *95% Confidence Interval of the Difference*, which is then divided into *Lower* (−.6698) and *Upper* (1.4958). The numbers −.6698 and 1.4958 represent the range of values of the sample statistic that is likely to contain the unknown (but estimated) difference value for the population. If we look at the column to the left, we see *Mean Difference*, which is the observed difference in our sample. We see that our difference of .41300 is within the given *confidence interval*. The *95% confidence level* informs us that in repeated sampling of the population, 95% of the samples would yield a difference value within the limits of our interval. Given that high level of confidence, we then use such sample values as estimates for the unknown population value.

12.5 Nonparametric Test: The Binomial Test of Equality   △

The *binomial test of equality* can be used as an alternative if the assumptions of normality and equal variability are not met. It should be noted that the binomial test is principally used with categorical data to test the equality of proportions. However, the jackrabbit scenario offers a viable opportunity for the application of the binomial test of equality. The reasoning is that if the speed of the Northern California rabbits is approximately the same as that of the Southern California rabbits, we would expect an equal number of speeds to fall above and below the mean speed (8.3 mph) of the Southern California rabbits. If the unknown speed is not the same, then the proportions of cases above and below the mean would be significantly different. The reader might also note that the binomial test requires that scale data be transformed into categorical data; this is done seamlessly by SPSS.

The alternative hypothesis (H_A) when using the nonparametric binomial test of equality is that the proportions of Northern California jackrabbits running above and below the average speed of the Southern California rabbits (8.3 mph) will *not* be equal. The null hypothesis (H_0) is that these proportions will be equal. Let's conduct the binomial test on our jackrabbit speed data and then compare the results with the findings of the parametric test.

For this particular example, we show two ways to conduct the nonparametric binomial test. The first method, using *Legacy Dialogs*, is compatible with earlier versions of SPSS. The second way utilizes the newer versions of SPSS.

When conducting the nonparametric binomial test using the *Legacy Dialogs* approach, follow the steps listed next.

- Start SPSS, and click **Cancel** in the SPSS opening window.
- Click **File**, select **Open**, and click **Data**.

- In the file list, locate and click **rabbits.sav**, and then click **Open**.
- Click **Analyze**, select **Nonparametric Tests**, select **Legacy Dialogs**, and then click **Binomial** (the *Binomial Test* window opens; see Figure 12.5).
- Click **Jackrabbit Speeds**, and then click the arrow.
- Click **Cut Point**, and type *8.3*.
- Check the *Test Proportion* box, making sure it reads *0.50* (see Figure 12.5).

Figure 12.5 *Binomial Test* Window for Jackrabbit Speed Data

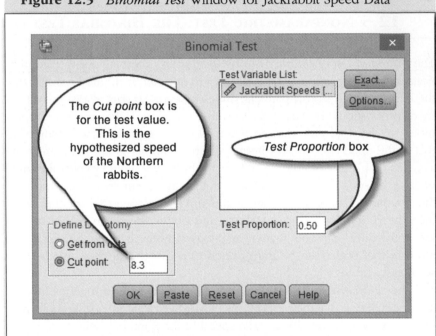

- Click **OK** (the Output Viewer opens; see Figure 12.6).

Figure 12.6 Binomial Test Results—Output Viewer

Binomial Test

		Category	N	Observed Prop.	Test Prop.	Exact Sig. (2-tailed)
Jackrabbit Speeds	Group 1	<= 8.3	9	.45	.50	.824
	Group 2	> 8.3	11	.55		
	Total		20	1.00		

When conducting the nonparametric binomial test using the nonparametric tests one-sample approach, follow the steps listed next.

- Assuming that SPSS is running and *rabbits.sav* is open, do the following
- Click **Analyze**, select **Nonparametric tests**, then click **One-sample** (the *One-Sample Nonparametric Tests* window opens). Click the Objective tab, then click **Customize analysis**.
- Click the Fields tab, and move *Jackrabbit Speeds* to the *Test Fields* panel.
- Click the Settings tab, click **Customize tests**, then check the box next to *Compare observed binary probabilities to hypothesized (Binomial test)* (see Figure 12.7 for the appearance of the window at this time).

Figure 12.7 The *One-Sample Nonparametric Tests* Window (Binomial)

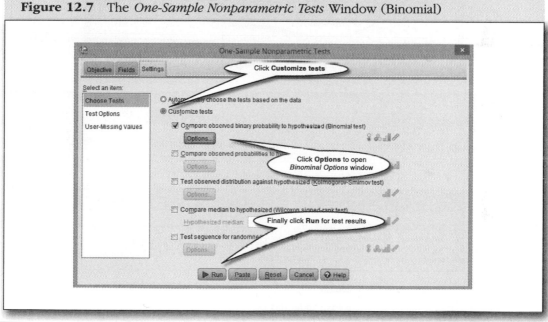

- Click **Options** (the *Binomial Options* window opens) in the *Define Success for Continuous Fields* panel; click **Custom cut point**, and then type *8.3* in the box (see Figure 12.8).
- In the *Binomial Options* window click **OK**, and then click **Run** (*Output Viewer* opens with test results—see Figure 12.9).

Figures 12.6 and 12.9 show that the test first separated the jackrabbit speeds into two groups (look at the column titled *Category* in Figure 12.6 and under *Null Hypothesis* in Figure 12.9): *Group 1* consisted of speeds less

Figure 12.8   Binomial Options for Nonparametric One-Sample Test

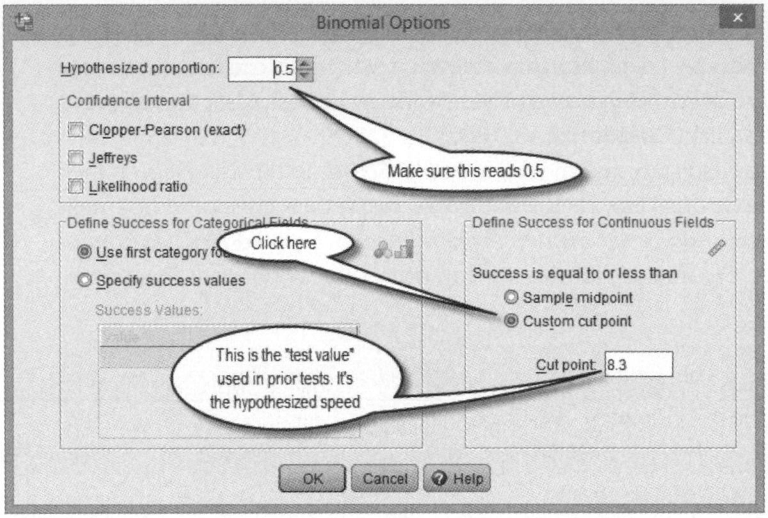

Figure 12.9   Hypothesis Test Summary for One-Sample Test

Hypothesis Test Summary

	Null Hypothesis	Test	Sig.	Decision
1	The categories defined by Jackrabbit Speeds <=8.300 and >8.300 occur with probabilities 0.5 and 0.5.	One-Sample Binomial Test	.824[1]	Retain the null hypothesis.

Asymptotic significances are displayed. The significance level is .05.

[1]Exact significance is displayed for this test.

than or equal to 8.3 mph, and *Group 2* consisted of speeds greater than 8.3. The test reasons that the proportions of observations for the Northern California rabbits (separated into two groups) should be the same (50% in each category). This would be true if the mean speed of the Northern California rabbits was approximately 8.3 mph. If the mean speed of our Northern California rabbits was significantly different, then the proportions would not be equal in the two groups. For instance, let's say that the Northern California rabbits sprinted across the highway at a blistering average speed of 24 mph. In this case, the binomial would be unlikely to indicate that 50% fell below and 50% above the test value of 8.3 mph.

The *Exact Sig. (2-tailed)* is reported as .824 in Figures 12.6 and 12.9. The value of .824 is greater than .05, indicating that the null hypothesis *cannot* be rejected. This finding agrees with the finding of the one-sample *t* test. We have insufficient evidence to support the researcher's idea that the Northern California and Southern California rabbits have significantly different highway crossing speeds.

12.6 SUMMARY △

In this chapter, we introduced the concept of hypothesis testing as a major component of inferential statistical analysis. Since this is the first chapter on hypothesis testing, time was spent on an explanation of the steps involved in the methods used to conduct such tests. Examples of writing the alternative and null hypotheses were provided. The hypothesis tests known as the *one-sample t test* and a nonparametric alternative, *the binomial test of equality*, were the particular tests presented. The one-sample *t* test basically relies on one random sample of a population, calculates the mean of that sample, and examines whether it equals some hypothesized population value. The binomial test separates all values into two groups and then uses the hypothesized value to test whether the observed values are equally distributed above and below the hypothesized value. If they are not so distributed, we have some evidence that the test value (hypothesized value) may be some other value.

In the following chapter, we present another form of the *t* test, but this time, it is to examine the mean differences for two independent samples. This test is known as the *independent-samples t test*. The nonparametric alternative is known as the *Mann-Whitney U test*.

12.7 REVIEW EXERCISES △

12.1 You are a seller of heirloom garden seeds. You have several machines that automatically load the seeds into packages. One of your machines is *suspected* of sometimes being inaccurate and of underloading or overloading the packages. You take a random sample of 20 packages and record their weights as follows: 3.09, 2.74, 2.49, 2.99, 3.22, 2.51, 2.28, 3.54, 2.52, 3.20, 3.09, 2.56, 3.43, 3.25, 2.69, 2.49, 3.30, 2.69, 2.89, and 3.57. In the past, it has been determined that the average weight should be 2.88 ounces. Write the null and research hypotheses and select the correct statistical test to determine if your sample evidence indicates that the machine is malfunctioning.

12.2   There is an annual race to reach the top of Kendall Mountain in the small town of Silverton, Colorado. A health scientist believes that the time to reach the top has significantly changed over the past 10 years. The average time for the first 12 runners to reach the summit in the 2005 race was 2.15 hours. He records the times for the top 12 runners in 2015 as follows: 1.43, 1.67, 2.13, 2.24, 2.45, 2.50, 2.69, 2.86, 2.92, 2.99, 3.35, and 3.36 hours, respectively. You must now write the null and alternative hypotheses in an attempt to produce evidence in support of your research hypothesis.

12.3   An instructor in a community college auto mechanics class asked his students if they thought that gas consumption, as measured in miles per gallon (mpg), had improved for the Ford Focus in the past 5 years. It was a team project, so they got together and found that on average the Ford Focus was rated at 24.3 mpg in 2010. No data for the current year (2015) were available, so they worked out a random sampling plan and collected the following current consumption data for 12 Ford Focus vehicles: 26.4, 26.5, 26.9, 26.9, 26.4, 26.4, 26.8, 26.9, 26.4, 26.9, 26.8 and 26.4 mpg. The plan was to somehow compare their sample data with the 5-year-old data—can you help these students?

CHAPTER 13

INDEPENDENT-SAMPLES *t* TEST AND MANN-WHITNEY *U* TEST

13.1 INTRODUCTION AND OBJECTIVES △

This chapter continues the theme of hypothesis testing as an inferential statistical procedure. In the previous chapter, we investigated whether there was a significant difference between the mean of one random sample of a population and some hypothesized mean for that population.

We now address the situation where we compare two sample means. This test is known as the *independent-samples t test,* and its purpose is to see if there is statistical evidence that the two population means are significantly different. You may recall that the sample means are estimates of the unknown means in the sampled population. Therefore, if a significant difference is detected in the sample means, we make an inference that the unknown means of the population are also different. In the following section, you are given details regarding the independent nature of the samples and how the samples may be obtained for the independent-samples *t* test. You are also given the data assumptions required to use this test.

The *Mann-Whitney U test* is the alternative nonparametric test that may be used when the data assumptions required of the independent-samples *t* test cannot be met. Rather than comparing means, which requires scale data, it uses the ranks of the values. Using ranks only requires that the data be measured at the *ordinal* level. However, the ultimate purpose of the Mann-Whitney U test is the same as that of the independent-samples t test—to search for statistical evidence that the sampled populations are significantly different.

OBJECTIVES

After completing this chapter, you will be able to

Describe the data assumptions appropriate for using the independent-samples t test

Write the research question and null hypothesis for the independent-samples t test

Input data for, conduct, and interpret the independent-samples t test using SPSS

Describe circumstances appropriate for the use of the Mann-Whitney U test

Conduct and interpret the Mann-Whitney U test using two different approaches

△   13.2 Research Scenario and Test Selection

The scenario involves an investigation meant to determine if two makes of automobiles obtain significantly different gas mileages. The dependent variable is the number of miles per gallon (mpg) for the Solarbird and the Ecohawk. Recall that the *dependent* variable is the variable that is subject to change as a result of the manipulation of the *independent* variable. In this example, the independent variable is the type of automobile, Solarbird or Ecohawk. The experiment will use 12 Solarbird and 12 Ecohawk automobiles, each driven over identical courses for 350 miles each. What would be the appropriate statistical test to determine if the Solarbird and the Ecohawk get significantly different average gas mileages (dependent variable)?

We know that miles per gallon is *scale* data, which is a requirement of the t test. We also understand that the test vehicles will be randomly selected from a wide range of dealerships in the western United States. Random selection is another requirement of the t test. Prior research has also shown that the values for miles per gallon follow a normal curve and that the variances are approximately equal. Given this information, we see that all data requirements for the *t test* have been met.

Only one question remains: Will the samples be independent? As the scenario is explained, there will be two samples taken from two independent populations of Solarbirds and Ecohawks. Thus, we will have two independent samples. Based on this information, we select the *independent-samples t test* for this investigation.

Before moving on to the next section, we address how samples in the independent-samples t test may be obtained. Often, the concern over how the two samples may be obtained is a source of confusion when attempting to select the appropriate t test.

In the scenario just presented, it is very clear that you have two populations; a random sample is obtained from each, and then the sample means are compared. However, the clarity presented in this exercise regarding the independence of the samples is not always the case. For example, another sampling method might require that you take two random samples from one population and then compare the means of the two samples. Yet another alternative would be to take one random sample and divide this sample into two groups, perhaps males and females, and compare the means of these two groups. Regardless of the sampling process, the major consideration for the independent-samples t test is that the measurements taken on the samples must be independent. Independence means that the measurement is taken on another individual or object (e.g., an automobile). This is in contrast with the hypothesis test known as the *paired-samples t test*, which is presented in Chapter 14. In the *paired-samples t-test*, the measurements are taken on the same individual or object but at different times and/or under different conditions.

13.3 RESEARCH QUESTION AND NULL HYPOTHESIS   △

Before reading the next sentence, it would be instructive for you to look away and attempt to visualize the researcher's question, or the reason for conducting the investigation on these two automobile models. The researcher's idea is that there are statistically significant differences in the average (mean) miles per gallon for the Solarbird and Ecohawk automobiles. For the purpose of the testing procedure, we refer to the researcher's idea as the alternative hypothesis, and we write it as H_A: $\mu_1 - \mu_2 \neq 0$. In plain language, this expression simply states that the difference between the population means (for *all* Solarbirds and Ecohawks) is not equal to zero. The alternative hypothesis agrees with the researcher's idea—that there are differences in average miles per gallon for the Solarbird and Ecohawk automobiles.

The null hypothesis states the opposite and is written as H_0: $\mu_1 - \mu_2 = 0$. Once again in plain language, the null hypothesis depicts the outcome that the difference between the average miles per gallon for the two populations of automobiles is equal to zero. Remember that if the null hypothesis is rejected, the researcher will have statistical evidence in support of the alternative hypothesis.

△ 13.4 DATA INPUT, ANALYSIS, AND INTERPRETATION OF OUTPUT

We begin this section by entering the variable information and the miles per gallon data for the 24 vehicles participating in this investigation. Solarbird and Ecohawk miles per gallon data are presented in Figure 13.1.

Figure 13.1   Miles per Gallon Data for the Independent-Samples *t* Test

Solarbird	34.5	36.2	33.2	37.0	32.7	33.1	30.5	37.2	33.5	32.0	36.2	35.7
Ecohawk	38.5	39.2	33.2	39.0	36.7	35.1	38.7	36.3	33.5	34.9	36.8	37.7

Next, we set up the SPSS database and then let the SPSS program do the hard work of analysis. The new database will consist of two variables: one for the miles per gallon data for all 24 vehicles and the other a *grouping* variable. The *grouping* (SPSS's term) variable simply labels the miles per gallon data as coming from a Solarbird or Ecohawk.

- Start SPSS, and click **Cancel** in the *SPSS Statistics* opening window.
- Click **File**, select **New**, and click **Data**.
- Click **Variable View**, type *mpg* (this is the name of the first variable), and then type *miles per gallon* in the cell below the *Label* column (select 2 decimal places and the *scale* level of measurement for this variable).
- Remain in the Variable View screen, and type *make* (this is the name of the second variable); then type *make of car* in the cell below the *Label* column (for this variable, set decimals to zero, and specify the *nominal* level of measurement in the *Measure* column).
- Click the right side of the cell below *Values*.
- The *Values* window opens; then type *1* in the *Value* box and *Solarbird* in the *Label* box, and click **Add**. Type *2* in the *Value* box and *Ecohawk* in the *Label* box, click **Add**, and then click **OK**.

- Click **Data View**, and type in all the "mpg" data (see Figure 13.1), beginning with *34.50* and ending with *37.70* (you should have data for all 24 cars—24 rows of data—entered in the first column of the Data View screen). Click the first cell below the "make" variable, and type *1* in the first 12 rows and *2* in the next 12 rows. (Now would be a good time to visually check the accuracy of your data entry.)
- Click **File**, then click **Save As** (the *Save Data As* window opens) in the *File name* box of the *Save Data As* window; type *miles_per_gallon* (note the underscores in the file name).
- Click **Save**.

You have now entered and saved the *miles_per_gallon* data for all vehicles. The data entry method just described, where you have all values for the dependent variable (miles per gallon) in one column, is required before SPSS will perform the test. Some may think that it would be more logical to enter two variables, one for Solarbird's miles per gallon and one for Ecohawk's miles per gallon, but SPSS does not work that way. Now let's do the fun part—the analysis—and see if we can discover the answer to the research question.

- Click **Analyze**, select **Compare Means**, and then click **Independent-Samples T Test**.
- Click **miles per gallon**, and then click the upper arrow (this moves the test variable to the right panel).
- Click **make of car**, and then click the lower arrow (this moves the *Grouping Variable* to the right panel, and the window should look like Figure 13.2; you may notice the question marks in the *Grouping Variable* box—don't worry as these will go away once you define the groups in the next step).
- Click **Define Groups** (the *Define Groups* window opens; see Figure 13.3).
- Click the *Group 1* box, and type *1* (1 is the *value* for the Solarbird).
- Click the *Group 2* box, and type *2* (2 is the *value* for the Ecohawk). (Completing this and the prior bullet point will eliminate the two *Grouping Variable* question marks seen in Figure 13.2.)
- Click **Continue**, and then click **OK** (the Output Viewer opens; see Figures 13.4 and 13.5).

The Output Viewer opens, displaying the tables shown in Figures 13.4 and 13.5. Figure 13.4 summarizes the descriptive statistics (*mean, standard deviation*, and *standard error*) for the "miles per gallon" variable. At first glance, we see that the Solarbird attained an average of 34.3167 mpg,

Figure 13.2   The *Independent-Samples T Test* Window

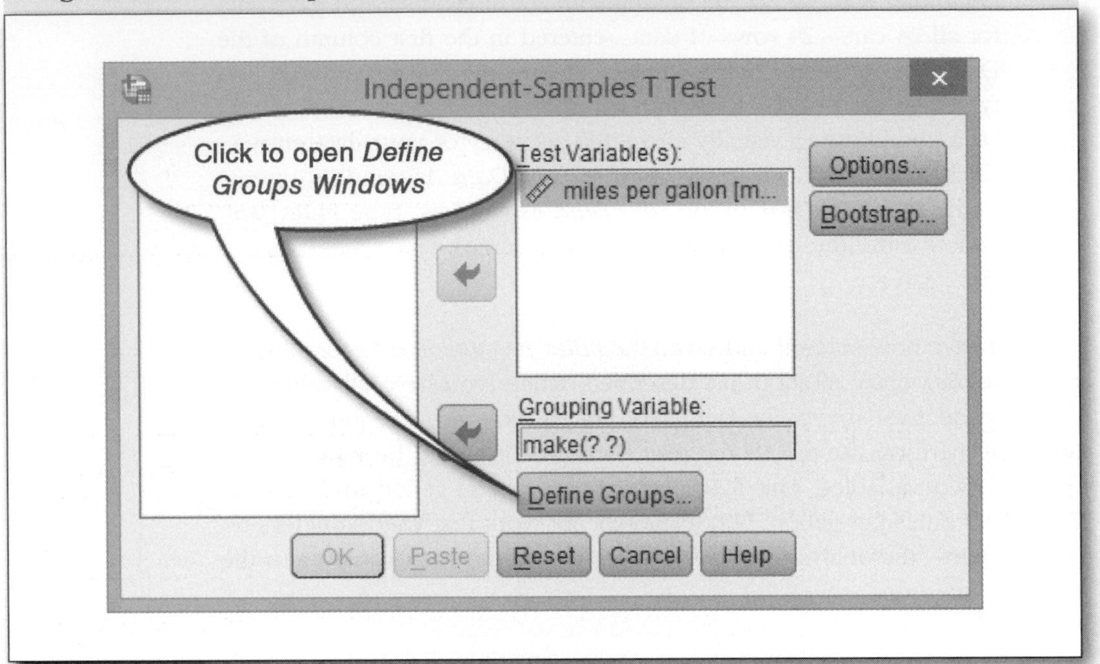

Figure 13.3   The *Define Groups* Window

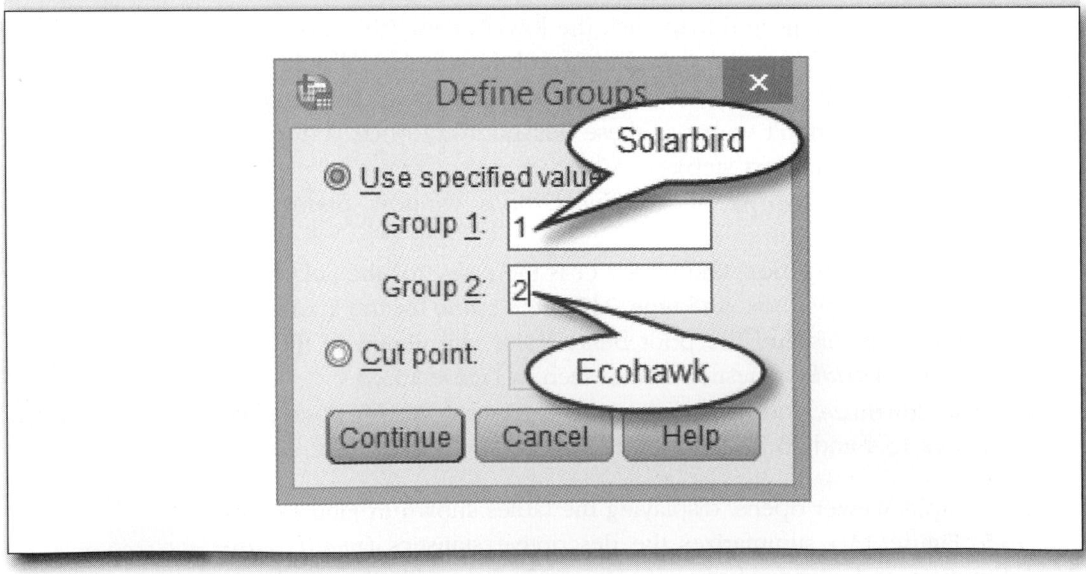

Figure 13.4   Group Statistics for Miles per Gallon

Group Statistics

	make of car	N	Mean	Std. Deviation	Std. Error Mean
miles per gallon	Solarbird	12	34.3167	2.14257	.61851
	Ecohawk	12	36.6333	2.09299	.60419

Figure 13.5   Independent-Samples *t* Test for Miles per Gallon

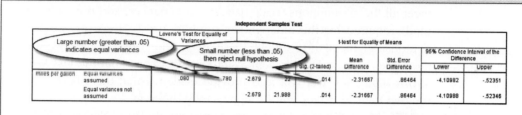

whereas the Ecohawk managed an average of 36.6333 mpg. The standard deviations for average miles per gallon are close to being identical (2.14257 and 2.09299, respectively).

Figure 13.5 provides the answer to the research question. The researcher's idea is that there is a significant difference in the average miles per gallon for the Solarbird and Ecohawk automobiles. What does the *independent-samples t test* tell us?

When looking at Figure 13.5, we see that the 95% confidence interval of the difference (last two columns) ranges from −4.10982 to −.52351 when *equal variances are assumed*, which is the case for these data (Sig. = .780). Furthermore, we note that our calculated difference of −2.31667 is within this range of the confidence interval.

The significance level of .014 informs us that it is very unlikely that the observed mean absolute difference of 2.31667 was due to chance. Specifically, we can state that there was a .014 probability that the observed difference was the result of chance and that the null hypothesis can be rejected. Another way to state this is that there is only a 1.4% (100 × .014) chance that the differences in gas mileage could be attributed to chance. The difference can be taken seriously, and there are significant differences in average miles per gallon attained by the Solarbird and Ecohawk. We further conclude that the Ecohawk gets superior gas mileage (36.63 mpg) when compared with the Solarbird's average of 34.32 mpg.

From the researcher's standpoint, we can say that the investigation was a success. The null hypothesis of no difference was rejected, and the researcher now has statistical evidence in support of the idea that these two makes of automobiles have significantly different rates of gas consumption.

△   13.5 NONPARAMETRIC TEST: MANN-WHITNEY *U* TEST

The miles per gallon data for the Solarbirds and Ecohawks were found to meet all the assumptions (scale data, equal variances, and normally distributed) required for the independent-samples *t* test. However, we wish to demonstrate the SPSS test when these assumptions are not met. We will use the same database for this demonstration. We expect that the less sensitive *Mann-Whitney U* test will also provide evidence that the gas mileage for the two vehicle makes will be significantly different.

The Mann-Whitney *U* test is the nonparametric test selected as the alternative to the independent-samples *t* test. The Mann-Whitney *U* test uses data measured at the *ordinal* level. Thus, SPSS *ranks* the miles per gallon scale data and then performs the statistical operations. Recall that it is a legitimate statistical manipulation to transform data from the higher levels of measurement (in this case *scale*) to lower levels (*ordinal*). The observations from both groups are combined and ranked, with the average rank assigned in the case of ties. If the populations are identical in location, then the ranks for miles per gallon should be randomly mixed between the two samples.

The *alternative hypothesis* is that the distributions of miles per gallon (ranks) are not equally distributed between the Solarbirds and Ecohawks. The *null hypothesis* is that the distributions of miles per gallon (ranks) are equal for both the Solarbirds and the Ecohawks.

Follow the bullet points to use the Mann-Whitney *U* test to discover if the null hypothesis can be rejected.

- Open SPSS, and open *miles_per_gallon.sav*.
- Click **Analyze**, select **Nonparametric Tests**, and then click **Independent Samples**.
- Click **Field**s in the *Nonparametric Tests: Two or More Independent Samples* window.
- Click **miles per gallon**, and then click the upper arrow.
- Click **make of car**, and then click the lower arrow (a window now appears as shown in Figure 13.6).

Figure 13.6 Nonparametric Tests: Two or More Independent Samples

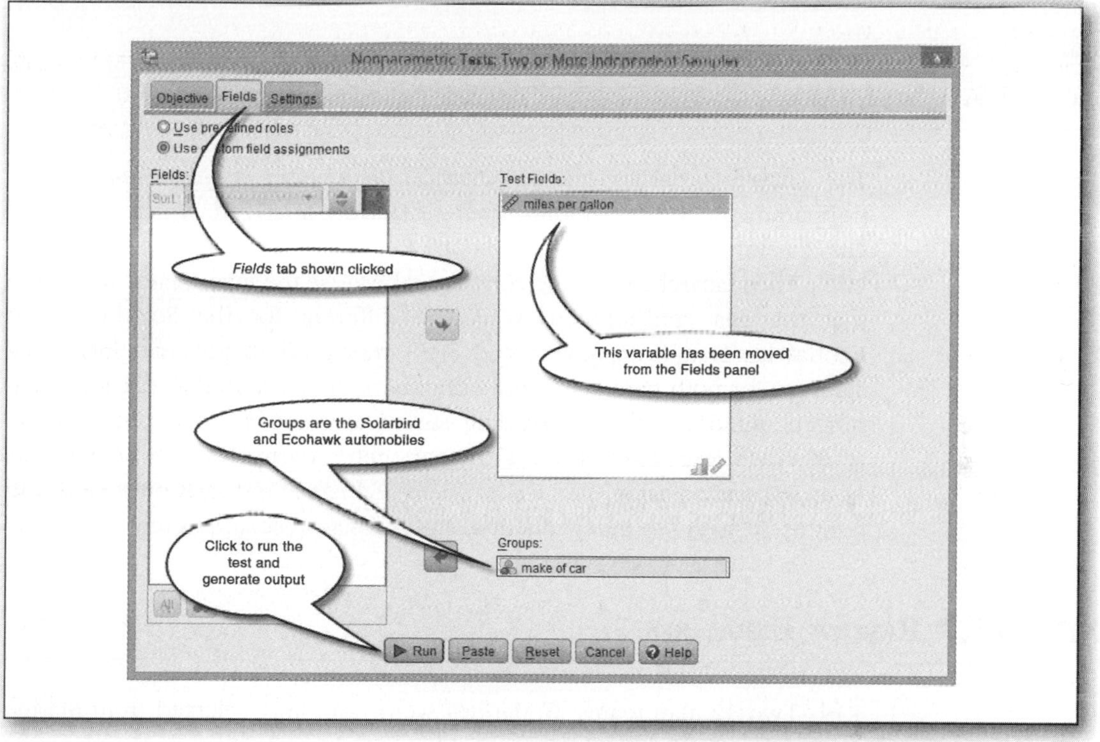

Figure 13.7 Hypothesis Test Summary for Miles per Gallon

Hypothesis Test Summary

	Null Hypothesis	Test	Sig.	Decision
1	The distribution of miles per gallon is the same across categories of make of car.	Independent-Samples Mann-Whitney U Test	.014[1]	Reject the null hypothesis.

Asymptotic significances are displayed. The significance level is .05.

[1]Exact significance is displayed for this test.

- Click **Run** (the Output Viewer window opens; see Figure 13.7).

Figure 13.7 shows the SPSS output from the less sensitive, nonparametric Mann-Whitney *U* test. This test also found a significant difference (level of .014) in the miles per gallon attained by these two automobiles. The null hypothesis of equality was rejected just as in the independent-samples *t* test.

△ 13.6 Summary

This chapter presented the parametric *independent-samples t test* and the nonparametric *Mann-Whitney U test*. Data were analyzed, which compared gas consumption (in miles per gallon) for two different makes of automobiles. The research was being conducted because the investigator suspected that the gas consumption would be different for the Solarbirds and Ecohawks. Data were given, and SPSS was used to generate inferential statistics for both parametric and nonparametric tests. Both indicated a significant difference, thus providing statistical evidence in support of the researchers' idea. In Chapter 14, the parametric *paired-samples t test* and its nonparametric analog, the *Wilcoxon signed-ranks* test, are presented and contrasted with the independent-samples tests presented here.

△ 13.7 Review Exercises

13.1 Two 12-man teams of Marines were randomly selected from Marine Corps Air Stations Miramar and Yuma to be compared on their Combat Fitness Test. Their scores ranged from a low of 263 to a perfect score of 300. Miramar scores: 267, 278, 295, 280, 268, 286, 300, 276, 278, 297, 298, and 279. Yuma scores: 263, 272, 286, 276, 267, 284, 293, 270, 272, 296, 279, and 274. The Yuma team leader and researcher had the idea that the scores were unequal. Can you help the Yuma team leader write the null and alternative hypotheses and select the appropriate test(s) to see if there is evidence in support his idea?

13.2 The local bank president had the idea that the money held in individual savings accounts would be significantly different for males and females. A random sample of the dollars in male and female savings accounts was recorded as follows. Males: 5,600, 5,468, 5,980, 7,890, 8,391, 9,350, 10,570, 12,600, 8,200, 7,680, 6,000, and 8,900. Females: 4,900, 5,200, 5,000, 7,000, 8,000, 9,050, 9,900, 12,000, 8,000, 7,500, 5,900, and 8,500. Write the null and alternative hypotheses, and select the correct test to seek evidence in support of the bank president's contention that male and female saving habits are significantly different.

13.3 For this review exercise, you will select and open the SPSS sample file called *bankloan.sav*. You will test for significant differences in the categories of education ("ed") and whether they have previously defaulted on a loan ("default"). There are five educational categories and two for the "default" variable. Write the alternative hypothesis and null hypothesis, and use the appropriate statistical test to see if the distribution of levels of education is the same for the categories that had previously defaulted.

CHAPTER 14

PAIRED-SAMPLES *t* TEST AND WILCOXON TEST

△ **14.1 INTRODUCTION AND OBJECTIVES**

We continue the theme of the previous two chapters by showing how SPSS is used to accomplish hypothesis testing. Chapters 12 and 13 covered the *one-sample t test* and the *two-sample independent t test*. Also addressed were their nonparametric alternatives, the *binomial test of equality* and the *Mann-Whitney U test*.

This chapter presents the *paired-samples t test*, which compares measurements (means) taken on the same individual or on the same object but at different times. First addressed is the *parametric* test—the *paired-samples t test*. The *nonparametric* alternative, the *Wilcoxon test*, compares ranks for the two measurements and is covered later in the chapter.

The purpose of the paired-samples test is to test for significant differences between the means of two related observations. Usually, the test is used when an individual or object is measured at two different times. Such an investigative approach is often described as a *pretest* and *posttest* research methodology. A mean value is determined, which is the *pretest* measurement; some action intervenes, such as an educational lecture, and another mean is calculated, which is the *posttest* measure. Using this methodology, we look for a significant difference between the means of our pretest and posttest. If the investigation was well designed, meaning that all other factors were controlled for, then we can attribute significant differences to the intervening action. In the current example, the intervening action is the educational lecture. A significant difference in scores (hopefully, the posttest scores were greater) between the means of the pretest and posttest would provide evidence that learning took place as a direct result of the lecture.

The paired-samples *t* test can be used in many different situations, not just when measuring human performance. One example might be to select a random sample of Harley Davidson motorcycles and record the miles per gallon for Brand X gasoline. You could then drain the tanks and refill with Brand Y, record the miles per gallon, and look for a significant difference between miles per gallon for Brand X and Brand Y.

When our distribution of differences fails to meet the criteria of normality, the Wilcoxon signed-ranks test is the nonparametric alternative. The Wilcoxon test demonstrated in this chapter compares two sets of values (scores) that come from the same individual or object. This situation occurs when you wish to investigate any changes in scores over a period of time. The difference between the paired-samples *t* test and the Wilcoxon test is that the latter does not require normally distributed difference data. The Wilcoxon test is able to use data measured at the *ordinal*, *interval*, or *ratio* level. Therefore, in our main example in this chapter, which measures blood pressure at the interval level, we can rank the data (which become ordinal) and use the Wilcoxon test to see if the median ranks are significantly different.

OBJECTIVES

After completing this chapter, you will be able to

Describe the data assumption requirements when using the paired-samples *t* test

Write a research question and alternative and null hypotheses for the paired-samples *t* test

Use SPSS to answer the research question for the paired-samples *t* test

Describe the circumstances appropriate for using the Wilcoxon nonparametric test

Use SPSS to conduct the Wilcoxon test

Interpret SPSS's output for the Wilcoxon test

△ 14.2 Research Scenario and Test Selection

Whenever human subjects are used in research, it is necessary that the principal investigator obtain approval to ensure that the participants are not harmed. We are not sure if the research scenario we are about to describe would pass a review by a Human Subjects Committee, but let's assume that it did pass. The basic concept is that an unexpected violent and chaotic event would significantly change human systolic blood pressure readings. In other words, the researcher intends to measure blood pressure (*pretest*), scare the heck out of human subjects, and then measure their blood pressure (*posttest*) once again.

Twelve men aged 20 to 25 years were randomly selected from a group of recruits who had volunteered for service in the U.S. Marine Corps. Early in their training, they were ordered to report to a barracks, and their individual systolic blood pressures were recorded by 12 technicians. In this research approach, the pretest measurements are the systolic blood pressure readings. Following the measurement procedure, the recruits were instructed to relax, and during this period of "relaxation," a realistic recording of nearby gunfire, explosions, and screaming was played from loudspeakers just outside the barracks. The simulated chaos is called the intervening variable or stimulus in this type of research methodology. The recruits were commanded to take cover and take no action. The simulated chaos continued for several minutes until the "all clear" was announced. The young recruits were then measured for systolic blood pressure levels by the same 12 technicians. This was the posttest measurement. What is the appropriate test for this research scenario?

The reasoning process for test selection should start with a consideration of the level of measurement. Since we have *scale* data, we are able to look for differences between means; therefore, you can consider one of the *t* tests. Since we have pairs of measurements that were taken on the same subject (pretest and posttest on each recruit), we select the *paired-samples t test*. For the paired-samples test, the differences between the pairs of observations are assumed to be distributed normally. The larger the sample size, the more likely that this assumption of normality will be met. In spite of our small sample ($n = 12$) in the current example, we assume normality and proceed with the test.

△ 14.3 Research Question and Null Hypothesis

As you may have guessed, the researcher's idea is that the pretest and posttest blood pressure readings will be significantly different. Stated in statistical terms, we write the alternative hypothesis as H_A: $\mu_1 - \mu_2 \neq 0$. This expression states that the difference between the population means for our two sample measurements is not equal to 0. If you discover that the means are not equal, then the data agree with the researcher's idea. You now have evidence of significant differences in the pretest and posttest blood pressure readings.

Note: The reader may notice that the researcher's idea would most likely be that there would be an increase in blood pressure readings following the simulated violent chaos. This is technically known as a directional or one-sided hypothesis. In most cases, when using SPSS, it is unnecessary to specify directionality. In the current example, if significance is found, we can look at whether there was an increase or decrease in blood pressure. We are then justified in stating that there is either a significant increase or decrease in blood pressure following the simulated chaos. Such directionality is illustrated when the analysis of our data is completed and SPSS's output is interpreted.

Remember that the null hypothesis states the opposite of the H_A. Therefore, the null hypothesis is written as H_0: $\mu_1 - \mu_2 = 0$. This expression states that the difference between the population means is equal to 0. The researcher is investigating the scenario to determine if the null hypothesis can be rejected. If the null hypothesis is rejected, then we have some evidence that the alternative hypothesis is true. In that, this would mean that there is a significant difference in the blood pressure readings.

14.4 DATA INPUT, ANALYSIS, AND INTERPRETATION OF OUTPUT △

We begin our demonstration of the *paired-samples t test* by entering our variable information and the two systolic blood pressure readings for the 12 recruits participating in this experiment.

- Start SPSS, then click **Cancel** in the *SPSS Statistics* opening window.
- Click **File**, select **New**, and click **Data**.
- Click **Variable View**; in Column 1, Row 1, type *bl_pres1*; then type *Blood Pressure (pretest)* in the *Label* column, set decimals to 0, and change the *Measure* column to read *scale*.
- Remain in *Variable View* and in Column 1, Row 2 type *bl_pres2* and then type *Blood Pressure (posttest)* in the *Label* column and set decimals to 0 and change the *Measure* column to read *scale*.
- Click **Data View**, click the cell for Row 1 and Column 1, and type all the data for the variable "bl_pres1," beginning with *122* and ending with *129*. You should have 12 values in Column 1 when finished (see Figure 14.1).

Figure 14.1   Data View for Recruits' Pretest and Posttest Blood
Pressure Readings

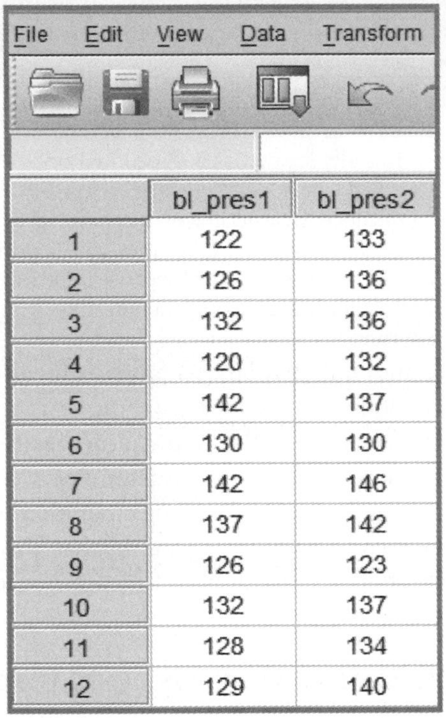

File	Edit	View	Data	Transform		
		bl_pres1	bl_pres2
1	122	133
2	126	136
3	132	136
4	120	132
5	142	137
6	130	130
7	142	146
8	137	142
9	126	123
10	132	137
11	128	134
12	129	140

- Click the cell for Row 1 and Column 2, and type all the data for the variable "bl_pres2," beginning with *133* and ending with *140*. You should have 12 values in Column 2 when finished. (Check to see if your data entry now looks exactly like Figure 14.1.)
- Click **File**, click **Save As** (the *Save As* window opens), and type *Blood Pressure* in the *File name* box of the *Save As* window.
- Click **Save**.

You have now entered and saved the blood pressure data that were observed for the 12 military recruits. We now do the analysis.

- Click **Analyze**, select **Compare Means**, and then click **Paired-Samples T Test** (the *Paired-Samples T Test* window opens; see Figure 14.2).
- Click **Blood Pressure (pretest)**, and then click the arrow (the variable is moved to the panel on the right).

Figure 14.2 Paired-Samples *t* Test—Blood Pressure Experiment

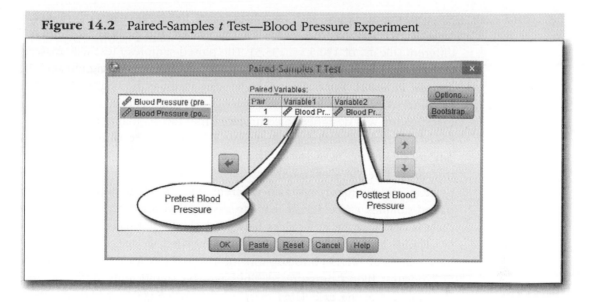

- Click **Blood Pressure (posttest)**, and then click the arrow (the variable is moved to the panel on the right).
- Click **OK** (the Output Viewer opens, displaying Figures 14.3 and 14.4).

Figure 14.3 Paired-Samples *t* Test—Pretest and Posttest Blood Pressure Statistics

Paired Samples Statistics

		Mean	N	Std. Deviation	Std. Error Mean
Pair 1	Blood Pressure (pretest)	130.50	12	7.026	2.028
	Blood Pressure (posttest)	135.50	12	5.916	1.708

Figure 14.4 Paired-Samples *t* Test for the Blood Pressure Experiment

Paired Samples Test

| | | Paired Differences | | | | | | | |
| | | | | | 95% Confidence Interval of the Difference | | | | |
		Mean	Std. Deviation	Std. Error Mean	Lower	Upper	t	df	Sig. (2-tailed)
Pair 1	Blood Pressure (pretest) - Blood Pressure (posttest)	-5.000	5.543	1.600	-8.522	-1.478	-3.125	11	.010

Reject the null hypothesis

Looking at Figure 14.3, which presents the basic statistics for each variable, we see that the means for the pretest and posttest are indeed different (means of 130.5 vs. 135.5). The paired-samples *t* test will answer the question whether the observed difference is actually "significant." If our test indicates significance, then we can attribute the change in blood pressure to the simulated violent chaos. The reader should note that the investigation was designed to control for other factors that may have influenced the recruits' blood pressure levels. If our test does not find a significant difference, then any change in blood pressure could be attributed to chance. The *Std. Deviations (standard deviations)* for both pretest and posttest measurements are also different. This is not a problem for the paired-samples *t* test. Recall that the assumption for the paired-samples *t* test is only that the differences between pairs be normally distributed.

Figure 14.4 presents the information needed to decide whether we have statistical evidence to support the researcher's idea that the pretest and posttest blood pressure readings will be significantly different. We first notice that the calculated sample mean of −5.000 is within the 95% confidence limits of −8.552 (lower) and −1.478 (upper). Then we look at the last column, *Sig. (2-tailed)*; we find the value .010. This value tells us that the probability is .010 and that the observed difference is due to chance. The probability, stated as a percentage (100 × .010), permits us to say that there is only a 1% chance that the observed difference resulted from chance. Based on the results of the paired-samples *t* test, we now have statistical evidence that the simulated violent chaos resulted in a significant blood pressure change—the investigation was a success. The direction of the blood pressure change (increase or decrease) can be easily determined by looking at the mean pretest and posttest readings given in Figure 14.3. It is evident that there was an increase in mean blood pressure readings (from 130.5 to 135.5) following the simulated violent chaos (stimulus). Since the *t* test found a significant difference, we may now say that there was a *significant* increase in systolic blood pressure following the simulated chaos.

Let's now examine the same data with the assumption that they do not meet the requirements of the parametric paired-samples *t* test.

△   14.5 Nonparametric Test: Wilcoxon Signed-Ranks Test

Recall that the *paired-samples t test* assumes that the differences between the pairs are normally distributed. If the differences between the pairs greatly differ from normality, it would be a good idea to use the *Wilcoxon signed-ranks test.* The Wilcoxon test converts the blood pressure *interval* data to the *ordinal* level of measurement. It then calculates the medians of the ranks for the pretest and posttest groups, after which it tests to see if there is a significant difference between the medians of the pretest and posttest categories.

The alternative and null hypotheses remain the same as with the paired-samples *t* test. The alternative hypothesis states that the difference is *not* equal to 0, whereas the null hypothesis states that it does equal 0.

Follow the bullet points to see what the Wilcoxon signed-ranks test informs us regarding the rejection of the null hypothesis.

- Start SPSS, click **File**, select **Open**, and click **Data**.
- Click **Blood Pressure.sav**, and then click **Open**.
- Click **Analyze**, select **Nonparametric Tests**, and then click **Related Samples** (the *Nonparametric Test: Two or More Related Samples* window opens; see Figure 14.5). (Note that the window shown in

Figure 14.5 The *Nonparametric Test: Two or More Related Samples* Window

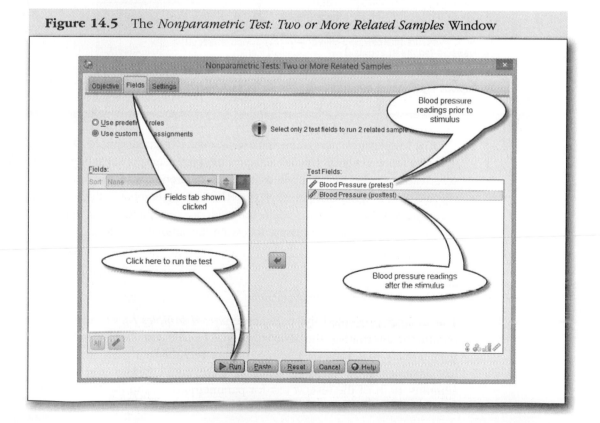

Figure 14.5 shows what you see after the Fields tab is clicked and the variables have been moved to the *Test Fields* panel—see the next two bullet points.)

- Click the Fields tab at the top of the window if it is not already done (the window does not appear as in Figure 14.5).
- Click **Blood Pressure (pretest)**, and then click the arrow. This moves the variable to the *Test Fields* panel.
- Click **Blood Pressure (posttest)**, and then click the arrow. This moves the variable to the *Test Fields* panel (the *Nonparametric Test: Two or More Related Samples* window should now appear as shown in Figure 14.5).
- Click **Run** (the Output Viewer opens; see Figure 14.6).

Figure 14.6   Hypothesis Test: Nonparametric Wilcoxon Signed-Rank Test

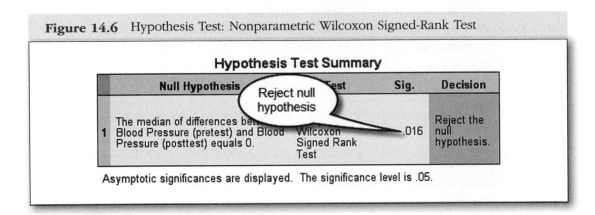

The less sensitive nonparametric test alternative (related-samples Wilcoxon test) also found a significant difference for the pretest and posttest blood pressure readings. The null hypothesis that the difference was equal to 0 was rejected with a significance level of .016. This is in agreement with the paired-samples *t* test. We now have additional evidence that there was a significant increase in systolic blood pressure following the simulated violent chaos.

△   14.6 SUMMARY

This chapter presented the parametric *paired-samples t test* and the non-parametric alternative, the *Wilcoxon signed-ranks test*. Data were derived from a study that subjected young military recruits to a frightening stimulus. Systolic blood pressure readings were taken before and after the stimulus. Both of the tests (*t* test for parametric data and the Wilcoxon for nonparametric data) found a significant change in blood pressure following the introduction of the stimulus. The data input, analysis, and interpretation of SPSS's output were discussed and explained in relationship to the researcher's reason for conducting the investigation. Chapter 15 continues with hypothesis testing but addresses the situation where you have three or more means to compare. To accomplish the testing of three or more means for significance, you will use *analysis of variance* and the *Kruskal-Wallis H test*.

△ 14.7 REVIEW EXERCISES

14.1 The researcher has the idea that listening to hard rock music can directly influence one's perception of the world. The professor randomly selects a group of 15 individuals from his "Introduction to Psychology" lecture hall class of 300 freshman students. He wishes to test his theory on his sample of 15 students by giving them the World Perception Test (WPT), having them listen to loud hard rock music, and then regiving the WPT. The test results are as follows. The WPT premusic scores are 16.34, 16.67, 17.18, 16.73, 16.37, 16.91, 17.32, 16.61, 16.67, 17.23, 17.26, 16.70, 16.43, 17.32, and 16.52. The WPT postmusic scores are 16.62, 16.49, 16.91, 16.61, 16.34, 16.85, 17.12, 16.43, 16.49 17.20, 16.70, 16.55, 16.25, 17.06, and 16.37. Write the null and alternative hypotheses, and see if you can produce statistical evidence in support of the professor's idea that listening to hard rock music can actually change your perception of the world.

14.2 Data were collected from 10 major oil well drilling operators that recorded the number of hours lost per week due to work-related accidents. A rigorous safety program was instituted, and the number of lost hours was once again recorded following the introduction of the safety program. The presafety program values are 41, 55, 63, 79, 45, 120, 30, 15, 24, and 24. The postsafety program values are 32, 49, 50, 73, 43, 115, 32, 9, 22, and 19.

A research consultant was hired to examine the data and determine if the safety program significantly changed the weekly hours lost from on-job injuries. Write the null and alternative hypotheses, and select and conduct the appropriate test to seek evidence that the program was successful.

14.3 The chemical engineer added a chemical to a fast-burning compound that changed the oxygen consumption once the reaction started. He did a pretest to measure the oxygen consumption index; then he added the chemical and recorded the posttest oxygen index. The pretest values are 120, 139, 122, 120, 124, 120, 120, 125, 122, 123, 126, and 138. The posttest values are 121, 140, 123, 121, 125, 122, 121, 126, 123, 124, 127, and 139. Write the null and alternative hypotheses, select the correct test, and look for differences between the pretest and posttest oxygen index values.

15

One-Way ANOVA and Kruskal-Wallis Test

15.1 Introduction and Objectives △

We continue with the theme of the prior three chapters in that we use SPSS to accomplish hypothesis testing. Let's list the hypothesis tests covered so far, with their nonparametric alternatives given in parentheses: *one-sample t test* (binomial), *two-sample independent t test* (Mann-Whitney), and *paired-samples t test* (Wilcoxon). You may have noticed that in all these tests we investigated various assertions about *two* means. The hypothesis test presented in this chapter also involves the testing of means, the difference being that we now have three or more independent group means that we wish to test for significant differences. The appropriate test for three or more means is the ANOVA, an acronym for *analysis of variance*. Don't let the name of the test, analysis of *variance*, mislead you into thinking that it does not represent another method of comparing the means of various groups. It is a procedure that tests for significant differences between three or more means. It determines significance via calculation of the F statistic. The value of F is calculated when the variance of the total group is compared with the variances of the individual groups. We refer the reader to any basic statistical text for details on the underlying mathematics used with ANOVA.

With ANOVA, you have one *independent* variable. Recall that the independent variable is thought to have an effect on the dependent variable. The *independent* variable may be measured at the *scale* (*interval/ratio*) or *categorical* (*nominal/ordinal*) level. The *dependent* variable must be measured at the *scale* level. The one independent variable may consist of any number of groups but at least three. Each group of the independent variable represents a unique treatment. These treatments are most often referred to as *levels* of the independent variable. Variables designated as string (alphabetic/nominal) in SPSS are permissible as the independent variable for the ANOVA procedure.

For example, we have one independent variable, "antibacterial spray," but we have four different brands of the spray, labeled A, B, C, and D. The *levels* of the independent variable are A, B, C, and D; they could also be referred to as four different treatment groups. The dependent variable could be the number of bacteria killed by an application of the various *levels* of the *one* independent variable, "antibacterial spray."

It is important to note that a significant *F* statistic, resulting from the ANOVA test, only tells you that there is a significant difference between at least two group means, while not identifying which two are different. To answer which of the pairs are significantly different, we must conduct *post hoc* analysis. Post hoc simply means "after the fact." Once significance is established, then we do additional work (after the fact) to identify which of the multiple pairs of means contributed to the significant *F* statistic. SPSS makes this additional work easy.

OBJECTIVES

After completing this chapter, you will be able to

Describe the data assumptions required for the ANOVA hypothesis test

Write a research question, alternative hypothesis, and null hypothesis for ANOVA

Use SPSS to answer the research question for the ANOVA procedure

Conduct post hoc analysis (Scheffe test) for ANOVA when significance is discovered

Describe the data characteristics required for the Kruskal-Wallis nonparametric test

Use SPSS to conduct the Kruskal-Wallis test

Interpret SPSS's output for the Kruskal-Wallis test

As in the previous chapters, we also present a nonparametric analog when data characteristics fail to meet the minimum data requirements for using the ANOVA test. The *Kruskal-Wallis* test is the nonparametric alternative often used when the level of measurement is *ordinal*. The Kruskal-Wallis approach is to rank the data and compare the median of the ranks for all groups with the individual group medians. The mathematical procedures are similar to those used in ANOVA, except that now we compare medians rather than means. If the Kruskal-Wallis test identifies overall significance, then SPSS can examine each pair for significance. This additional analysis is known as *pairwise comparisons* and is a type of post hoc study, as mentioned in the previous paragraph.

15.2 RESEARCH SCENARIO AND TEST SELECTION △

The Municipal Forest Service (MFS) had the idea that the monies spent to maintain five disparate mountain hiking trails in the Santa Lucia Mountains, California, were significantly different. Random samples of such expenditures were taken from the records for the past 50 years. Six years were selected at random and the expenditures for trail maintenance recorded. The MFS wished to use the findings of their analysis to help them allocate resources for future trail maintenance.

The dependent variable is the amount of money spent on each of the five hiking trails. The independent variable, "hiking trails," consists of five separate trails. Since the data were measured at the *scale* level, we may calculate means and standard deviations for the amount of money spent. The mean expenditures for each trail were based on the random samples of size six ($n = 6$). We assume that the distributions of trail expenditures are approximately normally distributed with equal variances. Which statistical test would you use to develop evidence in support of the MFS's belief that the amount of money spent on trail maintenance was significantly different for the five trails?

We have scale data for our dependent variable ("expenditures in dollars"); therefore, we are able to look for the differences between means. We can eliminate *t* tests from consideration since we have more than two means. At this point, the ANOVA, designed to compare three or more means, seems like a worthy candidate. To use ANOVA, there must be random selection for sample data—this was accomplished. The distributions of expenditures appear to approximate the normal curve, and their variances approximate equality. Based on this information, we select the ANOVA. If the ANOVA results in significance, the *Scheffe post hoc* analysis will be used to identify which pairs of means contributed to the significant *F* value.

△ 15.3 Research Question and Null Hypothesis

The researcher's idea, and reason for conducting this investigation, is the belief that there is a significant difference in maintenance expenditures for the five hiking trails. Recall that the alternative hypothesis (H_A) is simply a restatement of the researcher's idea. We write the following expression for the alternative hypothesis:

H_{A1}: One or more of the five hiking trails have unequal maintenance expenditures.

The null hypothesis (H_0) states the opposite and is written as

$$H_{01}: \mu_1 = \mu_2 = \mu_3 = \mu_4 = \mu_5.$$

The expression H_{01} states that there are no differences between the means of the populations. The MFS researcher would prefer to reject the null hypothesis, which would then provide statistical evidence for the idea that maintenance expenditures for the five trails are indeed significantly different.

If there is evidence of overall significance, leading to the rejection of the null hypothesis (H_{01}), the researcher would then wish to identify which of the five groups are different and which are equal. The following null and alternative hypotheses will facilitate that task.

If the null hypothesis (H_{01}) is rejected, then the following null hypotheses should be tested:

$$H_{02}: \mu_1 = \mu_2,\ H_{03}: \mu_1 = \mu_3,\ H_{04}: \mu_1 = \mu_4,\ H_{05}: \mu_1 = \mu_5,\ H_{06}: \mu_2 = \mu_3,$$
$$H_{07}: \mu_2 = \mu_4,\ H_{08}: = \mu_2 = \mu_5,\ H_{09}: \mu_3 = \mu_4,\ H_{10}: \mu_3 = \mu_5,\ H_{011}: \mu_4 = \mu_5.$$

The alternative hypotheses for these new null hypotheses follow:

$$H_{A2}: \mu_1 \neq \mu_2,\ H_{A3}: \mu_1 \neq \mu_3,\ H_{A4}: \mu_1 \neq \mu_4,\ H_{A5}: \mu_1 \neq \mu_5,\ H_{A6}: \mu_2 \neq \mu_3,$$
$$H_{A7}: \mu_2 \neq \mu_4,\ H_{A8}: \mu_2 \neq \mu_5,\ H_{A9}: \mu_3 \neq \mu_4,\ H_{A10}: \mu_3 \neq \mu_5,\ H_{A11}: \mu_4 \neq \mu_5.$$

△ 15.4 Data Input, Analysis, and Interpretation of Output

We begin by setting up a new database and then entering the expenditures and canyon data recorded by the MFS. SPSS requires that the data be entered as two variables, one listing the expenditures and the other that specifies the canyon (grouping variable). The procedure to accomplish the data entry is provided in the bullets following Figure 15.1. The amount of money, in thousands of dollars, spent to maintain each of the five trails is given in Figure 15.1.

Figure 15.1 Municipal Forest Service Trail Maintenance in Thousands of Dollars (Santa Lucia Mountains)

| Trail | Six Randomly Selected Years | | | | | |
	Year 1	Year 2	Year 3	Year 4	Year 5	Year 6
Eaton Canyon (1)	20	11	17	23	29	19
Grizzly Flat (2)	11	3	8	13	11	12
Rattlesnake (3)	7	5	3	9	8	14
Bailey Canyon (4)	3	3	1	3	8	14
Millard Canyon (5)	16	21	16	11	6	16

Let's next create a new database in a format compatible with SPSS's ANOVA procedure.

- Start SPSS, and click **Cancel** in the *SPSS Statistics* opening window.
- Click **File**, select **New**, and click **Data**.
- Click **Variable View**, and type *Dollars* in the cell found in Row 1, Column 1; then type *Dollars (in thousands) spent on trail maintenance* in the cell beneath the *Label* column, set decimals to 0, and set *Measure* to read *scale*.
- Type **Trail** in the cell found in Row 2, Column 1; then, type *Name of Trail* in the cell beneath the *Label* column heading.
- Click the right side of the cell in Row 2 in the *Values* column (the *Value Labels* window opens).
- Add the trail information in the *Value Labels* window as follows: Begin by typing *1* in the *Value* box and *Eaton* in the *Label* box, then click **Add**. Repeat this process for the remaining trails, where *2 = Grizzly Flat, 3 = Rattlesnake, 4 = Bailey Canyon,* and *5 = Millard Canyon* (the *Value Labels* window should now appear as shown in Figure 15.2).
- Once all the trail information is entered, click **OK**; set decimals to 0, and change *Measure* to read *nominal* for this "Trail" variable.

Figure 15.2   *Value Labels* Window for the ANOVA Example

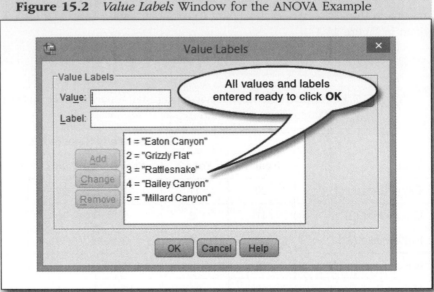

- Click **Data View**, and click the cell found in Row 1, Column 1; then type data for *Eaton Canyon* in Rows 1 through 6: *20*, *11*, *17*, *23*, *29*, and *19*. Repeat this process for the remaining data that are shown in Figure 15.3. Enter data in the cell found in Row 7, Column 1 for *Grizzly Flat*, and continue until all the "Dollar" variable data are entered. When finished, there should be 30 values in Column 1 of this new database.
- Click the cell beneath the "Trail" variable, and type *1* for the six rows of this variable, *2* for the next six rows, *3* for the next six rows, *4* for the next six rows, and finally *5* for the last six rows of data (see Figure 15.3 for the data entry procedure for the first 10 cases).
- Click **File**, click **Save As**, and then type *trail*.
- Click **Save**.

You have now entered and saved the MFS expenditure data for the five trails of interest. Let's proceed to the analysis using the power of the SPSS ANOVA procedure.

- Click **Analyze**, select **Compare Means**, and click **One-way ANOVA**.
- Click **Dollars (in thousands)**, and then click the upper arrow (which moves the variable to the *Dependent List* box).
- Click **Name of Trail (Trail)**, and then click the lower arrow (this click moves the variable to the *Factor* box, and the window should look like Figure 15.4).

Figure 15.3   Data View for One-Way ANOVA (First 10 of 25 Cases Shown)

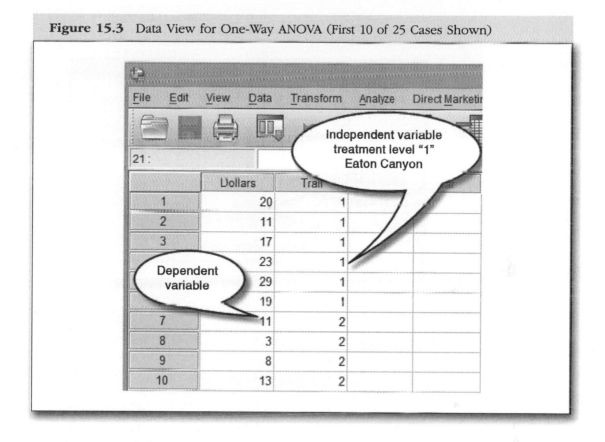

- Click **Post Hoc** (the *One-Way ANOVA: Post Hoc Multiple Comparisons* window opens, as shown in Figure 15.5).
- Click **Scheffe**, click **Continue**, and then click **OK** (the Output Viewer opens; see Figures 15.6 and 15.7).

The most important thing to note in Figure 15.6 is that SPSS found a significant difference between at least one of the 10 pairs of means. This is shown by the .000 found in the *Sig.* column. In other words, we now have statistical evidence in support of the researcher's idea that the expenditures for these trails were not equal. As in the prior tests for differences between means, significance is indicated by the small value (.000) shown in the *Sig.* column in Figure 15.6. (*Note:* SPSS rounds the *Sig.* number to three places; therefore, the probability that the observed differences are due to chance is actually not 0 but less than .0005.)

If the researcher were doing this analysis by hand, the real work of identifying which trails were significantly different would now begin.

Figure 15.4   The *One-Way ANOVA* Window

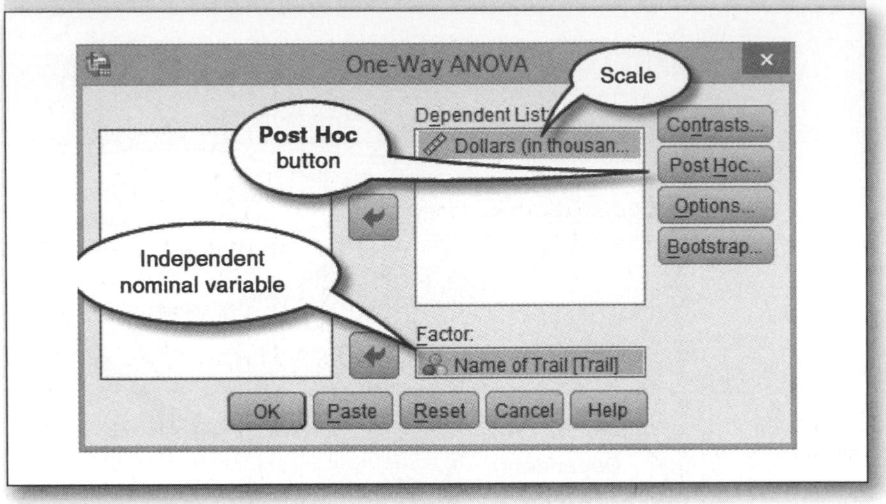

Figure 15.5   Post Hoc Analysis Selection Window

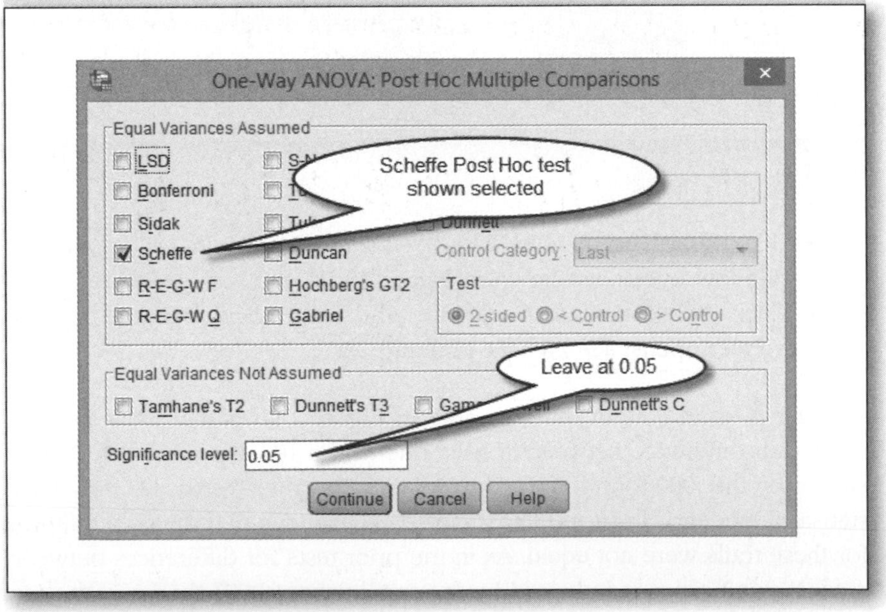

Fortunately, we clicked **Scheffe**, which instructed SPSS to compare all 10 possible combinations of mean values for these five hiking trails. This comparison is presented in Figure 15.7.

Figure 15.6 ANOVA Table for Trail Expenditure

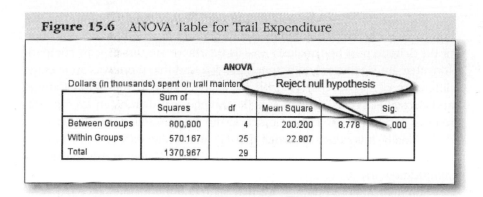

ANOVA

Dollars (in thousands) spent on trail mainten[ance]

	Sum of Squares	df	Mean Square		Sig.
Between Groups	800.800	4	200.200	8.778	.000
Within Groups	570.167	25	22.807		
Total	1370.967	29			

Reject null hypothesis

Figure 15.7 Post Hoc Analysis: Scheffe Test of Significance (in Thousands of Dollars)

Multiple Comparisons

Dependent Variable: Dollars (in thousands) spent on trail maintenance

Scheffe

(I) Name of Trail	(J) Name of Trail	Mean Difference (I-J)	Std. Error	Sig.	95% Confidence Interval Lower Bound	Upper Bound
Eaton Canyon	Grizzly Flat	10.167*	2.757	.024	[ ].33	[ ]9.33
	Rattlesnake	12.167*	2.757	.005	3.01	21.33
	Bailey Canyon	14.500*	2.757	.001	5.34	23.66
	Millard Canyon	5.500	2.757	.429		14.66
Grizzly Flat	Eaton Canyon	-10.167*	2.757	.024		.01
	Rattlesnake	2.000	2.757	.969	-7.16	11.16
	Bailey Canyon	4.333	[ ]	.654	-4.83	13.49
	Millard Canyon	[ ].667	[ ].757	.589	-13.83	4.49
Rattlesnake	Eaton Canyon			.005	-21.33	-3.01
	Grizzly Flat			.969	-11.16	7.16
	Bailey Canyon			.947		[ ]9
	Millard Canyon			.244	-[ ]	[ ]
Bailey Canyon	Eaton Canyon	-14.500*	2.757	.001	-23.66	-5.34
	Grizzly Flat	-4.333	2.757	.654	-13.49	4.83
	Rattlesnake	-2.333		.947	-11.49	6.83
	Millard Canyon			.056	-18.16	.16
Millard Canyon	Eaton Canyon			.429	-14.66	3.66
	Grizzly Flat			.589	-4.49	13.83
	Rattlesnake	6.667	2.757	.244	-2.49	15.83
	Bailey Canyon	9.000	2.757	.056	-.16	18.16

*. The mean difference is significant at the 0.05 level.

significant

significant

significant

Grizzly Flat and Eaton Canyon same as Eaton Canyon and Grizzly Flat

same as above

same as above comparison

Note: Figure 15.7 may seem a little overwhelming because of its sheer length. Be aware that it presents 20 comparisons for two means when there are actually only 10. This is because, for instance, *Eaton Canyon* is compared with *Grizzly Flat,* and once again *Grizzly Flat* is compared with *Eaton Canyon.* The same probabilities are assigned to both comparisons. The table in Figure 15.7 repeats this for the remaining trails, which adds considerable complexity to the table. The callouts in Figure 15.7 attempt to point out where the SPSS output introduces these complexities.

The Scheffe test is another test of significance, but this time the test compares each possible combination of means one at a time. Examination of the Scheffe post hoc analysis reveals that there are three mean comparisons that are significantly different. First, we find that there was an average difference of $10,167 per year on the maintenance of the Eaton Canyon and Grizzly Flat trails. Second, it is determined that there is an average difference of $12,167 for expenditures on the Eaton Canyon and Rattlesnake trails. Finally, it is determined that there is an average difference of $14,500 for the Eaton and Bailey Canyon trails. All these differences are statistically significant at the .05 level.

Summarizing our findings, we may state that the maintenance expenditures for these five mountain trails are significantly different. There is now statistical support for the researcher's hypothesis as well as additional details regarding which of the trail pairs are significantly different.

Let's now examine the same data but with the belief that they do not meet the data assumptions required for the ANOVA procedure.

△   15.5 NONPARAMETRIC TEST: KRUSKAL-WALLIS TEST

The application of the nonparametric alternative for the one-way ANOVA, the *Kruskal-Wallis* test, is now demonstrated using the same trail maintenance data. The same data are used, but this time let's assume that the data severely differ from the normal distribution.

The Kruskal-Wallis test is similar to ANOVA in many ways. One similarity is that the null hypotheses for both parametric and nonparametric tests assume that the random samples are drawn from identical populations. The alternative hypotheses for both ANOVA and Kruskal-Wallis assume that the samples come from nonidentical populations. Presented another way, we can state that the alternative hypothesis looks for statistically significant differences between the groups. One thing to be kept in mind is that the Kruskal-Wallis test is not as powerful as ANOVA—it may miss the significance.

Assumptions for the use of the Kruskal-Wallis test are that (1) the samples are random, (2) the data are at least measured at the ordinal level, and (3) the scaled distributions are identically shaped. Next, let's demonstrate the application of the Kruskal-Wallis procedure.

- Start SPSS, click **File**, select **Open**, and click **Data**.
- Click **trail.sav**, and then click **Open**.

- Click **Analyze**, select **Nonparametric Tests**, and then click **Independent Samples** (the *Nonparametric Test: Two or More Related Samples* window opens).
- Click **Fields**, click **Dollars**, and then click the upper arrow.
- Click **Name of Trail**, and then click the lower arrow.
- Click **Run** (the Output Viewer opens showing the *Hypothesis Test Summary*, as shown in Figure 15.8).

Figure 15.8   Kruskal-Wallis—*Hypothesis Test Summary*

Hypothesis Test Summary

	Null Hypothesis	Test	Sig.	Decision
1	The distribution of Dollars (in thousands) spent on trail maintenance is the same across categories of Name of Trail.	Independent-Samples Kruskal-Wallis Test	.003	Reject the null hypothesis.

Asymptotic significances are displayed. The significance level is .05.

The test determines that maintenance costs are significantly different with a *Sig.* value of .003. Recall that this value simply states that the probability that the noted differences are attributable to chance alone is only .003. Since we have rejected the null hypothesis, that the number of dollars spent on the maintenance of each trail is the same, we might wish to do comparisons between all pairs of mountain trails to establish which trail pairs are indeed significantly different. You will recall that a similar procedure was followed when conducting the *Scheffe* test in the ANOVA procedure. Follow the steps below to accomplish this task.

- Double click on **Hypothesis Test Summary** in the Output Viewer, as shown in Figure 15.8. This opens the *Model Viewer* window, as shown in Figure 15.9. Pay special attention to the lower right portion of the *Model Viewer*, which is enlarged and shown in Figure 15.10.
- Click the small black arrow to the right of *Independent Samples Test View* in the *View* box (this action is shown in Figures 15.9 and 15.10 with the callouts).

Figure 15.9 *Model Viewer*—Trail Expenditure Data

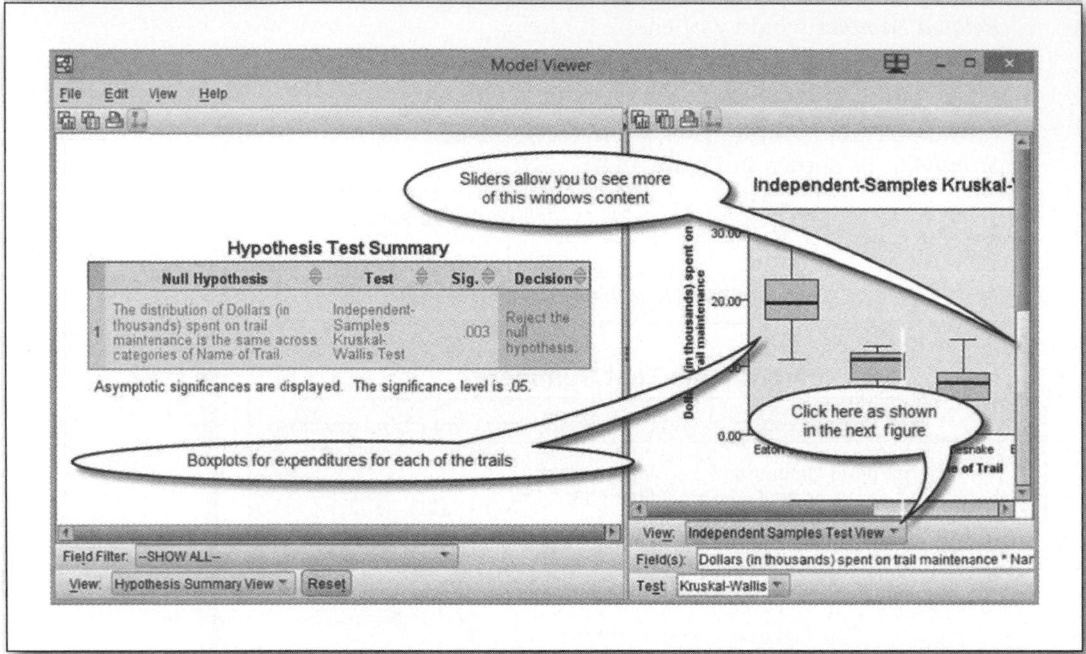

Figure 15.10 Enlarged Lower-Right Corner of the *Model Viewer* Window

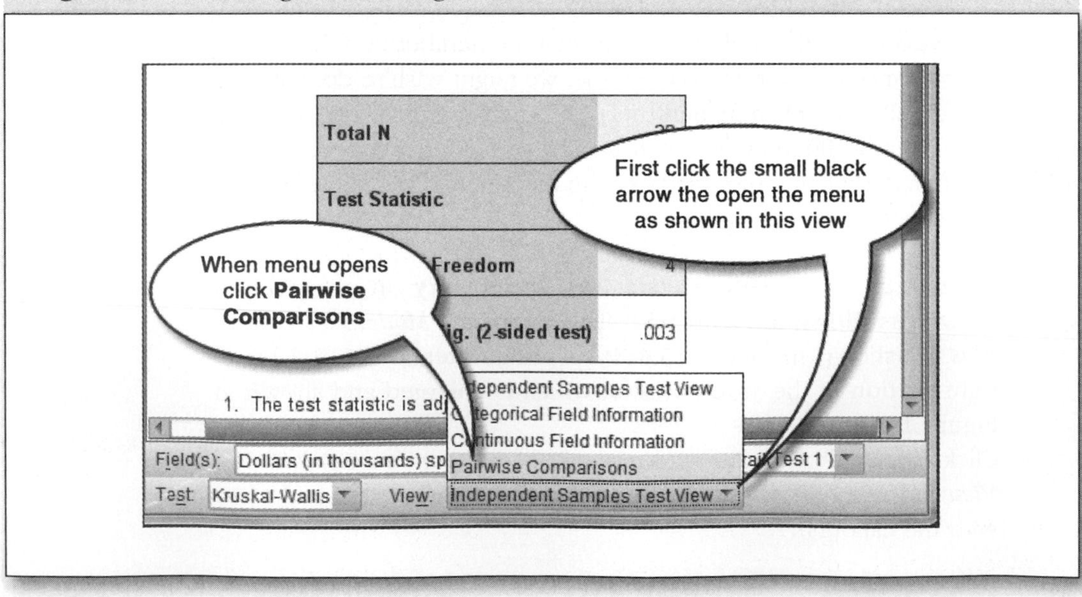

- From the new menu, select and click **Pairwise Comparisons** (once clicked, Figure 15.11 appears in the Output Viewer, which compares all possible pairs of the five mountain trails)

Figure 15.11   Pairwise Comparisons for the Kruskal-Wallis Test

Sample1-Sample2	Test Statistic	Std. Error	Std. Test Statistic	Sig.	Adj.Sig.
Bailey Canyon-Rattlesnake	3.333	5.061	.659	.510	1.000
Bailey Canyon-Grizzly Flat	6.417	5.061 (Significant)		.205	1.000
Bailey Canyon-Millard Canyon	-12.667	5.061	-2.503	.012	.123
Bailey Canyon-Eaton Canyon	18.000	5.061	3.557	.000	.004
Rattlesnake-Grizzly Flat	3.083	Significant	.609	.542	1.000
Rattlesnake-Millard Canyon	-9.333	5.061	-1.844	.065	.651
Rattlesnake-Eaton Canyon	14.667	5.061	2.898	.004	.038
Grizzly Flat-Millard Canyon			-1.235	.217	1.000
Grizzly Flat-Eaton Canyon		5.061	2.289	.022	.221
Millard Canyon-Eaton Canyon	5.333	5.061	1.054	.292	1.000

Callouts: "Significant", "Significant", "Not significant but ANOVA detected a significant difference"

Each row tests the null hypothesis that the Sample 1 and Sample 2 distributions are the same.
Asymptotic significances (2-sided tests) are displayed. The significance level is .05.

The Kruskal-Wallis test is less sensitive than the ANOVA. Although both the ANOVA and the Kruskal-Wallis tests found significance between Eaton and Bailey and Eaton and Rattlesnake, the nonparametric test did detect differences between Eaton and Grizzly Flat. This finding is shown in Figure 15.11.

15.6 SUMMARY △

In this chapter, we presented the procedures used to check for significant differences between three or more means. The ANOVA test was used for data meeting the parametric requirements. The ANOVA test requires that data for each of the randomly sampled populations be approximately normally distributed, that they have equal variances, and that the dependent variable be measured at the *scale* level. Once these conditions are met and the ANOVA test is applied, an *F* statistic is used to determine if any of the pairs of means are significantly different. In our example, significance was found, and the data were subjected to further analysis. The purpose of further analysis was to identify those pairs of means that contributed to the significant *F* statistic. This type of analysis is known as *post hoc* analysis. We selected the *Scheffe* critical value test to compare all mean combinations.

When the data fail to meet the parametric requirements, the *Kruskal-Wallis* nonparametric test may be used. This procedure was demonstrated on the same data used to show the ANOVA procedure. The application of the Kruskal-Wallis test resulted in the determination that there was a significant difference between 2 of the pairs of means out of the 10 possible comparisons. The ANOVA test indicated three significant comparisons owing to its greater sensitivity compared with the Kruskal-Wallis test. In the next chapter, we double our fun by addressing the situation where there are two or more independent variables—with the *two-way ANOVA*.

△   15.7 Review Exercises

15.1   An El Salvadorian pig farmer, Jose, had the idea to add a by-product from the production of cane sugar to his pig feed. The idea was that the pigs would eat more, gain weight, and be worth more at market time. He had 24 weaner pigs weighing from 20 to 40 pounds. He randomly divided the pigs into three groups of eight. He concocted three different feed types, each containing different levels of the cane sugar by-product (*low* sugar, *medium,* and *high*). The farmer decided to record the pounds of feed consumed by each pig for 1 week. The pigs fed the low-sugar feed consumed 8.5, 8.0, 13.2, 6.8, 6.45, 6.0, 9.12, and 9.75 pounds. The pigs fed the medium-sugar feed consumed 10.99, 10.5, 9.67, 8.61, 10.92, 12.8, 9.03, and 9.45 pounds. The pigs fed high-sugar feed consumed 10.39, 9.97, 13.78, 12.69, 12.8, 9.67, 9.98, and 10.67 pounds.

Your task is to seek evidence that there is a significant difference in consumption for the three different feed types. Write the null and alternative hypotheses. If statistical significance is determined, then identify which groups contribute to overall significance. Once you complete the analysis, answer Jose's question about whether he should add the cane sugar by-product to his pig feed and which of the three feeds is the best.

15.2 A chemical engineer had three different formulas for a gasoline additive that she thought would significantly change automobile gas mileage. She had three groups of 15 test standard eight-cylinder engines that simulated normal driving conditions. Each group received a different gasoline formulation (A1, A2, and A3) and was run for several hours. Simulated mileage for the A1 group was 35.60, 34.50, 36.20, 33.10, 36.10, 34.80, 33.90, 34.70, 35.20, 35.80, 36.60, 35.10, 34.90, 36.00, and 34.10. Mileage for A2 was 36.80, 35.30, 37.00, 32.90, 36.80, 35.60, 35.10, 35.80, 36.90, 36.60, 36.80, 36.60, 35.80, 36.30, and 36.00. Mileage for the A3 group was 37.79, 36.29, 38.01, 33.80, 37.79, 36.58, 36.03, 36.79, 37.89, 37.57, 37.79, 37.59, 36.78, 37.29, and 37.01.

Your job is to investigate the mileage numbers in an effort to provide evidence in support of her contention that the groups would have significantly different gas mileage. Write the *null* and *alternative* hypotheses. If you find a difference, therefore rejecting the null, you must identify the groups contributing to the significant F statistic with *post hoc* analysis. Can you provide evidence in support of the chemical engineer's contention that her formulas will significantly alter gas mileage for these test engines?

15.3 Bacteria counts were taken at the four Southern California beaches of Santa Monica, Malibu, Zuma, and Ventura. The researcher's idea was that the different beaches would yield significantly different bacteria counts. The Santa Monica beach count was 16.2, 12.0, 16.4, 15.5, 16.5, 22.0, and 23.0. The Malibu count was 18.3, 18.2, 18.3, 17.4, 18.4, 24.1, and 25.2. The Zuma count was 17.2, 17.3, 17.2, 16.4, 17.3, 23.0, and 24.3. The Ventura count was 20.2, 20.9, 21.1, 20.3, 20.2, 26.1, and 28.4. Check the distributions for normality—just by looking, you would suspect that they don't approximate the normal curve.

Select the correct testing approach based on your normality findings, and write the null and alternative hypotheses. If you find significant differences in the bacteria counts at the four beaches, do additional work to identify the specific beaches that contribute to the overall finding. What is the answer to the researcher's idea that the beaches have different bacteria counts?

CHAPTER **16**

TWO-WAY (FACTORIAL) ANOVA

△ 16.1 INTRODUCTION AND OBJECTIVES

We continue with the theme of the prior three chapters in that we use SPSS to accomplish hypothesis testing. In Chapter 15, you were introduced to *one-way analysis of variance* (ANOVA), which tested for significant differences between three or more means. Recall that when using *one-way* ANOVA you have *one* independent variable that can have three or more groups, called *treatments* or *levels of the independent variable*.

This chapter presents the circumstances where you have at least *two* independent variables with any number of groups (levels) in each. The *two-way* ANOVA is sometimes referred to as *factorial* ANOVA. The use of *two-way* ANOVA permits the evaluation of each independent variable's effects on the dependent variable, which are known as the **main** effects. But it also does more—it tests for *interaction* effects. Interaction means that the factorial analysis looks for significant changes in the dependent variable as a result of *two or more* of the independent variables working together.

For example, you are interested in investigating how tall corn grows under various conditions—the height of the corn, measured in inches, is the dependent variable. You are interested in studying two independent variables: The first variable is seed type (two levels of Pearl and Silver), and the second independent variable is fertilizer type (two levels of seaweed and fish). To succinctly describe such a study's design, we state that we are using a 2×2 *factorial* ANOVA. In other words, we have two independent variables: The first has two groups (seed types), and the second also has two groups (fertilizer types). If you had three seed types and four fertilizer types, it would be described as a 3×4 *factorial* ANOVA. Or some statisticians might call it a 3×4 *two-way* ANOVA. In this chapter, we will refer to this research design as *factorial* ANOVA and sometimes as *two-way* ANOVA.

Let's use the 2×2 design and the hypothetical summarization of data in Figure 16.1 to explain the concept of *main* and *interaction* effects. The *main* effects investigated would be seed and fertilizer types. Are there significant differences in the height of corn (measured in inches) when grown from Pearl or Silver seeds? Looking at the column means, found in Figure 16.1, we find that the value is 50 inches for both types of seed. Therefore, there is no difference in height, and we conclude that there is *no* main effect due to seed type. The other main effect would be fertilizer type. Are there significant differences in the height of corn when the seed is fertilized with seaweed or fish? The data indicate *no* differences in the height of corn when fertilized with these two fertilizer types. This is shown by the row mean heights of 50 inches for both fertilizer types. However, there is an interaction effect between seed and fertilizer type. Pearl seed corn grows taller when seaweed fertilizer is used (60 inches) rather than fish fertilizer (40 inches). Silver seed corn grows taller when using fish fertilizer (60 inches) rather than with seaweed fertilizer (40 inches). This example may appear to be an oversimplification, but we feel it does make it easier to conceptualize the difference between *main* and *interaction* effects.

The *two-way* ANOVA is specifically designed to investigate data to determine whether there are *main* and/or *interaction* effects.

Figure 16.1 Hypothetical Data for Corn-Growing Study—Shows Height in Inches (Main and Interaction Effects)

	Pearl Seed	Silver Seed	Row Means (inches)
Seaweed fertilizer	60	40	50
Fish fertilizer	40	60	50
Column means (inches)	50	50	

OBJECTIVES

After completing this chapter, you will be able to

Describe the data assumptions required for two-way ANOVA

Input data for the two-way ANOVA procedure

Use SPSS to conduct two-way ANOVA

Describe the main effects in two-way ANOVA

Describe the interaction effects in two-way ANOVA

Write alternative and null hypotheses for two-way ANOVA

Interpret the SPSS output to identify main and interaction effects

△ 16.2 Research Scenario and Test Selection

An investigation that would study two independent variables, detergent type and water temperature—each consisting of two groups—was proposed; therefore we have a 2×2 *factorial design*. The investigators suspected that the type of detergent and water temperature may have some impact on the whiteness of the clothes (main effects). Those conducting the study also wish to determine if detergent type and water temperature interact in a manner that may affect the whiteness level (*interaction effects*).

Thus, we have one dependent variable, whiteness, measured at the *scale* level and two independent variables. The two independent variables used in this scenario are type of detergent (Mighty Good and Super Max) and water temperature (hot and cold). Both of the independent variables are measured at the *nominal* level. Since each load of clothing will be washed independently, a repeated measures design is inappropriate. If we begin with the assumption of normality and equal variances of the populations, what statistical test would you suggest for this investigation?

The reasoning for deciding which statistical test is appropriate might start with a consideration of levels of measurement for the selected variables. We have a dependent variable measured at the scale level, which informs us that we can calculate means and standard deviations for our groups. We also have two independent variables measured at the nominal level. *Two-way* ANOVA procedures will work with nominal independent variables as long as the dependent variable is measured at the scale level. We will use a 2×2 *factorial design* since we have two groups for each of our two independent variables. We also assume that the dependent variable is approximately normally distributed and that the groups have equal variances. The assumption of normality and equality of variances will be empirically tested as we proceed through the analysis.

16.3 RESEARCH QUESTION AND NULL HYPOTHESIS △

The researchers wished to investigate whether there is a difference in the whitening power of Mighty Good and Super Max detergent. For the detergent main effect research question, we have the following null and alternative hypotheses:

$$H_{01} : \mu_{\text{Mighty\_Good}} = \mu_{\text{Super\_Max}}$$

$$H_{A1} : \mu_{\text{Mighty\_Good}} \neq \mu_{\text{Super\_Max}}.$$

They also wished to investigate if there is a significant main effect on the whiteness of clothes when washed in hot or cold water. For the water temperature main effect research question, we have the following null and alternative hypotheses:

$$H_{02} : \mu_{\text{hot}} = \mu_{\text{cold}}$$

$$H_{A2} : \mu_{\text{hot}} \neq \mu_{\text{cold}}.$$

The researchers also wished to determine whether detergent type and water temperature may work together to change the whiteness of the clothing (*interaction effects*). Figure 16.6 shows the interaction effects by presenting the mean whiteness levels for (Mighty Good + cold water), (Mighty Good + hot water), (Super Max + cold water), and (Super Max + hot water). Our final SPSS output will answer all these questions regarding the level of whiteness as it may relate to both main and interaction effects.

In the next section, we input, analyze, and interpret the findings from our clothes-washing experiment.

△   16.4 Data Input, Analysis, and Interpretation of Output

Your new database for this example will be named *Whiteness Study*. We are confident that your skill level is such that detailed bullet points are not required to enter the variable information. Instead of bullet points, we have provided Figure 16.2, which shows how your Variable View screen should look once the variable information has been entered. As shown in the call-outs in Figure 16.2, the *Values* windows for detergent and water should be completed as follows: values for detergent should be 1 = *Mighty Good* and 2 = *Super Max*, and values for water should be 1 = *Hot* and 2 = *Cold*.

Figure 16.2   Variable View Screen for the Whiteness Study

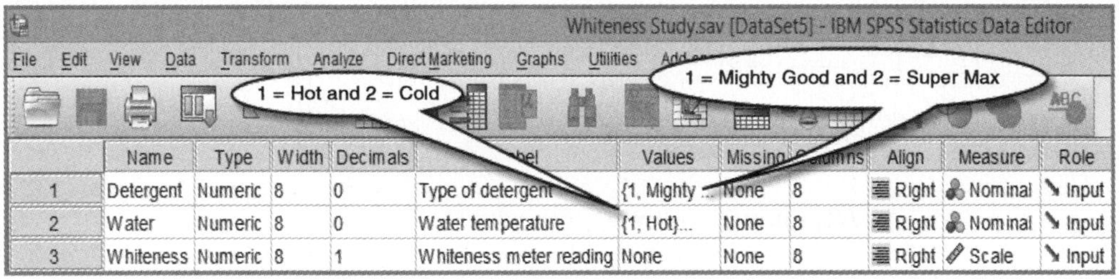

Enter the variable information, and then carefully enter all the data (40 cases) shown in Figure 16.3. Note that you should not enter the data in the column titled *Case #* as SPSS assigns these numbers automatically when each row of data is numbered. We have included these numbers in our table for reference purposes only.

The following list presents a much-abbreviated procedure for entering the variable information (see Figure 16.2) and the actual data (see Figure 16.3). This is followed by a more detailed explanation of how to get SPSS to do the work required for factorial analysis. Once the Output Viewer displays the results, the various tables are explained.

- Start SPSS, and click **Cancel** in the *SPSS Statistics* opening window.
- Click **File**, select **New**, and click **Data**.
- Click **Variable View** (enter all the variable information as represented in Figure 16.2).
- Click **Data View** (carefully enter all the data for the three variables as given in Figure 16.3).

Figure 16.3 Data for the 2×2 ANOVA Whiteness Study (40 Cases)

Case #	Detergent	Water	Whiteness	Case #	Detergent	Water	Whiteness
1	2	1	68.7	21	2	2	65.2
2	2	1	50.1	22	2	2	79.3
3	2	1	58.1	23	2	2	60.2
4	2	1	56.0	24	2	2	67.8
5	2	1	42.3	25	2	2	65.7
6	2	1	52.0	26	2	2	51.9
7	2	1	27.8	27	2	2	61.7
8	2	1	53.4	28	2	2	62.5
9	2	1	56.3	29	2	2	63.1
10	2	1	57.7	30	2	2	66.6
11	1	1	60.9	31	1	2	65.2
12	1	1	61.3	32	1	2	75.9
13	1	1	42.2	33	1	2	57.3
14	1	1	49.8	34	1	2	64.5
15	1	1	47.7	35	1	2	60.9
16	1	1	33.9	36	1	2	50.8
17	1	1	43.7	37	1	2	59.3
18	1	1	19.5	38	1	2	58.4
19	1	1	45.1	39	1	2	58.8
20	1	1	53.6	40	1	2	62.3

- Click **File** and then **Save As**, type **Whiteness Study** in the *File name* box, and then click **OK**.
- Click **Analyze**, select **General Linear Model**, and then click **Univariate** (the *Univariate* window opens; see Figure 16.4).
- Click **Whiteness meter reading**, and then click the arrow by the *Dependent Variable:* box.
- Click **Type of detergent**, and then click the arrow for the *Fixed Factor(s):* box.
- Click **Water temperature**, and then click the arrow for *Fixed Factor(s):* (your window should look like Figure 16.4).
- Click **Options** (the *Univariate: Options* window opens; Figure 16.5).
- Click **Descriptive statistics** and **Homogeneity tests**.
- Click **Continue** (the *Univariate* window opens), and then click **OK** (the Output Viewer opens).

Figure 16.4   *Univariate Window* for the 2×2 ANOVA Study

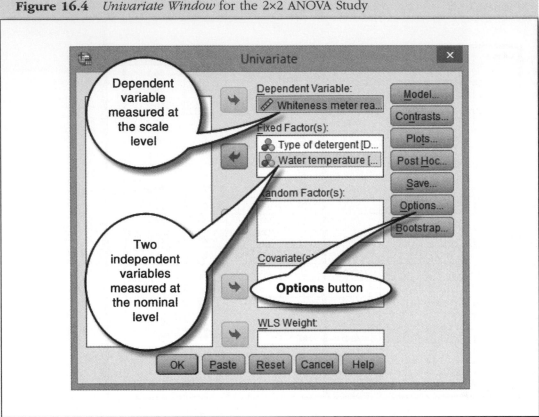

The reader is advised that once **OK** is clicked, a considerable amount of SPSS output is produced. We begin by looking at the *Descriptive Statistics* table (see Figure 16.6), which presents the means and standard deviations for all groups. We can learn much about our whiteness study from the data presented in Figure 16.6.

The table presented in Figure 16.6 is both informative and important—so let's take a closer look at the information it contains. It presents the means of the various groups receiving the different treatments. These means provide an insight into the outcome of our experiment—they make differences obvious but do not address whether those differences are significant. For that we must look at Figure 16.8—not now but in a couple of minutes.

We first look at the *main effects* of detergent type and water temperature. The means for detergent type are in the ovals as shown in Figure 16.6. Looking at the top oval, you see 53.555; notice that this is the average of the means of that group (i.e., average of 45.77 and 61.34 = 53.555). By getting this average, we are cancelling out any effect that water temperature may

Figure 16.5 *Univariate: Options* Window

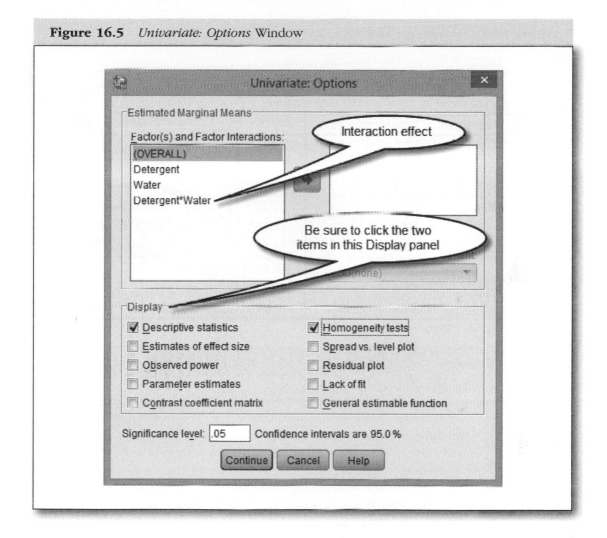

have on the whiteness value. Thus, we have the *main effect* of *Mighty Good* detergent on whiteness. The same process is duplicated for the cells below—for *Super Max* detergent showing a mean of 58.32. The question becomes "Is the difference between 53.555 and 58.32 significant? Is the *main effect* of detergent type significant?" We'll find out soon!

Next, we look at the main effect of water temperature on the whiteness level of clothing. For these values, we basically cancel out the influence of detergent type by following the same process just shown for water temperature. Look at the rectangular outline in Figure 16.6, and look more closely at the value for Hot of 49.005. Can you guess how this value was calculated? It is the average of the hot water values in the two cells above, 45.77 and 52.24. As in the explanation above, this has the effect of cancelling out any

Figure 16.6 *Descriptive Statistics* Window for the Whiteness Study

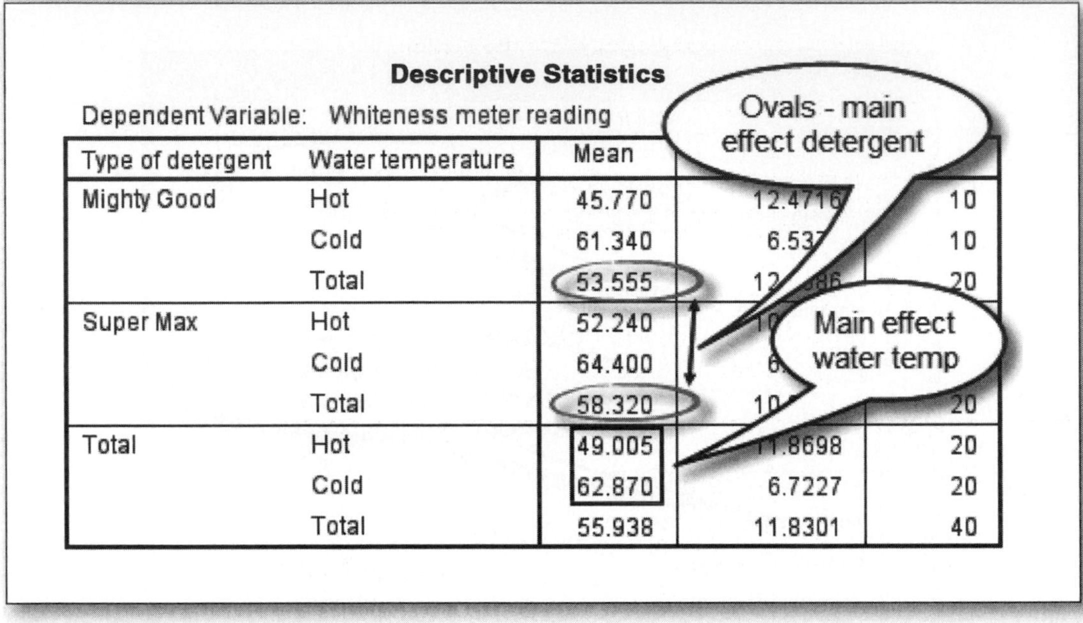

Descriptive Statistics

Dependent Variable: Whiteness meter reading

Type of detergent	Water temperature	Mean		
Mighty Good	Hot	45.770	12.4716	10
	Cold	61.340	6.53	10
	Total	53.555	12.86	20
Super Max	Hot	52.240	10	
	Cold	64.400	6	
	Total	58.320	10	20
Total	Hot	49.005	11.8698	20
	Cold	62.870	6.7227	20
	Total	55.938	11.8301	40

Ovals - main effect detergent

Main effect water temp

influence the type of detergent might have on the whiteness level. It leaves us with mean values for hot and cold water. We can test these means, representing whiteness levels, for hot water (49.005) and cold water (62.87) to see if they are significantly different (not yet but soon).

Before looking at Figure 16.8 and answering our significance questions, let's examine the table shown in Figure 16.7. The physical appearance of the table is small, but its importance is large when conducting our factorial analyses. When reading this table, you are interested in whether the null hypothesis is rejected. If the null hypothesis is rejected, then any significant difference identified by factorial analysis would be suspect, and it would be a good idea to use another approach.

The null hypothesis for *Levene's* test is one of equal error variances for all groups. Therefore, *Levene's* test is one of those cases where rejecting the null hypothesis would indicate that further analysis using *two-way* ANOVA is inappropriate. *Two-way* ANOVA requires equal error variances for all groups. Figure 16.7 presents a significance level of .360, informing us that by failing to reject the null hypothesis, we have statistical evidence that the error variances are equal across all groups. We may proceed with an increased level of confidence in any findings of our factorial analysis.

Figure 16.7   Levene's Test of Equality for Variances

Levene's Test of Equality of Error Variances[a]

Dependent Variable: Whiteness meter reading

F	df1	df2	Sig.
1.105	3	36	.360

Tests the null hypothesis that the error variance of the dependent variable is equal across groups.

a. Design: Intercept + Detergent + Water + Detergent * Water

Figure 16.8 presents the most relevant information regarding whether the researcher can reject or fail to reject the null hypothesis. Differences in whiteness levels when using Mighty Good and Super Max detergents (53.555 vs. 58.320) were found to be *not* significant, the level being .123. Since .123 is larger than .05, it can be said that the differences are due to chance. We can say that the *main effect* of detergent was not significant. The main effect for water temperature on whiteness levels was found to be significant (49.005 vs. 62.870) at .000. Therefore, we can say that water temperature does significantly influence whiteness levels. Nonsignificance was identified when the interaction effects between detergents and water temperatures (detergent * water) were compared. See the *Sig.* of .576 in Figure 16.8. Once again, any interaction differences could only be attributed to chance.

What did we learn from our analysis of the data resulting from the clothes-washing experiment? What did the *two-way (factorial)* ANOVA reveal? We can state that when clothes are washed in Mighty Good or Super Max, the level of whiteness does not change significantly. We can also state that when clothes are washed in cold water, the level of whiteness increases significantly (see Figures 16.6 and 16.8). No interaction effect was identified for these two types of detergents and the temperature of water. In other words, we may say that detergent type and water temperature did not work together to change the level of whiteness by the end of the wash cycle. The reader should understand that the data are hypothetical; therefore, one should not struggle for some plausible explanation for the observed increase in whiteness when using cold water.

Figure 16.8   ANOVA Table Showing Main and Interaction Effects

Tests of Between-Subjects Effects

Dependent Variable: Whiteness meter reading

Source	Type III Sum of Squares	df	Mean		Sig.
Corrected Model	2178.5	8	7.9		.000
Intercept	12516			1373.898	.000
Detergent	227.052			2.492	.123
Water	1922.382	1	1922.382	21.102	.000
Detergent * Water	29.070	1	29.070	.319	.576
Error	3279.549	36	91.000		
Total	1306				
Corrected Total	5458.054				

Main effect detergent (53.555 vs. 58.320)

Main effect water (49.005 vs. 62.870)

Interaction – not significant

a. R Squared = .399 (Adjusted R Squared = .349)

△   16.5 Summary

This chapter expanded on the method of *one-way* ANOVA by permitting the investigator the latitude to examine multiple independent variables. When using *two-way* ANOVA, multiple independent variables can consist of many individual groups. Such study designs can be extremely complex. An example that used two independent variables, each consisting of two groups, was used to demonstrate how SPSS analyzes such data. Chapter 17 introduces the reader to another form of the ANOVA test—repeated measures. The repeated measures procedure is used when the same subject or object is measured several times, usually over a period of time.

△   16.6 Review Exercises

16.1   A corn farmer is interested in reducing the number of days it takes for his corn to silk. He has decided to set up a controlled experiment that manipulates the *nominal* variable "fertilizer," having two categories: 1 = *limestone* and 2 = *nitrogen*. Another *nominal* variable is "soil type," with two categories: 1 = *silt* and 2 = *peat*. The dependent variable is *scale* and is the "number of days until the corn begins to silk." The data are given below; note that they must be entered into Data View in one continuous string of 40 cases. Once the data are entered, you must look for significant differences between the four study groups. You will also look for any interaction effects between fertilizer and soil and any influence they may have on the number of days to the showing of silk. Write the null and alternative hypotheses, select and conduct the correct test(s), and interpret the results.

	fertilizer	soil	silk										
1	1	1	54	21	1	2	55	31	2	2	50
2	1	1	56	22	1	2	54	32	2	2	49
3	1	1	60	23	1	2	57	33	2	2	55
4	1	1	65	24	1	2	52	34	2	2	53
5	1	1	67	25	1	2	57	35	2	2	57
6	1	1	67	26	1	2	60	36	2	2	51
7	1	1	63	27	1	2	64	37	2	2	55
8	1	1	62	28	1	2	58	38	2	2	57
9	1	1	66	29	1	2	60	39	2	2	52
10	1	1	59	30	1	2	63	40	2	2	53

16.2 A psychologist had the idea that different types of music and room temperature would influence performance on simple math tasks. She had two independent variables measured at the nominal level: (1) "music type," hard rock and classical, and (2) "room temperature," comfortable and hot. The dependent variable was a series of minimally challenging mathematical problems that were scored on a 0 to 100 scale. She randomly selected 24 students and then once again

	roomtemp	music	math
1	1	1	95
2	1	1	100
3	1	1	85
4	1	1	75
5	1	1	95
6	1	1	87
7	2	1	76
8	2	1	76
9	2	1	65
10	2	1	100
11	2	1	54
12	2	1	78
13	1	2	58
14	1	2	76
15	1	2	95
16	1	2	56
17	1	2	79
18	1	2	100
19	2	2	65
20	2	2	73
21	2	2	82
22	2	2	65
23	2	2	97
24	2	2	76

randomly assigned them to one of four groups. The data that resulted from her experiment are presented in the following table. Your task is to select the correct test, write the null and alternative hypotheses, and then interpret the results. Was there any significance on task performance as a result of music type or room temperature, or did these two variables act together to cause change?

16.3 The inspector general for a large state's motor vehicle department decided to collect some data on recent driving tests. The idea was to see if scores on the driving test (dependent *scale* variable) were significantly different for male and female (*nominal* independent variable) instructors. He also wanted to know if the time of day the test was given might also influence the scores. He first randomly picked two instructors and then collected data on recent tests they had administered. Time of day that the test was given was categorized as either early morning or late afternoon (the second nominal independent variable). He decided to randomly select six morning and six afternoon tests for each of his picked instructors. In the end, he had four unique groups consisting of six test takers each. You must write the null and alternative hypotheses and then select the correct test, interpret the results, and answer the inspector's questions. The data are given below.

	gender	timeday	test
1	1	1	87
2	1	1	76
3	1	1	93
4	1	1	89
5	1	1	74
6	1	1	87
7	2	1	73
8	2	1	71
9	2	1	81
10	2	1	69
11	2	1	75
12	2	1	63
13	1	2	63
14	1	2	51
15	1	2	52
16	1	2	61
17	1	2	52
18	1	2	51
19	2	2	89
20	2	2	91
21	2	2	89
22	2	2	83
23	2	2	77
24	2	2	76

ONE-WAY ANOVA REPEATED MEASURES TEST AND FRIEDMAN TEST

17.1 INTRODUCTION AND OBJECTIVES △

The one-way ANOVA (analysis of variance) repeated measures test is used when you wish to compare scale-level measurements taken on the same individuals at different times. An example might be the vitamin B-12 levels of six individuals taken for three (or more) consecutive months. In this situation, you would use the repeated measures test to compare the mean B-12 levels for each of the 3 months. The test could also be used when you need to take measurements on the same object under different conditions. An example would be baking vegan cookies under three (or more) different climate conditions (in the same stove and with the same ingredients) and then comparing the cookies' moisture content for the three different conditions. You may remember that you used this same general approach when you used the paired-samples t test in Chapter 14 but for two measurements only. The difference here is that you use the one-way repeated measures ANOVA when you have *three or more* times or conditions under which you take the measurements.

The Friedman test is a nonparametric test used as an alternative to the ANOVA repeated measures test when the assumptions for the parametric test cannot be assumed. It is used to compare the means of the rankings of three or more matched groups.

OBJECTIVES

After completing this chapter, you will be able to

Describe the purpose of the ANOVA repeated measures test

Use SPSS to conduct an ANOVA repeated measures test

Describe the assumptions related to an ANOVA repeated measures test

Describe the purpose of the Friedman test

Use SPSS to conduct the Friedman test

Describe the assumptions related to the Friedman test

△ 17.2 Research Scenario and Test Selection

A statistics instructor wished to test his conjecture that the test scores of the 15 students in his class significantly improved during the semester. The students in the class took three major tests during the standard 15-week semester. The professor wished to study the test performance for each of his 15 students. He was specifically interested to see if the test scores significantly changed during the semester.

We have *scale* data, and therefore we can look for differences between the means. We can assume equal variances for this type of data based on prior research. We can also test our assumption of equal variances as we proceed with the data analysis and terminate or alter the test method if the results so indicate. The paired-samples *t* test is not appropriate because there are more than two groups (there are *three* tests). Based on these considerations, the one-way ANOVA repeated measures will be the best test to determine if there are significant differences between the mean scores for the three tests. We must use a repeated measures test because we are *not* dealing with three different groups of students. In contrast, we are dealing with the same group of students who have repeated measures, namely, three exams over time during the semester.

17.3 RESEARCH QUESTION AND NULL HYPOTHESIS △

It may be expected that the students' scores on the three exams would improve over time as they become more acclimated to the subject matter and the instructor's approach to teaching. In contrast, it may be that the scores might decrease as the students become less interested in the subject or become overwhelmed by the increased complexity of the subject matter. The instructor assumes that the average test scores will not be equal and so chooses the .05 level of significance. Assumptions of normal distributions and equivalent variances apply to the repeated measures test. Also, sphericity is assumed. Sphericity refers to the equality of the variances of the differences between levels of the repeated measures factor—in our case, the three exams.

The alternative hypothesis (H_{A1}) is simply a restatement of the researcher's idea. We write the following expression for the alternative hypothesis:

H_{A1}: One or more of the mean test scores for the three tests are unequal.

The null hypothesis (H_{01}) states the opposite and is written as

$$H_{01}: \mu_1 = \mu_2 = \mu_3.$$

In this example, H_{01} states that there are no differences between the means of the populations. The professor would prefer to reject the null hypothesis, which would provide statistical evidence for the idea that the scores on the three tests were significantly different.

If there is evidence of overall significance, leading to the rejection of the null hypothesis (H_{01}), the researcher would most likely wish to identify which of the three groups are different and which are equal. The following null and alternative hypotheses will facilitate that task.

If the first null hypothesis (H_{01}) is rejected, then we may test the following:

$$H_{02}: \mu_1 = \mu_2, \ H_{03}: \mu_1 = \mu_3, \ H_{04}: \mu_2 = \mu_3.$$

The alternative hypotheses for these new null hypotheses are as follows:

$$H_{A2}: \mu_1 \neq \mu_2, \ H_{A3}: \mu_1 \neq \mu_3, \ H_{A4}: \mu_2 \neq \mu_3.$$

△   17.4 DATA INPUT, ANALYSIS, AND INTERPRETATION OF OUTPUT

Figure 17.1 shows the SPSS Data View screen as it should appear once you have entered the data for the three statistical tests. Follow the bullet points given below to also enter the variable information into the Variable View screen. These procedures are presented in a truncated manner since we believe that your ability to enter data and variable information has advanced considerably.

- Start SPSS, click **File**, select **New**, and click **Data**.
- In the Variable View screen, type in the following three variable names: "examone," "examtwo," and "examthree."
- Set *type* to *numeric*, *Decimals* to 0 (zero), *Align* to *Center*, and *Measure* to *scale* on these three variables.
- In the Data View screen, enter the data for each variable, as shown in Figure 17.1. (Visually check all data entries for accuracy.)

Figure 17.1   SPSS Data View for Repeated Measures Example

	examone	examtwo	examthree
1	49	59	62
2	79	88	86
3	69	76	80
4	58	67	72
5	63	78	81
6	72	85	84
7	70	77	87
8	48	58	60
9	77	85	93
10	60	72	74
11	69	76	81
12	62	74	78
13	54	68	71
14	71	82	86
15	61	74	81

- Click **File**, select **Save As**, type *repeat* in the *File Name* box, and then click **OK**.

Now that you have entered the variables and the data and saved these entries, you can proceed to the analysis. Assuming that the *repeat.sav* database is open, follow the steps given below:

- Click **Analyze**, select **General Linear Model**, and then click **Repeated Measures** (a window titled *Repeated Measures Define Factor(s)* will open; see Figure 17.2).
- Type *time* to replace *factor 1* in the *Within-Subject Factor Name:* box.
- Type *3* in the *Number of Levels* box (a window now appears as shown in Figure 17.2).

Figure 17.2 *Define Factors* Window

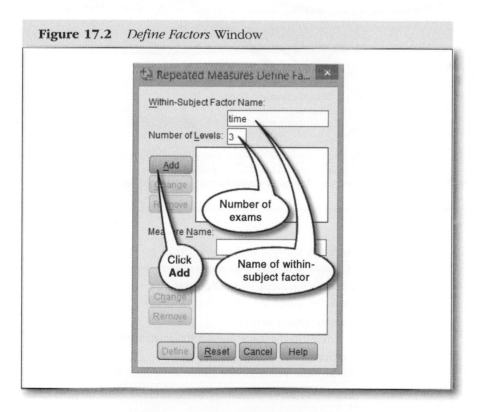

- Next, click **Add**, then click **Define** (a window titled *Repeated Measures* will open; see Figure 17.3).
- Click **examone(1)**, and then click the right arrow to place it in the *Within-Subjects Variables (time)* box. Do the same for **examtwo(2)** and **examthree(3)**.

Figure 17.3 *Repeated Measures* Window

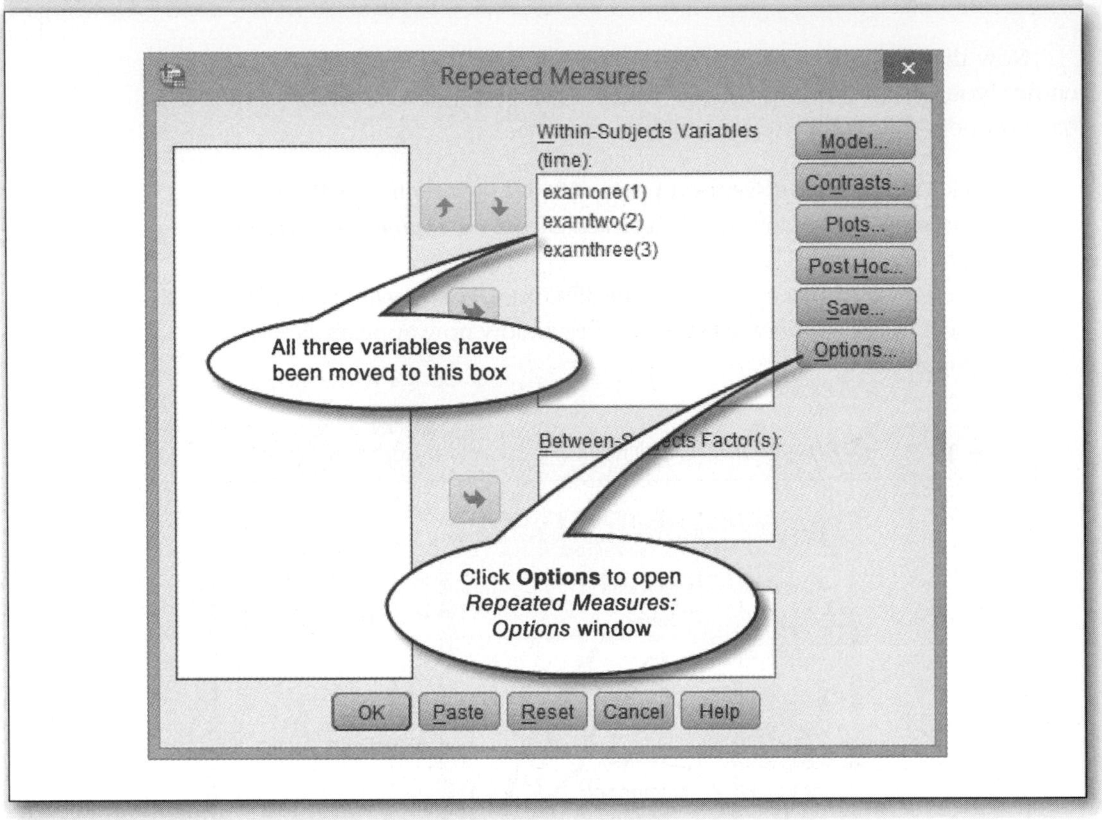

- Click **Options** (as shown in Figure 17.3). A window titled *Repeated Measures: Options* will open, as shown in Figure 17.4.
- Click **time** in the *Factorial(s) and Factor Interaction* box, and then click the arrow to move it to the *Display Means for:* box.
- Check **Compare main effects**.
- Select **LSD(none)** in the *Confidence interval adjustment* box.
- Click **Descriptive statistics**.
- Click **Continue**. You are returned to the *Repeated Measures* window.
- Click **OK**.

Following your final click of **OK**, a generous amount of output is produced by SPSS. We will not explain all the information provided but will certainly address important aspects that point out the validity and usefulness of the one-way repeated measures test. We will also emphasize those outputs that provide the information needed to answer our original research

Figure 17.4 *Repeated Measures: Options* Window

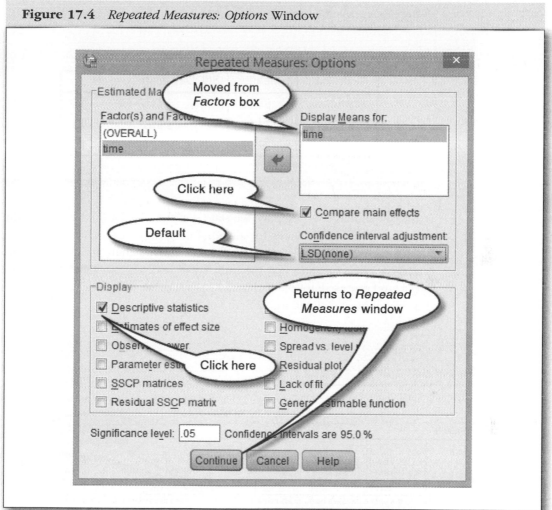

questions. To start with, we direct your attention to Figure 17.5, which shows the means and standard deviations for each of the three exams. When looking at the means, you should notice the consistent increase in mean scores from *examone* to *examthree* while also seeing that the variances are approximately equal. This was probably the information that piqued the professor's interest in the first place.

Figure 17.6 shows the results of Mauchly's test for equality of variances. This initial test of your data is very important, and it is one of those cases where you hope to *fail* to reject the null hypothesis. If you read the small print in the footnote below the table in Figure 17.6, you may get lost in its

Figure 17.5   Means and Standard Deviations for the Three Exams

Descriptive Statistics

	Mean	Std. Deviation	N
examone	64.13	9.410	15
examtwo	74.60	8.862	15
examthree	78.40	9.179	15

statistical language. To simplify, we state that the null for this test is that the variances of the differences for all possible groups are equal. Given this, we are happy to see a level of significance greater than .05, namely, .680. Therefore, we conclude that the null is *not* rejected, we have equal variances, and our assumption of sphericity is valid. We do not need to conduct the Greenhouse-Geisser and Huynh-Feldt tests.

Figure 17.6   Mauchly's Test for Equality of Variances of Differences Between all the Pairs

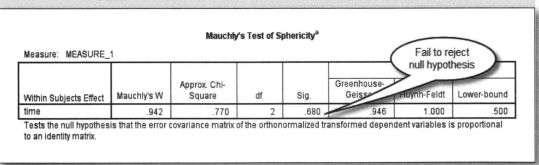

Mauchly's Test of Sphericity[a]

Measure: MEASURE_1

Within Subjects Effect	Mauchly's W	Approx. Chi-Square	df	Sig.	Greenhouse-Geisser	Huynh-Feldt	Lower-bound
time	.942	.770	2	.680	.946	1.000	.500

Fail to reject null hypothesis

Tests the null hypothesis that the error covariance matrix of the orthonormalized transformed dependent variables is proportional to an identity matrix.

Figure 17.7 shows the results of the tests of within-subjects effects. We are concerned with the row indicating time and sphericity assumed as this gives the level of significance, which indicates whether the means of the three exams are significantly different. The *F* value is 180, and the level of significance is .000. Since .000 is less than .05, we reject the null hypothesis that the means are equal. That is, we now have statistical evidence that exam scores did significantly change with the passage of time.

Figure 17.7 Tests of Within-Subjects Effects

Tests of Within-Subjects Effects

Measure: MEASURE_1

Source		Type III Sum of Squares	df	Mean Square	F	Sig.
time	Sphericity Assumed	1637.644	2	818.822	180.496	.000
	Greenhouse-Geisser	1637.644	1.891	865.930	180.496	.000
	Huynh-Feldt	1637.644	2.000	818.822	180.496	.000
	Lower-bound	1637.644		1637.644	180.496	.000
Error(time)	Sphericity Assumed		28	4.537		
				4.797		
				4.537		
	Lower-bound		14.000	9.073		

This row of data is what should be looked at in this table

Our final output is shown in Figure 17.8. This table presents the tests of significance for all possible pairs of the three exams—that is, Time 1 with Time 2, Time 1 with Time 3, and finally Time 2 with Time 3. You might first notice that the table repeats itself, as when the time of Exam 1 is compared with the time of Exam 2 and then again when the time of Exam 2 is compared with the time of Exam 1. Both of these comparisons yield exactly the same results.

Figure 17.8 Pairwise Comparisons for All Possible Combinations

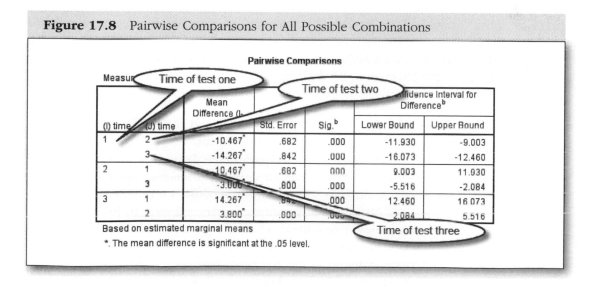

Pairwise Comparisons

(I) time	(J) time	Mean Difference (I-J)	Std. Error	Sig.[b]	Lower Bound	Upper Bound
1	2	-10.467*	.682	.000	-11.930	-9.003
	3	-14.267*	.842	.000	-16.073	-12.460
2	1	10.467*	.682	.000	9.003	11.930
	3	-3.800*	.800	.000	-5.516	-2.084
3	1	14.267*	.842	.000	12.460	16.073
	2	3.800*	.800	.000	2.084	5.516

Based on estimated marginal means

*. The mean difference is significant at the .05 level.

Time of test one

Time of test two

Time of test three

Inspection of the significance levels for all comparisons shows a value of .000, which is less than .05, which permits the rejection of the null hypothesis and provides evidence that the mean scores for these three exams are significantly different. Figure 17.8 indicates that the mean difference of −10.467 for Exam 1 and Exam 2 is significant. The mean difference of −14.267 for Exam 1 and Exam 3 is also significant. And the final comparison of Exam 2 and Exam 3 shows a significant mean difference of −3.8000. Given this information, we can return to Figure 17.5 and understand that the increase in the test score means over time (64.13, 74.60, and 78.40, respectively) represents statistically significant increases. We now have statistical evidence that supports the professor's original idea that the students did improve as the semester progressed. We could offer plausible explanations for such improvement but will leave that to the reader.

△   17.5 Nonparametric Test: Friedman Test

The data listed in Figure 17.1 can be shown to meet all the assumptions required for a parametric test (scale data, normally distributed, and equal variances). However, we wish to demonstrate the Friedman test available in SPSS when these assumptions are not met. The Friedman test is based on ranking of data. The student rankings are obtained by rating their scores on each of the three tests as follows: a rank of 1 for the lowest score of the three tests, a rank of 2 for the middle score, and a rank of 3 for the highest score. An example would be Student 2, who scored 79 (ranked 1) on "examone," 88 (ranked 3) on "examtwo," and 86 (ranked 2) on "examthree." The mean ranks are computed by summing the ranks for all students and for each test and dividing by 15. SPSS does these calculations and then summarizes the results as presented in the *Model Viewer* of the Output Viewer in Figure 17.11.

Follow these steps to perform the Friedman analysis:

- Start SPSS, click **File**, select **Open**, and click **Data**.
- In the file list, locate and click **repeat**.
- Click **Analyze**, click **Nonparametric Tests**, and then click **Related Samples**. A window titled *Non Parametric Tests: Two or More Related Samples* will open (this window is not shown).
- Click **Automatically compare observed data to hypothesized**.
- Click **Fields**, and then click **Use Custom Field Assignments**.
- Click **examone** in the *Fields* panel, then click the right arrow to place *examone* in the *Test Fields* panel. Click **examtwo**, then click the arrow. Click **examthree**, then click the arrow. All exams should now be in the *Test Fields* panel (this window is not shown).

- Click **Settings**, and then click **Customize Tests** (see Figure 17.9).
- Click **Friedman's 2-way ANOVA by ranks (k samples)**.
- Below this, in the *Multiple comparisons:* box, click the black arrow, and then select and click from the menu **All pairwise** if not already selected.

Figure 17.9   The *Nonparametric Tests: Two or More Related Samples* Window

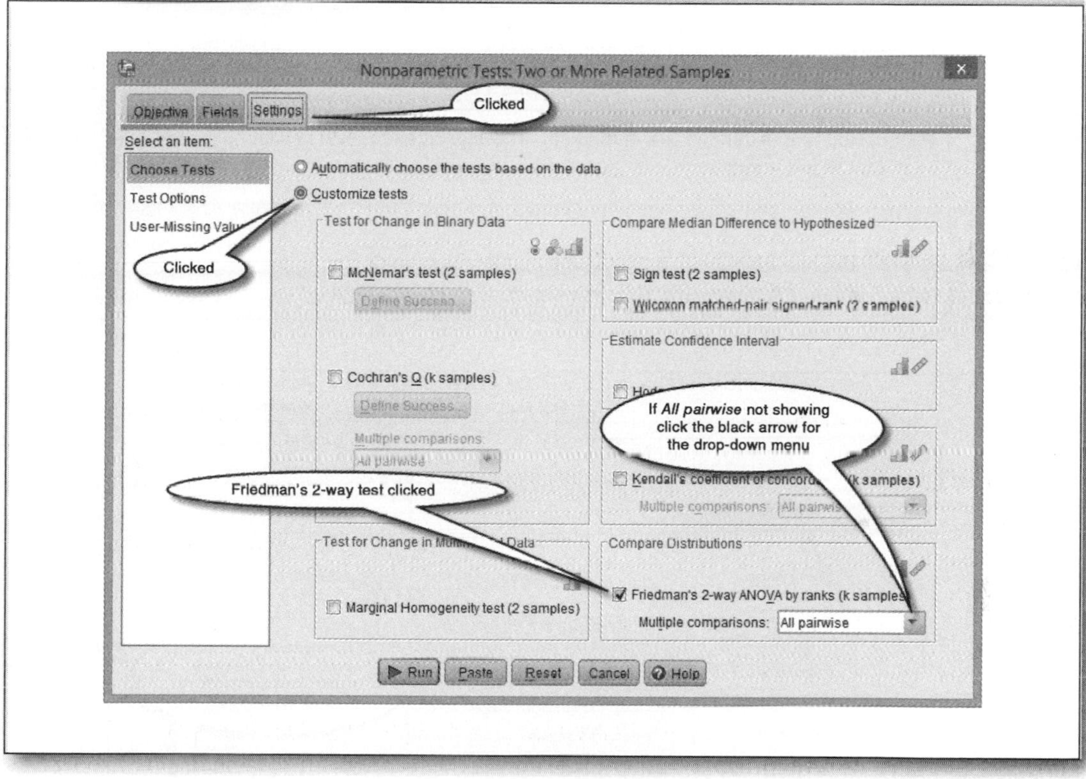

- Click **Run**.
- A table titled *Hypothesis Test Summary* will appear in the Output Viewer (see Figure 17.10).
- Double click on **Hypothesis Test Summary**, and a window will open consisting of two panels, showing the expanded *Model Viewer* panel and the *Hypothesis Test Summary* table on which you just double clicked. (Note that Figure 17.11 only shows the right panel of this window.)
- At the bottom of the expanded *Model Viewer* panel, click **View**, and then select **Pairwise Comparisons**. This click brings up Figure 17.12, which represents our final output for this demonstration.

Figure 17.10 Hypothesis Test Summary for Friedman's Two-Way Test

Hypothesis Test Summary

	Null Hypothesis	Test	Sig.	Decision
1	The distributions of examone, examtwo and examthree are the same.	Related-Samples Friedman's Two-Way Analysis of Variance by Ranks	.000	Reject the null hypothesis.

Asymptotic significances are displayed. The significance level is .05.

Figure 17.11 Model Viewer for the Friedman Two-Way Test

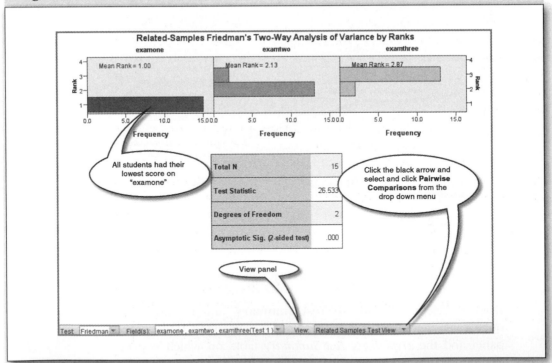

Let's examine each of these outputs individually and see how they compare with our findings when using the one-way ANOVA repeated measures test.

Figure 17.12 Pairwise Comparisons for the Friedman Two-Way Test

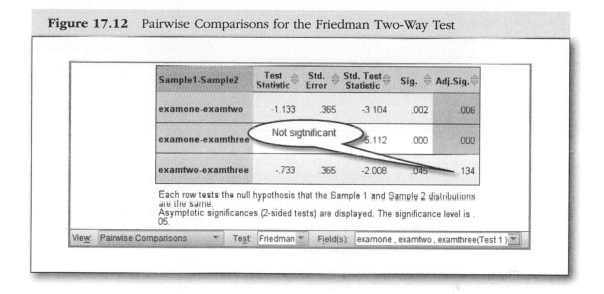

Sample1-Sample2	Test Statistic	Std. Error	Std. Test Statistic	Sig.	Adj.Sig.
examone-examtwo	-1.133	.365	-3.104	.002	.006
examone-examthree		Not sigtnificant	-5.112	.000	.000
examtwo-examthree	-.733	.365	-2.008	.045	.134

Each row tests the null hypothesis that the Sample 1 and Sample 2 distributions are the same.
Asymptotic significances (2-sided tests) are displayed. The significance level is .05.

View: Pairwise Comparisons ▼ Test: Friedman ▼ Field(s): examone , examtwo , examthree(Test 1) ▼

Figure 17.10: The null hypothesis in the Friedman test is that the distributions of the ranks of the exam scores are the same. In Figure 17.10, since .000 is less than .05, we reject the null hypothesis of no difference, indicating that there is a significant difference in the mean ranks of the three exams. This result agrees with that of the earlier one-way ANOVA repeated measures parametric test.

Figure 17.11: This output window provides more informative statistics, including the mean ranks as follows: "examone" = 1.00, "examtwo" = 2.13, and "examthree" = 2.87. Inspection of the results shown in Figure 17.11 gives a test statistic of 26.53 and a significance level of .000. Since .000 is less than .05, we reject the null hypothesis that there is no difference in the rank means of the three exams. Figure 17.11 also points out the clicking point (at the bottom of the window) needed to obtain the final output, which compares all possible combinations of the three mean ranks.

Figure 17.12: The final output presents three rows of data, which compare the distributions of ranks for "examone" with "examtwo," "examone" with "examthree," and "examtwo" with "examthree." When Exams 1 and 2 are compared, we find that the null can be rejected (significance of .006), and the ranks are not equal. When Exams 1 and 3 are compared, we find a significance level of .000; therefore, the null hypothesis of equal ranks can once again be rejected. When looking at the final comparison, we find a significance level of .134 (greater than .05); therefore, we fail to reject the null hypothesis of equality. For this comparison, the Friedman test fails to detect any significant difference in the ranks when Exam 2 is compared with Exam 3. This finding simply indicates that the power of the Friedman test to detect differences is less than that of the one-way ANOVA repeated measures approach.

Δ   17.6 Summary

The one-way ANOVA repeated measures test is appropriate when comparing three or more groups when the subjects undergo repeated measures. It is equivalent to a paired-samples t test, where only two groups are involved. If the assumptions for the ANOVA cannot be confirmed, then the nonparametric Friedman test can be used. However, it should be remembered that the power of a nonparametric test is less than that of a parametric test. In the next chapter, we will present the ANCOVA (analysis of covariance) procedure, which permits statistical control of a variable that might be negatively affecting the dependent variable.

Δ   17.7 Review Exercises

17.1   The high school track team coach recorded the time taken for the 100-yard dash by eight team members for three consecutive track meets during the regular season. His past experience informed him that they would improve their times throughout the season as they grew stronger and smarter. He had the idea that their level of improvement would qualify for statistical significance. Can you help the coach write the null and alternative hypotheses, select the correct test(s), interpret the analysis, and then answer his question? The data follow:

	runtime1	runtime2	runtime3
1	11.80	11.20	10.70
2	11.90	11.70	10.96
3	11.30	10.80	10.50
4	11.20	10.97	10.65
5	11.10	10.89	10.00
6	10.90	10.70	10.45
7	10.60	10.50	10.31
8	11.00	10.90	10.69

17.2 A farm manager was interested in studying several first-time straw-berry pickers over a period of 4 weeks. He felt that there was a signifi-cant difference in the number of pints picked per hour from one week to the next. Can you help him write the null and alternative hypoth-eses, input the data, select the correct tests, interpret the results, and answer his question concerning significant changes in the number of pints picked? The data follow:

	time1	time2	time3	time4
1	10	11	13	16
2	9	10	12	15
3	10	11	13	13
4	8	9	10	11
5	11	12	12	14
6	8	9	11	12
7	11	13	15	17
8	10	11	12	14
9	11	12	14	16
10	8	8	10	13

17.3 For this exercise, you must open the SPSS sample file called *bank-loan.sav*. A bank president is interested in comparing the last three variables: "preddef1," " preddef2," and "preddef3." These three variables were three different models created to predict whether a bank customer would default on a bank loan. Since they were all created from the same basic information, we can treat them as the same object and ana-lyze them using some type of repeated measures method. Can you help the bank president in determining if the three models are the same? Also, write the null and alternative hypotheses that you will be testing.

CHAPTER 18

ANALYSIS OF COVARIANCE

Δ **18.1 INTRODUCTION AND OBJECTIVES**

The analysis of covariance (ANCOVA) statistical test is a method of statistically equating groups on one or more variables and for increasing the power of a statistical test. Basically, ANCOVA allows the researcher to remove the effect of a known concomitant or control variable known as a *covariate*. We include a covariate in the analysis to account for the effect of a variable that does affect the dependent variable but that could not be accounted for in the experimental design. In other words, we use a *covariate* to adjust scores on a dependent variable for initial differences in another variable. For example, in a pretest–posttest analysis, we can use the pretest as a covariate to statistically control for the pretest scores, meaning that we statistically equate all the participants on their pretest scores, after which we examine their posttest scores. ANCOVA essentially tests whether certain factors have an effect on the dependent variable after removing the variance accounted for by the covariate.

There are some rather strict assumptions regarding the use of ANCOVA. All assumptions regarding the analysis of variance (ANOVA) are required. In addition, the covariate should have a reasonable correlation with the dependent variable, meaning that there is a linear relationship between the covariate and the independent variable. The ANCOVA must also satisfy the additional assumption of homogeneous regression slopes, meaning that the slopes of the regression lines representing the relationship between the covariate and the dependent variable are similar. We realize that we have presented some challenging concepts for the beginning statistician, but these are necessary for a variety of reasons, one being that ANCOVA is often used inappropriately because the assumptions have not been met, especially the assumption of homogeneity of regression slopes. With that said, there are occasions when ANCOVA seems to offer the only viable method of control, albeit statistical, open to the researcher.

OBJECTIVES

After completing this chapter, you will be able to

Describe the assumptions required for the use of the ANCOVA

Write the null and alternative hypothesis for a covariate problem

Describe the purpose of a covariate in the ANCOVA

Use SPSS to perform an ANCOVA

Interpret the results of an ANCOVA

18.2 RESEARCH SCENARIO AND TEST SELECTION △

A researcher wished to make a preliminary determination regarding which of the four novel methods of teaching reading available to a school district might be more effective for third-grade students. He randomly selected 24 third-grade students and then randomly assigned 6 of them to each of the four methods of reading instruction. A pretest in reading was administered to the 24 students. After several months of reading instruction, a posttest was administered.

A one-way ANOVA might have been considered a legitimate procedure as the dependent variable ("posttest score") was measured at the scale level. The independent variable was "teaching method," which was a nominal variable. The problem with this approach was that the pretest scores were thought to influence the posttest score independently of the teaching method. Therefore, we needed a design that could statistically control for any influence that the student's pretest score might have on the posttest score.

The ANCOVA was the answer as it would provide the analyst with the ability to control for any influence that the pretest scores might have on the posttest values.

△   18.3 Research Question and Null Hypothesis

The researcher had the idea that the four unique teaching methods would result in significantly different posttest scores on reading ability. The researcher also believed that the pretest values had the potential to influence the posttest values. When this possibility (pretest influence) was controlled for, then the investigator was convinced that the data would show that the mean scores for the four teaching methods were significantly different.

We state the null hypotheses as follows:

$$H_{01}: \mu_1 = \mu_2 = \mu_3 = \mu_4.$$

We state the alternative hypothesis as

H_{A1}: One or more of the four groups have mean test scores that are not equal.

If the null hypothesis (H_{01}) is rejected, the researcher would then wish to identify which of the four groups are different and which are equal. The following null and alternative hypotheses will facilitate that task.

If the first null hypothesis (H_{01}) is rejected, then we may test the following:

$$H_{02}: \mu_1 = \mu_2, H_{03}: \mu_1 = \mu_3, H_{04}: \mu_1 = \mu_4, H_{05}: \mu_2 = \mu_3, H_{06}: \mu_2 = \mu_4, H_{07}: \mu_3 = \mu_4.$$

The alternative hypotheses for these new null hypotheses are as follows:

$$H_{A2}: \mu_1 \neq \mu_2, H_{A3}: \mu_1 \neq \mu_3, H_{A4}: \mu_1 \neq \mu_4 \ H_{A5}: \mu_2 \neq \mu_3, H_{A6}: \mu_2 \neq \mu_4, H_{A7}: \mu_3 \neq \mu_4.$$

△   18.4 Data Input, Analysis, and Interpretation of Output

The variable and data information that must be inputted into SPSS for this ANCOVA example is provided in Figures 18.1 and 18.2. By now, you should feel comfortable with entering both data and variable information. The only thing we mention here is that the *values* and their *labels* are not directly shown in Figure 18.1. They are, however, listed in the callout balloon, which shows the information that must be entered in the *Value Labels* window.

If you choose, you may follow the bullet points that are given below; they will give you a rough guide for variable information and data entry, while also providing detailed instructions for the analysis.

Figure 18.1 SPSS Variable View Screen for Teaching Methods and Test Scores

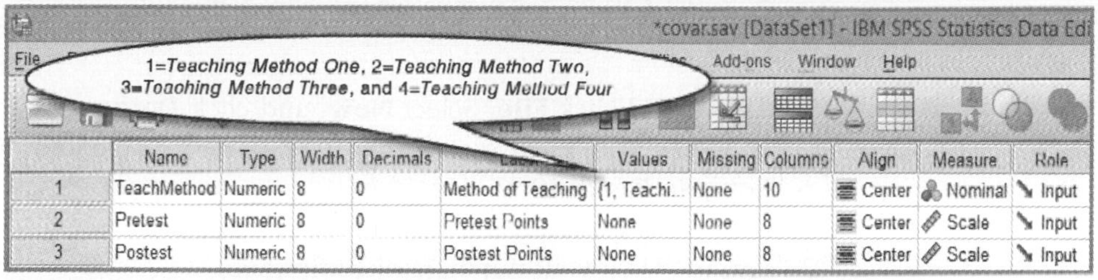

	Name	Type	Width	Decimals	Label	Values	Missing	Columns	Align	Measure	Role
1	TeachMethod	Numeric	8	0	Method of Teaching	{1, Teachi...	None	10	Center	Nominal	Input
2	Pretest	Numeric	8	0	Pretest Points	None	None	8	Center	Scale	Input
3	Postest	Numeric	8	0	Postest Points	None	None	8	Center	Scale	Input

1=Teaching Method One, 2=Teaching Method Two, 3=Teaching Method Three, and 4=Teaching Method Four

Figure 18.2 SPSS Data View Screen for Teaching Methods and Test Scores

	TeachMethod	Pretest	Postest
1	1	25	160
2	1	26	165
3	1	18	125
4	1	20	151
5	1	28	162
6	1	24	146
7	2	27	175
8	2	27	164
9	2	22	166
10	2	21	157
11	2	27	175
12	2	22	165
13	3	19	151
14	3	27	184
15	3	20	133
16	3	30	185
17	3	21	155
18	3	24	167
19	4	36	196
20	4	27	168
21	4	32	195
22	4	31	188
23	4	19	137
24	4	31	184

Let's proceed with covariate analysis using the ANCOVA approach. Just start the bullet points at the appropriate stage of your work. Note that the bullets assume that you have *not* entered the data and variable information into SPSS's Data Editor.

- Start SPSS, and click **File**, Select **New**, and click **Data**.
- Click the **Variable View** screen, and type in the variable information as shown in Figure 18.1.
- Click the Data View tab, and type in the data as shown in Figure 18.2. Be sure to check your numbers once finished.
- Once data entry has been completed, click **File**, then click **Save As**. The *Save Data As* window opens. In the *File name* box, type *covar*.
- Click **Save**.

You have now entered all the variable information and data values, so you are ready to begin the exciting part of the analysis—answering the questions about the data. We begin with a section that will tell us how to determine if the regression slopes are homogeneous.

Testing for the Homogeneity of Regression Slopes

The assumption of "homogeneity of regression slopes" is important when it comes to interpreting the results of an ANCOVA. Saying that one assumes homogeneity of regression slopes simply means that one assumes that the relationship between the dependent variable ("posttest scores") and the covariate ("pretest scores") is the same in each of the treatment groups (the four methods of teaching reading). For example, if there is a positive relationship between the pretest scores and the posttest scores in Teaching Method 1, we assume that there is a positive relationship between the pretest scores and the posttest scores for Teaching Methods 2, 3, and 4. Another way to state this is as follows: The regression lines when pretest scores are plotted against posttest scores for each of the four methods of teaching reading are essentially parallel, meaning that each has the same slope.

To have SPSS test for the homogeneity of regression slopes, we will conduct the ANCOVA test but in a fashion that will enable us to request a test of the interaction between teaching methods and pretest scores. The results of the interaction between pretest scores and posttest scores will answer the question regarding homogeneity of regression slopes. When the test is finished, and if the regression slopes are homogeneous, we will proceed with the standard ANCOVA procedure.

At this point, we assume that SPSS is up and running and that *covar.sav* is open and waiting for analysis.

- Click **Analyze**, select **General Linear Model**, and then click **Univariate** (a window titled *Univariate* will open).
- Click **Posttest Points**, and then click the right arrow to move it to the *Dependent Variable:* box.
- Click **Teaching Method**, and then click the right arrow to move it to the *Fixed Factor(s):* box.
- Click **Pretest Points**, and then click the right arrow to move it to the *Covariate(s):* box (see Figure 18.3).
- Click **Model**.
- Click **Custom**.
- Make sure the *Interaction* option is showing in the *Build Term(s)* box.

Figure 18.3 *Univariate* Window for the Test of Homogeneity of Regression Slopes

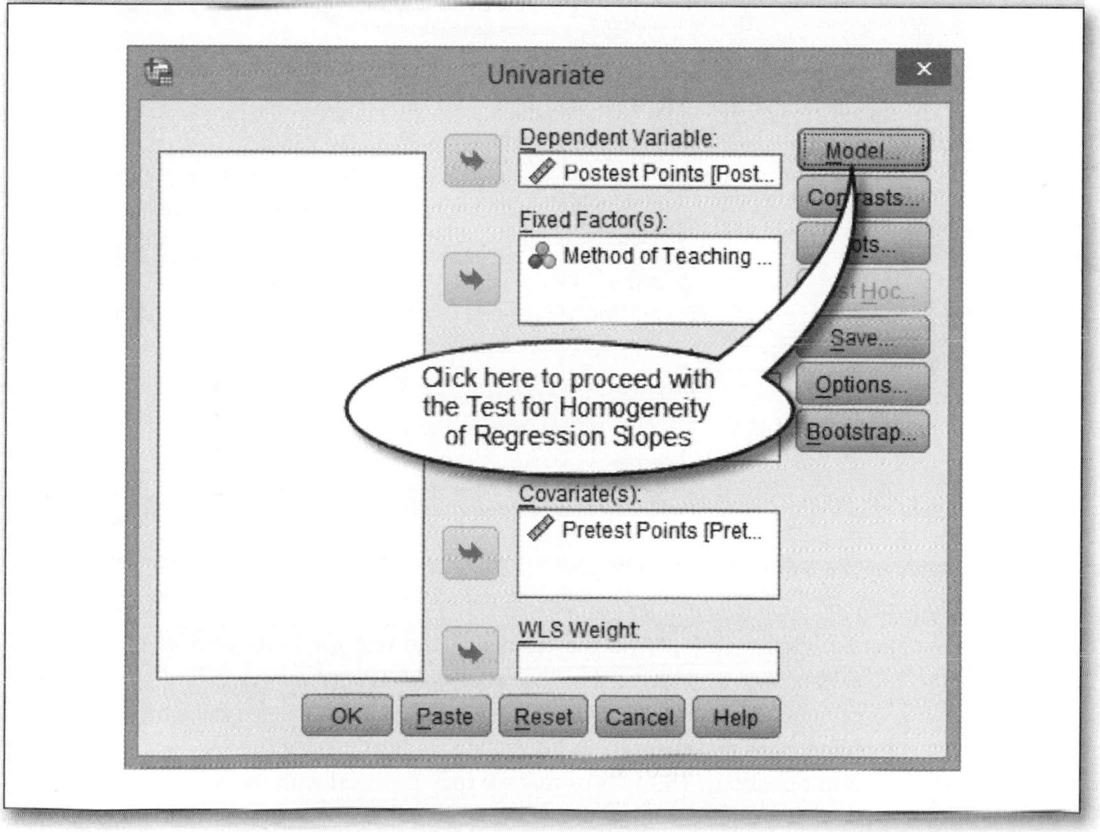

- Click **TeachMethod** in the *Factors & Covariates:* box, and then click the arrow to move it to the *Model:* box.
- Click the covariate **Pretest**, then click the arrow to also move it to the *Model:* box.
- Return to the *Factors & Covariates:* box, and click **TeachMethod**. While it is highlighted, hold down your computer's Ctrlkey, and click **Pretest**. Finally, click the arrow to move both to the *Model:* box. Carefully examine Figure 18.4, which shows how the *Univariate* window should appear at this point. This procedure is a little different from any done before, so make sure it is done correctly.
- Click **Continue**, and then click **OK**.

Figure 18.4 *Univariate Model* Window for Homogeneity of Regression Slopes

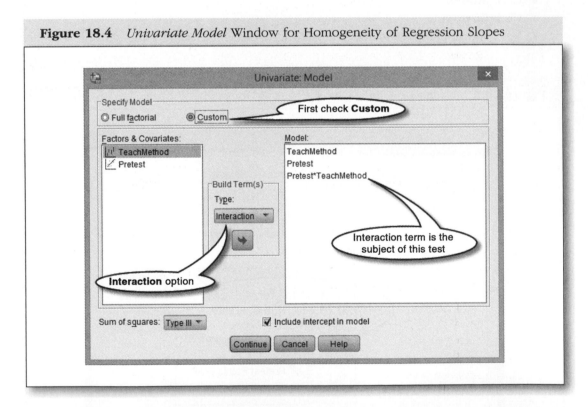

Figure 18.5 shows the results of the test for homogeneity of regression slopes. On inspecting the *TeachMethod * Pretest* row, we find that the level of significance is .340. Since .340 is greater than .05, we fail to reject the null hypothesis that there is no interaction, indicating that the regression slopes are homogeneous. This tells us that we may proceed with the ANCOVA procedure.

Figure 18.5   Homogeneity of Regression Slopes Test Results

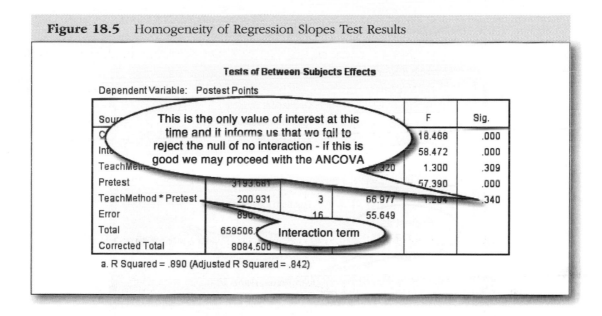

a. R Squared = .890 (Adjusted R Squared = .842)

Main Analysis for ANCOVA

We will assume that you are ready to continue. So with SPSS running and *covar.sav* open, follow the bullets given below to start the actual ANCOVA test.

- Click **Analyze**, select **General Linear Model**, and then click **Univariate**. A window titled *Univariate* will open.
- If you are continuing from the test for homogeneity of regression slopes, the *Univariate* window will open with your prior work, as shown in Figure 18.3. It is important that you click **Reset** at the bottom of the window before proceeding.
- Click **Posttest Points**, and then click the right arrow to move it to the *Dependent Variable:* box.
- Click **Teaching Method**, and then click the right arrow to move it to the *Fixed Factor(s):* box.
- Click **Pretest Points**, and then click the right arrow to move it to the *Covariate(s):* box. Figure 18.6 shows what the *Univariate* window should look like just before clicking the **Options** button in the next bullet.
- Click **Options**, and the *Univariate: Options* window opens. Click **Descriptive Statistics**, and then click **Homogeneity Tests**.

Figure 18.6 *Univariate* Window for the ANCOVA Test

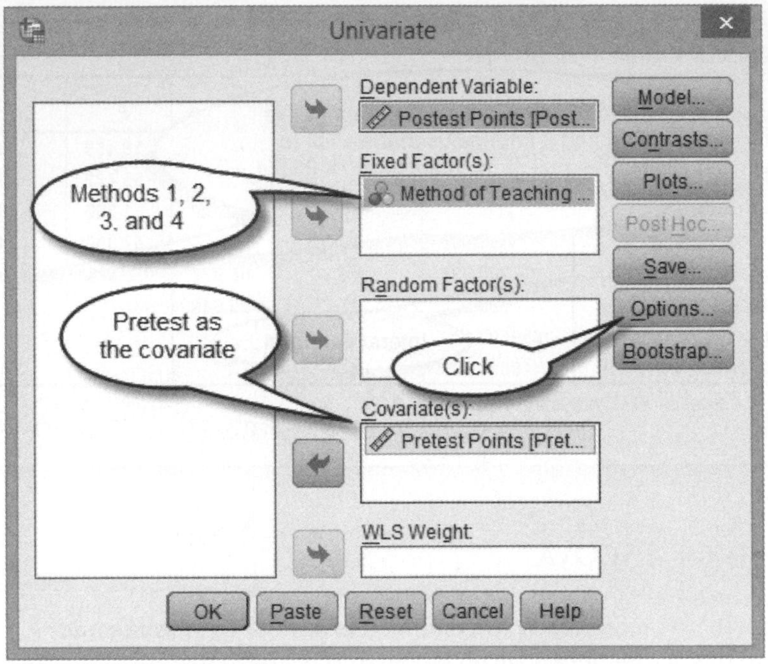

- In the *Factor(s) and Factor Interactions:* panel of the *Univariate: Options* window, click **Overall**, and then click the arrow to move it to the *Display means for:* box. Click **TeachMethod**, and then click the arrow to move it to the *Display means for:* box. Check the *Compare main effects* box, and then select **LSD** (see Figure 18.7 for the appearance of the *Univariate: Options* window at this time).

Click **Continue**, and then click **OK**. The Output Viewer now opens with the requested analysis. We next discuss each of the SPSS output tables to answer the researcher's questions concerning the various teaching methods. Figure 18.8 displays the mean posttest scores (dependent variable) for the four methods of teaching reading to third-grade students. A cursory inspection of the table of means reveals differences as great as 26.5 (Teaching Methods 1 and 4) and as little as 4.5 (Teaching Methods 2 and 3). Two additional comparisons show a difference of 15.5 points (Teaching Methods 1 and 2, Teaching Methods 3 and 4). Further analysis will show whether these values result in an overall significant difference.

Figure 18.7 *Univariate: Options* Window for "Teaching Method"

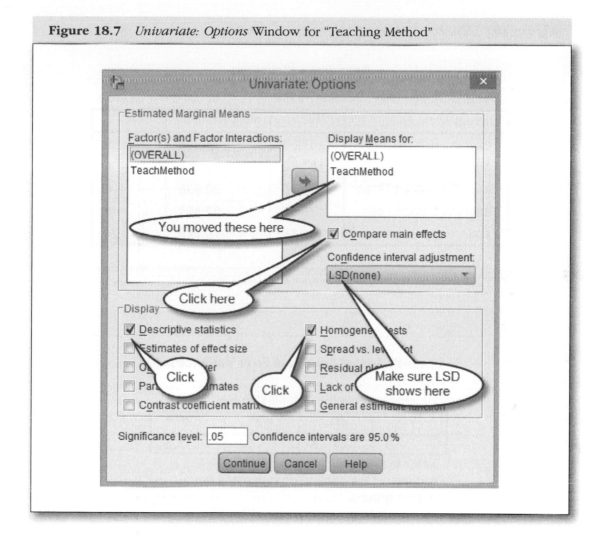

Levene's test results, shown in Figure 18.9, indicate that the assumption of equality of variances has *not* been violated, because .728 is greater than .05, meaning that we fail to reject the null hypothesis, which states that the variances are equal. Another important requirement for the ANCOVA test has been met.

The main results from the ANCOVA analysis are presented in the table titled *Tests of Between-Subjects Effects*, shown in Figure 18.10. We first want to see if our four teaching methods result in significantly different scores on the dependent variable ("Posttest"). To answer this question, we look at the row in the table for the independent variable group ("TeachMethod") and

Figure 18.8   Descriptive Statistics for Teaching Methods

Descriptive Statistics

Dependent Variable: Postest Points

Method of Teaching	Mean	Std. Deviation	N
Teaching Method One	151.50	14.816	6
Teaching Method Two	167.00	6.957	6
Teaching Method Three	162.50	20.236	6
Teaching Method Four	178.00	22.494	6
Total	164.75	18.748	24

Figure 18.9   Levene's Test of Equality of Error

Levene's Test of Equality of Error Variances[a]

Dependent Variable: Postest Points

F	df1	df2	Sig.
.438	3	20	.728

Tests the null hypothesis that the error variance of the dependent variable is equal across groups.

a. Design: Intercept + Pretest + TeachMethod

read across to the *Sig.* column. In our analysis, we find that *Sig.* is .04—it is less than .05; therefore, we can reject the null hypothesis of equal means. We now have evidence that the four teaching methods did influence the reading scores on the posttest after controlling for the effect of the pretest.

Another important finding is found in the row that presents results for the covariate ("Pretest"). This row will tell us whether there is a significant association between the dependent variable ("Posttest") and the covariate when the independent variable group ("TeachMethod") is statistically controlled. The *Sig.* for our covariate is .000; therefore, it is significant.

Figure 18.10 Main ANCOVA Results for Tests of Between-Subjects Effects

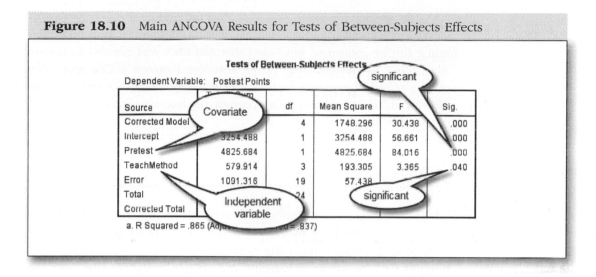

Tests of Between-Subjects Effects

Dependent Variable: Postest Points

Source	Type III Sum of Squares	df	Mean Square	F	Sig.
Corrected Model		4	1748.296	30.438	.000
Intercept	3254.488	1	3254.488	56.661	.000
Pretest	4825.684	1	4825.684	84.016	.000
TeachMethod	579.914	3	193.305	3.365	.040
Error	1091.316	19	57.438		
Total		24			
Corrected Total					

a. R Squared = .865 (Adjusted R Squared = .837)

(callouts: "significant", "Covariate", "Independent variable", "significant")

Now that overall significance has been established, we examine the *Pairwise Comparisons* table to discover which of the teaching methods are significantly different. This information is shown in Figure 18.11. As you can see, Methods 1 and 2 and 1 and 3 are significantly different. Therefore, we may reject H_{02}: $\mu_1 = \mu_2$ and H_{03}: $\mu_1 = \mu_3$, which provides statistical evidence in support of the alternative hypotheses (H_{A2}: $\mu_1 \neq \mu_2$ and H_{A3}: $\mu_1 \neq \mu_3$).

Figure 18.11 Pairwise Comparisons for Teaching Methods

Pairwise Comparisons

Dependent Variable: Postest Points

(I) Method of Teaching	(J) Method of Teaching	Mean Difference (I-J)	Std. Error	Sig.[b]	95% Confidence Interval for Difference[b] Lower Bound	Upper Bound
Teaching Method One	Teaching Method Two	-12.521*	4.388	.010	-21.705	-3.338
	Teaching Method Three	-11.000*		.021	-20.158	-1.842
	Teaching Method Four	-5.648		.266	-15.970	4.674
Teaching Method Two	Teaching Method One	12.521*	4.388	.010		21.705
	Teaching Method Three	1.521	4.388	.733		
	Teaching Method Four	6.873		.108		
Teaching Method Three	Teaching Method One	11.000*	4.376	.021	1.842	20.158
	Teaching Method Two	-1.521	4.388	.733	-10.705	7.662
	Teaching Method Four	5.352	4.932	.291	-4.970	15.674
Teaching Method Four	Teaching Method One	5.648	4.932	.266	-4.674	15.970
	Teaching Method Two	-6.873	4.790	.168	-16.899	3.154
	Teaching Method Three	-5.352	4.932	.291	-15.674	4.970

Based on estimated marginal means

*. The mean difference is significant at the .05 level.

(callouts: "Teaching methods 1 and 2 are significant", "Same as above", "Same as above", "Methods 1 and 3 are significant")

As an additional project, you may wish to conduct a one-way ANOVA (we did) to find out if it also detects significant differences. It did not. This finding further confirmed our belief that ANCOVA was needed to statistically control for the influence of the covariate ("Pretest"). The researcher now has statistical evidence that the four teaching methods resulted in significantly different means on the posttest reading scores for these third-grade children.

△   18.5 SUMMARY

The ANCOVA is used to statistically control for factors that cannot be, or were not, controlled for methodologically by methods such as random assignment. If the researcher has evidence that a factor or factors cannot be controlled for methodologically and might have an effect on the dependent variable, he or she should statistically control for these factors via an ANCOVA. Such factors should have a linear relationship with the independent variable.

In the following chapter, we present a descriptive statistic known as *correlation* as a way of introducing another family of similar statistical procedures. These descriptive procedures will emphasize the exploration of the strength and direction (positive or negative) of the relationships between variables. Correlation will then provide a foundation for additional chapters on various types of regression. Regression adds the rather exciting ability to predict unknown values based on observed data.

△   18.6 REVIEW EXERCISES

18.1   A metallurgist has designed a way of increasing the strength of steel. She has discovered a chemical that is added to samples of molten metal during the manufacturing process that have already been measured for strength. These pre-additive values are recorded as the variable called "preadd" in the data table below. She believes that the "preadd" values may influence the "postadd" measure of the steel's strength. She is looking for significant differences in strength for the four different manufacturing methods. If differences are found, she wishes to identify which ones contribute to the overall significance. Can you help her select the correct statistical procedure? She also needs help in writing the null and alternative hypotheses.

	method	preadd	postadd
1	1	5.00	12.65
2	1	5.10	12.85
3	1	4.24	11.18
4	1	4.47	12.29
5	1	5.29	12.73
6	1	4.90	12.08
7	2	5.20	13.23
8	2	5.20	12.81
9	2	4.69	12.88
10	2	4.50	12.53
11	2	5.20	13.23
12	2	4.69	12.85
13	3	4.36	12.29
14	3	5.20	13.56
15	3	4.47	11.53
16	3	5.10	13.00
17	3	4.58	12.45
18	3	4.90	12.92
19	4	6.00	14.00
20	4	5.20	12.96
21	4	5.66	13.96
22	4	5.57	13.71
23	4	4.36	11.70
24	4	5.57	13.56

18.2 A botanist measured the 3-day growth, in inches, of his marijuana plants at two different times (variables: "pregrowth" and "postgrowth") under four different growing conditions (variable: "peatsoil"). He felt that the initial growth rate influenced the second rate of growth. The

	peatsoil	pregrowth	postgrowth
1	4	1.40	2.20
2	4	1.41	2.22
3	4	1.26	2.10
4	4	1.30	2.18
5	4	1.45	2.21
6	4	1.38	2.16
7	3	1.43	2.24
8	3	1.43	2.21
9	3	1.34	2.22
10	3	1.32	2.20
11	3	1.43	2.24
12	3	1.34	2.22
13	2	1.28	2.18
14	2	1.43	2.26
15	2	1.30	2.12
16	2	1.48	2.27
17	2	1.32	2.19
18	2	1.38	2.22
19	1	1.56	2.29
20	1	1.43	2.23
21	1	1.51	2.29
22	1	1.49	2.27
23	1	1.28	2.14
24	1	1.49	2.26

scientist's main concern was the effect of soil type on growth rate during the second growth period. The problem was that he somehow wanted to statistically account for any differences in the second growth period that might be related to the first rate of growth. His ultimate quest was to identify any significant differences in the four samples that were grown in soils containing different percentages of peat. Select the correct statistical method, write the null and alternative hypotheses, do the analysis, interpret the results, and answer the botanist's questions.

18.3   An epidemiologist/psychologist was interested in studying the effects of early-childhood vaccinations and cognitive ability. He obtained records on randomly selected children who had received three levels of vaccinations during their first year of life. He randomly placed them in three groups defined by rates of vaccination ("vaccinated" is the nominal variable), where 1 = *high*, 2 = *low*, and 3 = *none*. The children had been tested for cognitive ability at 5 years of age ("precog" is the scale variable) and again at 10 years of age ("postcog" is another scale variable). The scientist's main reason for conducting the investigation was to search for any differential effects that the levels of vaccination might have on the children's cognitive ability. However, he was concerned about the potential effect that the "precog" scores might have on the "postcog" values. His major research question was whether the three levels of vaccination affected the children's cognitive ability at 10 years of age.

Can you help this scientist pick the appropriate statistical test that would offer a way to control for differences in the "precog" values? If you can, then write the null and alternative hypotheses, run the analysis, interpret the results, and answer his questions. The data are given in the table below:

	vaccinated	precog	postcog
1	1	50	49
2	1	63	55
3	1	45	46
4	1	78	69
5	1	54	53
6	1	81	80
7	2	43	44
8	2	65	66
9	2	43	42
10	2	75	73
11	2	42	40
12	2	80	74
13	3	34	45
14	3	35	43
15	3	67	87
16	3	41	55
17	3	78	82
18	3	53	65

PEARSON'S CORRELATION AND SPEARMAN'S CORRELATION

19.1 INTRODUCTION AND OBJECTIVES △

The preceding chapters on hypothesis testing have presented methods intended to give SPSS users tools that will assist them in fulfilling one of the major goals in statistics—the study of the relationships between variables. That goal was pursued by comparing group means, variances, and medians. We tested the null hypothesis and looked for statistically significant differences between various groups. We learned ways to develop statistical evidence in support of alternative hypotheses and thereby answer various research questions.

There are many situations in statistics where you are not comparing the scores of groups and cannot calculate means and variances. An example might be that you are interested in studying the relationship between hours of study and points earned on a test within a single sample of individuals. A situation where you are comparing two variables calls for the use of a descriptive statistic known as a *correlation coefficient*—the subject of this chapter.

We define a correlation coefficient as a numerical value and descriptive statistic that indicates the degree (strength) to which two variables are related and whether any detected relationship is positive or negative (direction). A positive relationship results if one variable increases and the other does also. A negative relationship is such that as one variable increases, the other decreases.

Two important descriptive purposes of both parametric and nonparametric correlation coefficients are to measure the strength and direction of the relationship between variables. To understand these two aspects of correlation, let's examine the hypothetical scale data presented in Figure 19.1.

Figure 19.1   Correlated Data for Two Variables

Name	Hours of Study	Points Earned
David	1	25
Julie	2	50
Mike	3	75
Maricela	4	100

A casual inspection of the data in Figure 19.1 reveals a relationship between the two variables. As the variable "Hours of Study" increases, so does the variable "Points Earned." We have evidence of a strong relationship since each 1 hour of study resulted in the student earning an additional 25 points. We say there is a linear relationship between hours of study and points earned. Furthermore, we say that the direction is positive since both variables increase together. The calculated correlation coefficient quantifies the strength and specifies the direction. If we used SPSS to calculate the correlation coefficient for these data, we would find its value to be +1.00. What does this value mean?

The plus sign indicates a positive correlation, whereas the value of 1.00 indicates a perfect correlation between hours of study and points earned on the exam. A calculated correlation coefficient can take on any number between 0 and positive or negative 1 (e.g. .43, .95, .88, −.35, −.99, or −.06). The closer the positive or negative correlation coefficient is to +1.00 or −1.00, the stronger the relationship. A correlation coefficient of 0 indicates that there is no relationship between the two variables.

There are many types of correlation coefficients that are designed for use with different levels of measurement. Two of the most commonly used correlation coefficients are Pearson's and Spearman's. These are the subject of our chapter on correlation. Let's take a brief look at these two types of correlation coefficients.

Pearson's product-moment correlation coefficient is used when both the variables are measured at the scale level (interval/ratio). The variables must also be approximately normally distributed and have a linear relationship. Thus, Pearson's coefficient is referred to as the parametric descriptive statistic. The data presented in Figure 19.1 fulfill these requirements; therefore, the appropriate correlation coefficient for the data would be Pearson's correlation coefficient.

When either variable is measured at the ordinal level or your scale data are not normally distributed, you should use the nonparametric Spearman correlation coefficient. For example, a researcher is interested in studying a survey instrument consisting of ranked responses. One variable ranks the importance of religion as very important, important, neutral, of little importance, and not at all important. The second variable ranks educational level as high school degree, associate degree, bachelor's degree, master's degree, and doctoral degree. To investigate the strength and direction of the relationship between religion and education, the nonparametric Spearman correlation coefficient is the correct statistic.

It must be mentioned that no matter how strong the calculated correlation coefficient might be, you cannot infer a causal relationship. Just because two variables move together in a linear manner, it does not indicate that one causes the change in the other. One of the most common examples of misuse of this statistic is when a high correlation coefficient is used to draw the conclusion that an increase (or decrease) in one variable is the cause of an increase (or decrease) in the other. The data in Figure 19.1 are a prime example—just by observing the strong relationship, we cannot say that increased study causes the increase in the number of points earned on the exam. From the information provided about hours of study and points earned, we cannot derive any information about the multitude of other variables that may have affected the points earned on the exams. We must not infer a causal relationship between the two variables.

Significance Test

SPSS automatically calculates a significance level for the correlation coefficient. In doing this, SPSS assumes that the cases were selected at random from a larger population. In this significance-testing scenario, the research question asks whether the calculated correlation coefficient can be taken seriously. When we say "taken seriously," we are simply asking whether the sample correlation coefficient is representative of the true population correlation coefficient or whether it is the result of a chance movement of the data.

As with prior chapters, we will first present the parametric statistic, Pearson's correlation coefficient, and finish the chapter with the nonparametric Spearman's.

OBJECTIVES

After completing this chapter, you will be able to

Define and describe the correlation coefficient

Describe the data assumptions required for Pearson's correlation coefficient

Explain the purpose of computing correlation coefficients

Use SPSS to calculate Pearson's and Spearman's correlation coefficients

Describe and interpret the results from Pearson's correlation coefficient

Use SPSS to conduct the test of significance for Pearson's correlation coefficient

Interpret the significance test for Pearson's correlation coefficient

Describe the data assumptions required for Spearman's correlation coefficient

Describe and interpret the results from Spearman's correlation coefficient

△ 19.2 Research Scenario and Test Selection

A statistics professor held the belief that there was a strong positive relationship between a student's scores on his or her first and second exams. Over her many years of teaching, it appeared that there was a consistent pattern—those who did well on the first exam also did well on the second. In other words, she hypothesized a linear relationship between these two variables. She wished to quantify the relationship in terms of strength and direction. Furthermore, this professor believed that the scores were normally distributed and that they represented a random sample of her population of students.

Since this professor sought to describe a relationship between variables measured at the scale level, she decided to use the descriptive statistic Pearson's product–moment correlation coefficient. Part of the study will also determine if the two exams have a linear relationship. Both distributions for the two exams will also be tested for normality.

19.3 RESEARCH QUESTION AND NULL HYPOTHESIS △

There are a number of questions that must be addressed prior to answering the professor's primary question regarding the direction and strength of any identified relationship between the two exams. First it must be determined if there is a linear relationship between the variables. Second, do both variables originate from a normally distributed population? If both of these questions are answered in the affirmative, we can then address the major question by the calculation of Pearson's correlation coefficient. This calculation will answer the professors' question regarding the strength and direction of the relationship between the two tests.

We will also use the SPSS program to seek evidence in support of the idea that the calculated coefficient can be taken seriously. The alternative hypothesis states that the population correlation coefficient does not equal 0, and we write H_A: $\rho \neq 0$. In the expression just given, ρ (rho) is the hypothesized population correlation coefficient. The null hypothesis is that the population has a true correlation coefficient of 0, H_0: $\rho = 0$. If the null hypothesis can be rejected, then we have some evidence that the calculated correlation is not due to chance—we can take it seriously.

19.4 DATA INPUT, ANALYSIS, AND INTERPRETATION OF OUTPUT △

You will use the *class_survey1.sav* database to demonstrate Pearson's correlation coefficient. In the class survey, there are two variables that recorded the points earned on two exams. We regard these data as scores for the professor's classes, as discussed in the preceding two sections. You will use SPSS to determine the strength and direction of the relationship between the student scores on these exams. Furthermore, we regard these exam scores as representative of a larger population. Therefore, we will interpret SPSS's test of significance in an effort to provide evidence that the sample correlation coefficient represents the unknown population correlation coefficient.

Check for a Linear Relationship Between the Variables

First, we check for a linear relationship between the two exams by using SPSS's *Chart Builder*. Then, we proceed with the test for normality for both variables.

- Start SPSS, then click **Cancel** in the *SPSS Statistics* opening window.
- Click **File**, select **Open**, and click **Data** (the *Open Data* window opens).
- In the *Open Data* window, locate and Click **class_survey1.sav**, and then click **Open**.
- Click **Graphs**, and then click **Chart Builder** (the *Chart Builder* window opens; see Figure 19.2).

Figure 19.2   *Chart Builder* Window Used to Check for a Linear Relationship

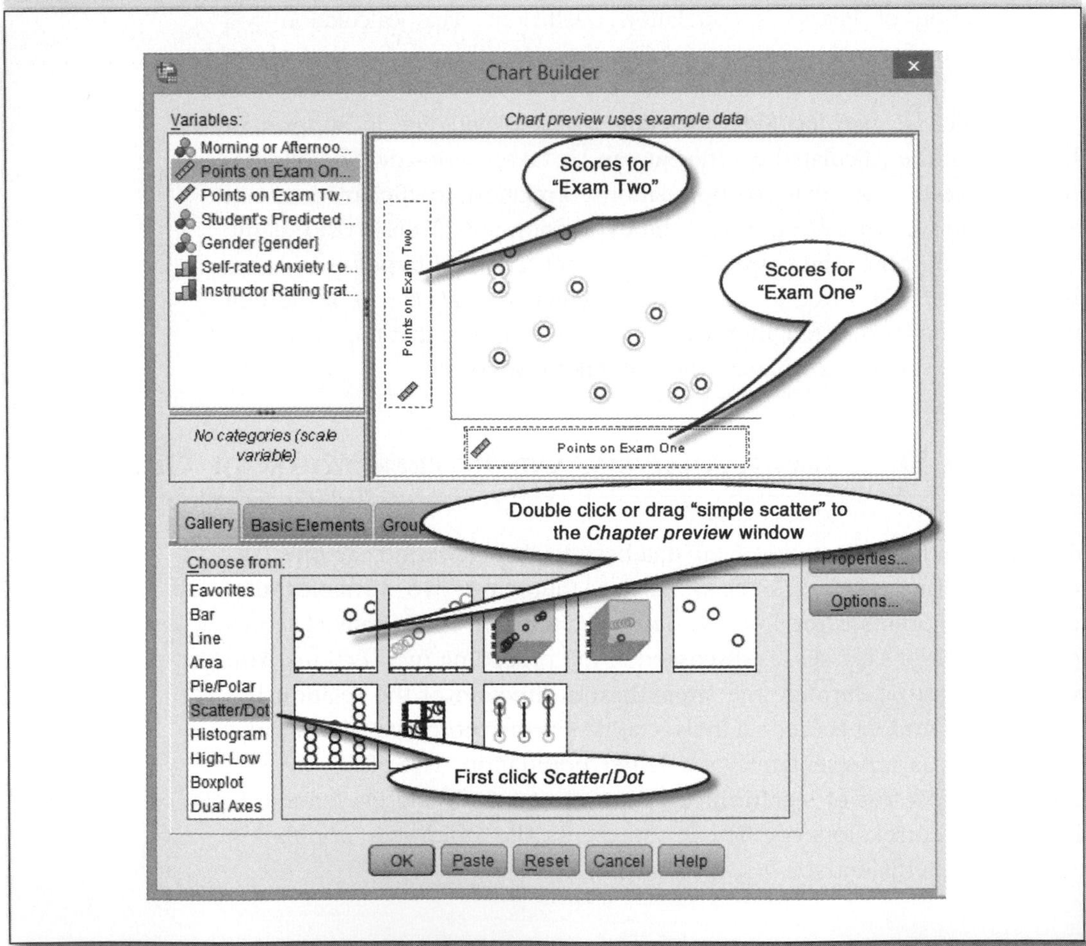

- Click **Scatter/Dot** (in the *Choose from:* panel), and then double click **Simple Scatter** (which moves this chart style to the *Chart preview* pane).
- Click and drag **Points on Exam One** to the *x*-axis in the *Chart preview* pane.

- Click and drag **Points on Exam Two** to the *y*-axis in the *Chart Builder* pane.
- Click **OK** (the Output Viewer opens; see Figure 19.3).

Figure 19.3 shows that there is a weak linear relationship between these two variables. We next check both variables to see if they are approximately normally distributed.

Figure 19.3   Scatterplot Showing a Weak/Positive Relationship Between Exams

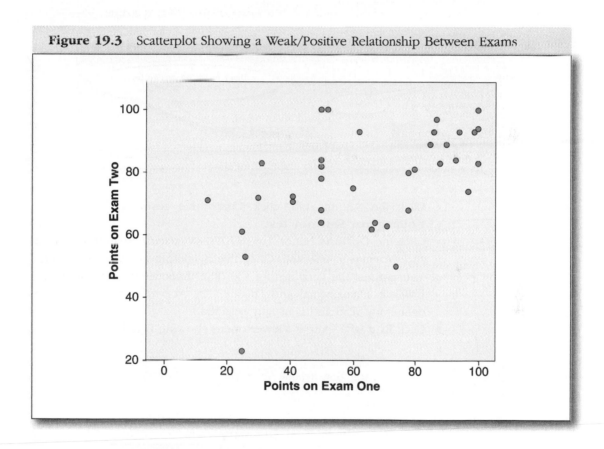

Check Both Variables for Normality

The checks for normality involve the use of the nonparametric Kolmogorov-Smirnov test as presented below.

- Click **Analyze**, select **Nonparametric Tests**, and then click **One-Sample** (the *One-Sample Nonparametric Tests* window opens—it has three tabs at its top: Objective, Fields, and Settings; see Figure 19.4).

- Click the Objective tab, and then click **Customize analysis**.
- Click the Fields tab (make sure that "Points on Exam One" and "Points on Exam Two" are the only variables in the *Test Fields:* panel as shown in Figure 19.4). (*Note:* You may have to click on the variables and use the arrow to add or remove variables in the *Test Fields:* panel to accomplish this task.)

Figure 19.4   Upper Portion of the *One-Sample Nonparametric Tests* Window

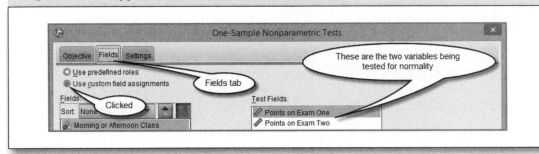

- Click the Settings tab, click **Customize tests**, and then click **Kolmogorov-Smirnov test**.
- Click the **Options** button, and in *Hypothesized Distributions* make sure **Normal** is checked. Once this is confirmed, click **OK**. (*Note:* You can test the variables for the other hypothesized distributions, *Uniform*, *Exponential*, and *Poisson*. The *Normal* distribution is the default for SPSS and is usually selected.)
- Click **Run** (the Output Viewer opens showing Figure 19.5).

Figure 19.5   Kolmogorov-Smirnov Test Results Showing Both Exam Distributions Are Normal

Hypothesis Test Summary

	Null Hypothesis	Test	Sig.	Decision
1	The distribution of Points on Exam One is normal with mean 65.946 and standard deviation 25.95.	One-Sample Kolmogorov-Smirnov Test	.199[1]	Retain the null hypothesis.
2	The distribution of Points on Exam Two is normal with mean 78.108 and standard deviation 16.55.	One-Sample Kolmogorov-Smirnov Test	.200[1,2]	Retain the null hypothesis.

Asymptotic significances are displayed. The significance level is .05.

The results of the Kolmogorov-Smirnov test are satisfactory as both distributions are determined to be normal. The significance levels are .199 (Exam 1) and .200 (Exam 2); both values exceed .05; therefore, we fail to reject the null hypothesis that the distributions are normal. We now feel confident that we have evidence supporting the use of the Pearson correlation coefficient and the hypothesis test for the exam data.

Calculation of Pearson's Correlation Coefficient and Test of Significance

- Click **Analyze**, select **Correlate**, and then click **Bivariate** (the *Bivariate Correlations* window opens; see Figure 19.6).
- Click **Points on Exam One**, and then click the arrow (which moves it to the *Variables* panel).

Figure 19.6 *Bivariate Correlations* Window—Exams Selected

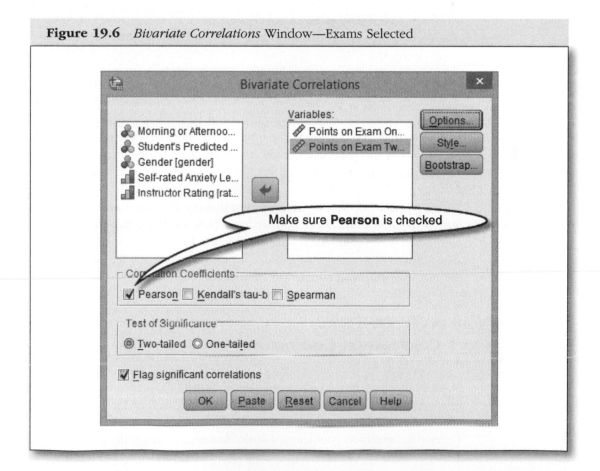

- Click **Points on Exam Two**, and then click the arrow (which moves it to the *Variables* panel).
- Click **Pearson** (if not already checked).
- Click **Flag significant correlations** (if not already checked).
- Click **OK** (the Output Viewer opens; see Figure 19.7).

Figure 19.7   Pearson's Correlation Output: Exams 1 and 2

Correlations

		Points on Exam One	Points on Exam Two
Points on Exam One	Pearson Correlation	1	.536**
	Sig. (2-tailed)		.001
	N	37	37
Points on Exam Two	Pearson Correlation	.536**	1
	Sig. (2-tailed)	.001	
	N	37	37

**. Correlation is significant at the 0.01 level (2-tailed).

Figure 19.7 shows a correlation of .536 when Exam 1 and Exam 2 are compared. We can say that they are moderately correlated in a positive direction. Figure 19.7 also indicates that the correlation coefficient of .536 is significant at the .001 level. The interpretation is that there is a very small probability (1 in 1,000) that the observed correlation coefficient was due to chance. Another way to think of this is that we now have statistical evidence that the population correlation coefficient does not equal 0. We reject the null hypothesis, H_0: $\rho = 0$ (remembering that ρ = the hypothesized population correlation coefficient), and we now have evidence in support of the alternative hypothesis, H_A: $\rho \neq 0$. We can take the correlation coefficient of .536 seriously (meaning that it did not occur by chance). We next address the use of Spearman's correlation coefficient.

△ 19.5 NONPARAMETRIC TEST: SPEARMAN'S CORRELATION COEFFICIENT

Spearman's correlation coefficient is a statistic that shows the strength of the relationship between two variables measured at the ordinal level. Spearman's correlation coefficient can also be used as the nonparametric alternative to Pearson's correlation coefficient if scale data are not normally distributed.

If the *class_survey1.sav* database is not open, locate and open it because it is used to demonstrate Spearman's correlation coefficient as applied to two ordinal variables.

- Start SPSS, and click **Cancel** in the *SPSS Statistics* opening window.
- Click **File**, select **Open**, and click **Data** (the *Open Data* window opens).
- In the *Open Data* window, find and click **class_survey1.sav**, and then click **Open**.
- Click **Analyze**, select **Correlate**, and then click **Bivariate** (the *Bivariate Correlations* window opens; see Figure 19.6 in the previous parametric section). (*Note:* You will unclick **Pearson** three bullets below.)
- Click **Self-rated Anxiety**, and then click the arrow.
- Click **Instructor Rating**, and then click the arrow.
- Unclick **Pearson**, and then click **Spearman** (leave *Flag significant correlations* checked).
- Click **OK** (the Output Viewer opens; see Figure 19.8).

Figure 19.8   Spearman's Correlation Output: Self-Rated Anxiety and Instructor Rating

Correlations

			Instructor Rating	Self-rated Anxiety Level
Spearman's rho	Instructor Rating	Correlation Coefficient	1.000	-.073
		Sig. (2-tailed)	.	.669
		N	37	37
	Self-rated Anxiety Level	Correlation Coefficient	-.073	1.000
		Sig. (2-tailed)	.669	.
		N	37	37

"Instructor Rating" and "Self-rated Anxiety Level" are the two ordinal variables chosen to demonstrate Spearman's correlation coefficient. We are interested in studying these data to determine if there was a relationship between how a student rated the instructor's performance and his or her self-rated level of anxiety.

Figure 19.8 reports that the Spearman correlation coefficient is $-.073$ with a significance level of .669. What does this mean for our investigation of instructor rating and self-rated anxiety level? First of all, the r_s of $-.073$ informs us that there is no relationship between these two ordinal variables. Knowing how a person rated the instructor would be of no help in predicting his or her level of anxiety, and vice versa. We also note that the null hypothesis, which stated that the population correlation coefficient equals 0, was not rejected. From this, we conclude that our finding of a low correlation coefficient could be attributed to chance. We are unable to take the calculated correlation seriously, and additional study would be advised.

△ 19.6 SUMMARY

This chapter placed an emphasis on the analysis of scale and ordinal data already collected. We studied the relationships between two variables and then answered questions regarding the strength and direction of any detected relationship. As a statistical technique, correlation is descriptive rather than inferential. However, there was an inferential aspect presented in our correlation chapter. Hypothesis tests were used to answer the question whether the calculated correlation coefficient is representative of what exists in the population. Stated succinctly, the question is "Can we take the observed sample correlation seriously, or could it be the result of chance?" In the following chapter, we extend these correlation concepts to regression analysis.

△ 19.7 REVIEW EXERCISES

19.1 Assume you have collected a random sample of first-year students at a local community college and given them a general survey that included a number of items. A series of questions results in self-esteem ratings, and part of their official record includes their IQ. You want to calculate a correlation coefficient for these two variables including a significance level and then chart the results and add the *Fit Line*. Select the correct correlation coefficient, write the null and alternate hypotheses, and interpret the results. A summary of the data follows:

	selfesteem	IQ
1	3.8	110
2	4.2	130
3	3.9	126
4	4.1	127
5	3.7	128
6	3.5	128
7	4.5	135
8	4.5	149
9	3.7	135
10	4.7	140
11	3.8	131
12	4.2	142

19.2 Let's say you live on a little used back road that leads to the ski slopes. Over the years, you have noticed that there seems to be a correlation between the number of inches of snowfall and traffic on your road. You collect some data and now wish to analyze them using correlation and a test of significance. You also wish to visualize the data on a graph that includes a *Fit Line*. Write the null and alternative hypotheses, calculate the coefficient and the significance level, and then build the graph. The data are as follows:

	snowfall	cars
1	12.7	23
2	13.0	16
3	6.0	10
4	23.0	32
5	10.0	12
6	20.0	28
7	15.0	17
8	24.0	32
9	16.0	18
10	11.0	10
11	14.0	14

19.3 Assume you own a furniture store and you decided to record a random sample of rainy days and the number of patrons on those days. Calculate the correlation coefficient, build a graph, and test the numbers for significance. The data are as follows:

	rainfall	storepat
1	1.20	446.00
2	2.95	235.00
3	5.12	123.00
4	6.20	72.00
5	4.10	174.00
6	2.21	347.00
7	4.46	156.00
8	5.49	97.00
9	2.49	293.00
10	6.48	46.00
11	1.45	393.00
12	3.49	197.00

Chapter 20

Single Linear Regression

△ 20.1 Introduction and Objectives

Linear regression is actually an extension of Pearson's (linear) correlation. In Chapter 19, we demonstrated how two variables could be described in terms of the strength and direction of their relationship. We also discussed how linear correlation could assist us in predicting the value of one variable through the knowledge of the other. An example of such a prediction would be that the manager of a public swimming pool could observe a day's temperature and record the number of patrons at the pool. Several days' temperature could be observed, and the information could be used to estimate the number of expected patrons for any day's temperature just by looking at the data in an informal way. Let's explain how this informal use of correlation might be expanded to *single linear regression*—the subject of this chapter.

Single linear regression is a statistical technique that describes the relationship between two variables by the calculation of a prediction equation. The prediction equation averages all prior observed relationships between two variables. These averages are then used to develop a precise equation to predict unknown values of the dependent variable.

The descriptive word *single* in *single linear regression* refers to the number of independent variables. In single linear regression, *one* independent variable is used to predict the value of the dependent variable.

Single-variable regression is contrasted with multiple regression, where you have two or more independent variables. Multiple regression is discussed in Chapter 21.

As mentioned in the first paragraph, once we have observed the number of swimmers and that day's temperature over a period of, say, 30 days, we can use those data to write a precise mathematical equation that describes the observed relationship. The equation is known as the prediction equation and is the basic component of regression analysis. Once you have written the prediction equation (using past observations), you can insert any day's temperature into the equation. For the purposes of our illustration, let's pick 105 degrees. The day's temperature is the independent variable. You could then solve the equation for 105 degrees and predict the value of the dependent variable. The dependent variable is the number of people seeking the cool water of the pool on a day having a temperature of 105 degrees. The techniques described also make it possible to determine the goodness of one's predictions. SPSS accomplishes this by applying the principles associated with inferential statistics, hypothesis testing, and significance levels. Therefore, we use the power of SPSS to calculate (write) the prediction equation, use the equation to make predictions, and then determine whether the prediction is statistically significant.

OBJECTIVES

After completing this chapter, you will be able to

Describe the purpose of single linear regression

Input variable information and data for single linear regression

Describe the data assumptions required for single linear regression

Interpret scatterplots concerning the data assumptions for regression

Interpret probability plots concerning the data assumptions for regression

Use SPSS to conduct single linear regression analysis

Describe and interpret the SPSS output for single linear regression

Interpret the coefficients table for single linear regression

Write the prediction equation and use SPSS to make predictions

Write the prediction equation and use a calculator to make predictions

△ 20.2 RESEARCH SCENARIO AND TEST SELECTION

The scenario used to explain the SPSS regression function centers on a continuation of the pool example presented in the introduction. You will enter hypothetical data for the "number of patrons" (dependent variable) at a public swimming pool and that "day's temperature" (independent variable). The research will investigate the relationship between a day's temperature and the number of patrons at the public swimming pool. The researcher also wishes to develop a way to estimate the number of patrons based on a day's temperature. It appears that the single linear regression method might be appropriate, but there are data requirements that must be met.

One data requirement (assumption) that must be met before using linear regression is that the distributions for the two variables must approximate the normal curve. There must also be a linear relationship between the variables. Also, the variances of the dependent variable must be equal for each level of the independent variable. This equality of variances is called homoscedasticity and is illustrated by a scatterplot that uses standardized residuals (error terms) and standardized prediction values. And yes, we must assume that the sample was random.

△ 20.3 RESEARCH QUESTION AND NULL HYPOTHESIS

The current research investigates the relationship between a day's temperature and the number of patrons at the swimming pool. We are interested in determining the strength and direction of any identified relationship between the two variables of "temperature" and "number of patrons." If possible, we wish to develop a reliable prediction equation that can estimate the number of patrons on a day having a temperature that was not directly observed. We also wish to generalize to other days having the same temperature and to specify the number of expected patrons on those days.

The researcher wishes to better understand the influence that daily temperature may have on the number of public pool patrons. The alternative hypothesis is that the daily temperature directly influences the number of patrons at the public pool. The null hypothesis is the opposite: The temperature has no significant influence on the number of pool patrons.

20.4 DATA INPUT △

In this section, we enter the hypothetical data for 30 summer days, selected at random, from the records of a public swimming pool in a Midwestern city. Patrons were counted via ticket sales, and temperature was recorded at noon each day. These observations are then analyzed using the single linear regression procedure provided in SPSS. As in the recent chapters, you are not given detailed instructions on entering the variable information and the data—most likely you are quite proficient at this by now. Begin by entering the variable information into the Variable View screen, as depicted in Figure 20.1.

Figure 20.1   Variable View Screen After Entering the Variable Information

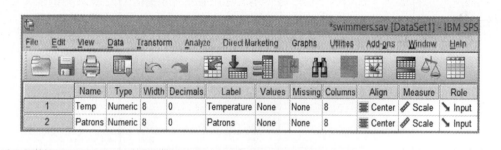

- Start SPSS, and then click **Cancel** in the opening window.
- Click **File**, select **New**, and click **Data**.
- Click the Variable View tab, and enter all the variable information as shown in Figure 20.1.
- Click the Data View tab, and enter all the data as shown in Figure 20.2 (please check your data for accuracy).
- Click **File**, then click **Save As**. Then type *swimmers* in the *File name* box.
- Click **Save** (your variable information and data are now saved as *swimmers.sav.*)

You have now entered the variable information and data for the 30 days for which the number of swimmers and day's temperature were recorded. Let's proceed with bullet points that will lead you through the regression analysis. Throughout these steps, you will be given explanations about the various outputs that are produced.

Figure 20.2 Daily Temperature and Patron Information for the *Swimmers* Database

	Temp	Patrons
1	99	281
2	78	199
3	89	241
4	108	321
5	86	250
6	69	170
7	95	279
8	96	254
9	78	200
10	95	281
11	80	201
12	85	245
13	85	201
14	92	279
15	70	159
16	92	265
17	83	199
18	97	281
19	86	201
20	87	250
21	115	360
22	103	319
23	109	321
24	104	319
25	76	161
26	99	279
27	77	180
28	96	239
29	65	121
30	95	241

△ 20.5 DATA ASSUMPTIONS (NORMALITY)

At this point, it is necessary to test both distributions for normality, which is easily accomplished by following the bullet points given next. It will take a few clicks; it may seem like more than a few, but with very little practice it easily becomes routine.

- We assume that the *swimmers* database is open; if not, open it at this time.
- Click **Analyze**, select **Nonparametric Tests**, and then click **One-Sample** (the *One-Sample Nonparametric Tests* window opens; you have seen it before—it has three tabs at the top: Objective, Fields, and Settings).
- Click the Objective tab, and then click **Customize analysis**.
- Click the Fields tab (if the variables "temperature" and "patrons" are not in the *Test Fields* pane, then move them by selecting and clicking the arrow).
- Click the Settings tab, click **Customize tests**, and then click **Kolmogorov-Smirnov test**.
- Click **Options** (the *Options* window opens), and look to see if *Normal* and *Use sample data* are checked; if not, do so.
- Click **Run** (the Output Viewer opens showing Figure 20.3).

Figure 20.3   Kolmogorov-Smirnov Test Results Showing Both Distributions Are Normal

Hypothesis Test Summary

	Null Hypothesis	Test	Sig.	Decision
1	The distribution of Temperature is normal with mean 89.633 and standard deviation 12.31.	One-Sample Kolmogorov-Smirnov Test	.200[1,2]	Retain the null hypothesis.
2	The distribution of Patrons is normal with mean 243.233 and standard deviation 57.34.	One-Sample Kolmogorov-Smirnov Test	.200[1,2]	Retain the null hypothesis.

Asymptotic significances are displayed. The significance level is .05.

[1]Lilliefors Corrected

[2]This is a lower bound of the true significance.

The results of the *Kolmogorov-Smirnov* test (think of vodka, although, in my opinion, Scotch is better) are satisfactory as both distributions are determined to be normal. This is another time when we hope that the data do not lead to a rejection of the null hypothesis. The null states that our distributions are normal, so we are hopeful that they do not depart from the normal to a great extent. In this case, the analysis provided by SPSS has provided evidence that the variables approximate a normal distribution. We may now move forward with our regression analysis.

△ 20.6 Regression and Prediction

When using SPSS to conduct regression analysis, it is easy to become over-whelmed by the number of options available for analysis and also by the resulting output. In this book, we will introduce just those options that will enable you to do basic regression analysis. As you become more proficient, you will go far beyond what is offered here.

Let's request that SPSS check for linearity, equal variances, and residuals (error terms) for normality. Assuming that the *swimmers* database is open, follow the procedure presented next. Once you have completed the bullet points, the output is explained in separate sections, which immediately follow.

As mentioned earlier, once you have finished with the bullet points below, a large volume of output is produced. We will explain the output tables and graphs one at a time. Some will be explained in an order different from the SPSS output. So please pay careful attention, and you should have no difficulty interpreting the output you generate.

- Click **Analyze**, select **Regression**, and then click **Linear**.
- Click **Patrons**, and then click the arrow next to the *Dependent:* box.
- Click **Temperature**, and then click the arrow next to the *Independent(s):* box (at this point, your screen should look like Figure 20.4).

Figure 20.4   *Linear Regression* Window for Swimmer Data

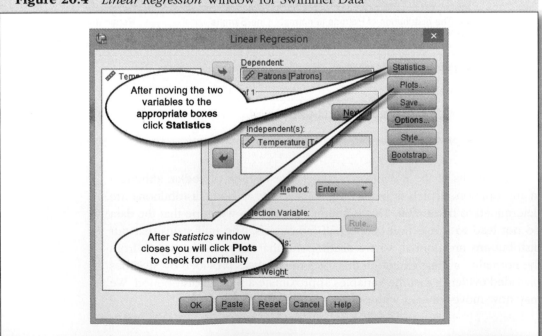

- Click **Statistics** (the *Linear Regression: Statistics* window opens; see Figure 20.5) (make sure that *Estimates* and *Model fit* are checked as shown).

Figure 20.5 The *Linear Regression: Statistics* Window

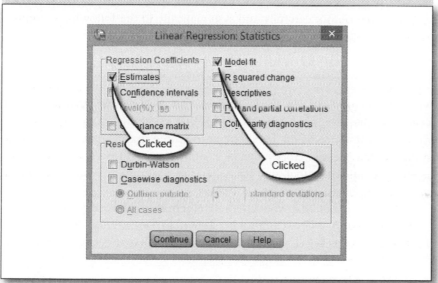

- Click **Continue**.
- Click **Plots** (the *Linear Regression: Plots* window opens; see Figure 20.6).
- Click ***ZPRED**, and then click the arrow next to the *Y:* box.

Figure 20.6 The *Linear Regression: Plots* Window

- Click ***ZRESID**, and then click the arrow next to the *X:* box.
- Click **Normal probability plot**.
- Click **Continue**, and then click **OK**.

Once you click **OK**, the Output Viewer opens with five tables and two graphs—please don't be dismayed. We will interpret and explain the relevant aspects of the output carefully and systematically in the following sections.

△   20.7 Interpretation of Output (Data Assumptions)

We continue our analysis and interpretation of the SPSS output by first looking at the tests intended to confirm that additional data requirements for regression are satisfied. Let's look at the three graphs that can be found in the Output Viewer. Hopefully, an examination of these graphs will give us more confidence that the data met the assumptions required when choosing the linear regression procedure.

You may recall that you checked **Normal probability plot** in the *Linear Regression: Plots* window, as shown in Figure 20.6. The result of this request is shown in Figure 20.7. The fact that the small circles are

Figure 20.7   Normal P-P Plot: Standardized Residuals (Error Terms)

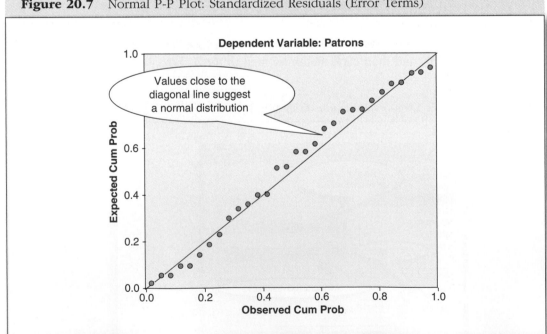

close to the straight line provides evidence that the error terms (the difference between the predicted and observed values) are indeed normally distributed. If the small circles stray far from the line, then we must assume nonnormal data and seek statistical procedures other than linear regression.

The final plot, Figure 20.8, results from the requests we made as shown in Figure 20.6. The scatterplot combines the standardized predicted values (*ZPRED*) with the values for the standardized residuals (*ZRESID*). For our regression to be accurate, there must be equal variability in the dependent variable for each level of the independent variable. Note that the plot you produce will not have the line titled "Reference line" for the *y*-axis, as shown in Figure 20.8. We have added the line to help you visualize how the bivariate variances are distributed along the reference line. These points represent the standardized residuals and the standardized predicted values. Looking at Figure 20.8, we see that the small circles are randomly dispersed, with a corresponding lack of pattern. Such an appearance indicates equal variances in the dependent variable for each level of the independent variable. This data characteristic is referred to as homoscedasticity and is a requirement for using linear regression. We can say that this requirement of linear regression is satisfied and may confidently continue with our regression analysis.

Figure 20.8 Standardized Residuals and Standardized Predicted Values

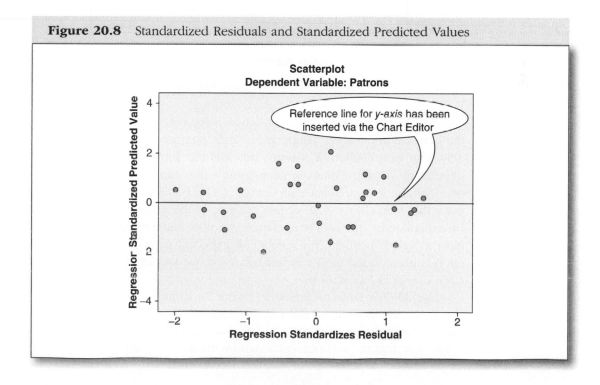

Now that the data assumptions required for using the single linear regression approach have been met, we will interpret the actual regression analysis.

△ 20.8 INTERPRETATION OF THE OUTPUT (REGRESSION AND PREDICTION)

Let's begin our interpretation of the regression output with the table titled *Model Summary* (shown in Figure 20.9) and see what can be learned about the relationship between our two variables. Most of the information in *Model Summary* deals with the strength of the relationship between the dependent variable and the model.

Figure 20.9   Model Summary for "Number of Patrons" (Dependent Variable)

Model Summary[b]

Model	R	R Square	Adjusted R Square	Std. Error of the Estimate
1	.959[a]	.920	.917	16.497

a. Predictors: (Constant), Temperature

b. Dependent Variable: Patrons

R is the correlation coefficient between the two variables; in this case, the correlation between "temperature" and "number of patrons" is high at .959. The next column, *R Square*, indicates the amount of change in the dependent variable ("number of patrons") that can be attributed to our one independent variable ("temperature"). The *R Square* value of .920 indicates that 92% (100 × .920) of the variance in the number of patrons can be explained by the day's temperature. We now begin to conclude that we have a "good" predictor for number of expected patrons when consideration is given to the day's temperature. Next, we will examine the ANOVA table shown in Figure 20.10.

The ANOVA table presented in Figure 20.10 indicates that the model can accurately explain variation in the dependent variable. We are able to say this since the significance value of .000 informs us that the probability is very low that the variation explained by the model is due to chance. The

Figure 20.10 ANOVA Table Indicating a Significant Relationship Between the Variables

ANOVA[a]

Model		Sum of Squares	df	Mean Square	F	Sig.
1	Regression	87713.492	1	87713.492	322.312	.000[b]
	Residual	7619.875	28	272.138		
	Total	95333.367	29			

a. Dependent Variable: Patrons
b. Predictors: (Constant), Temperature

Significant F statistic

conclusion is that changes in the dependent variable resulted from changes in the independent variable. In this example, changes in daily temperature resulted in significant changes in the number of pool patrons.

Prediction

Now comes an interesting (and, for many, exciting) part of the analysis, where the *Transform* and *Compute Variable* functions of SPSS and the regression output are used to define and then write the prediction equation. Using this equation, we will predict unknown values based on past observations.

The *Coefficients* table presented in Figure 20.11 is most important when writing and using the prediction equation. Please don't glaze over it; but we must present some basic statistics before you can use SPSS to do the tedious work involved in making predictions. The prediction equation takes the following form:

$$\hat{y} = a + bx,$$

where $\hat{y}$ is the predicted value, a the intercept, b the slope, and x the independent variable.

Let's quickly define a couple of terms in the prediction equation that you may not be familiar with. The *slope* (b) records the amount of change in the dependent variable ("number of patrons") when the independent variable ("day's temperature") increases by one unit. The *intercept* (a) is the value of the dependent variable when $x = 0$.

In simple words, the prediction equation states that you multiply the slope (b) by the values (x) of the independent variable ("temperature")

Figure 20.11   Coefficient Values for the Prediction Equation

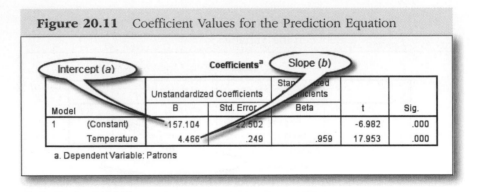

Model		Unstandardized Coefficients		Standardized Coefficients		
		B	Std. Error	Beta	t	Sig.
1	(Constant)	-157.104	22.502		-6.982	.000
	Temperature	4.466	.249	.959	17.953	.000

a. Dependent Variable: Patrons

and then add the result of the multiplication (bx) to the intercept (a)—not too difficult. But where (in all our regression output) do you find the values for the intercept and the slope? Figure 20.11 provides the answers. (*Constant*) is the intercept (a), and *Temperature* is the slope (b). The x values are already recorded in the database as *temp*—you now have everything required to solve the equation and make predictions. Substituting the regression coefficients, the slope and the intercept, into the equation, we find the following:

$$\hat{y} = -138.877 + (4.113x).$$

The x value represents any day's temperature that might be of interest and each of those temperatures recorded during our prior data collection.

Let's put SPSS to work and use our new prediction equation to make predictions for all the observed temperature values. By looking at the observed numbers of patrons and the predicted number of patrons, we can see how well the equation performs.

The following procedure assumes that SPSS is running and that the *swimmer* database is open.

- Click **Transform**, and then click **Compute Variable** (the *Compute Variable* window opens; see Figure 20.12, which shows the upper portion of the window with the completed operations as described in the following bullet points).
- Click the *Target Variable* box, and then type *pred_patrons*.
- Click the *Type & Label* box (the *Compute Variable: Type and Label* window opens), and then type *Predicted number of patrons* in the box titled *Label*.
- Click **Continue** (this returns you to the *Compute Variable* window).
- Click the Numeric Expression box.

- Click the minus button, and then use the keypad or type *157.104.* After this, click the plus button, then click the parenthesis button, then click inside the parentheses, and use the keypad or type *4.466.* Next, click the asterisk button, then select the variable "Temperature," and click the arrow, which moves it into the *Numeric Expression* box as part of the prediction equation (if you prefer, you can directly type the equation into the *Numeric Expression* window). Carefully check your completed equation with the one shown in Figure 20.12.
- Click **OK** at the bottom of the *Compute Variable* window (the bottom portion of this window is not shown in Figure 20.12).

Figure 20.12   Upper Portion of the Completed *Compute Variable* Window Showing the New Variable as It Appears in the Database

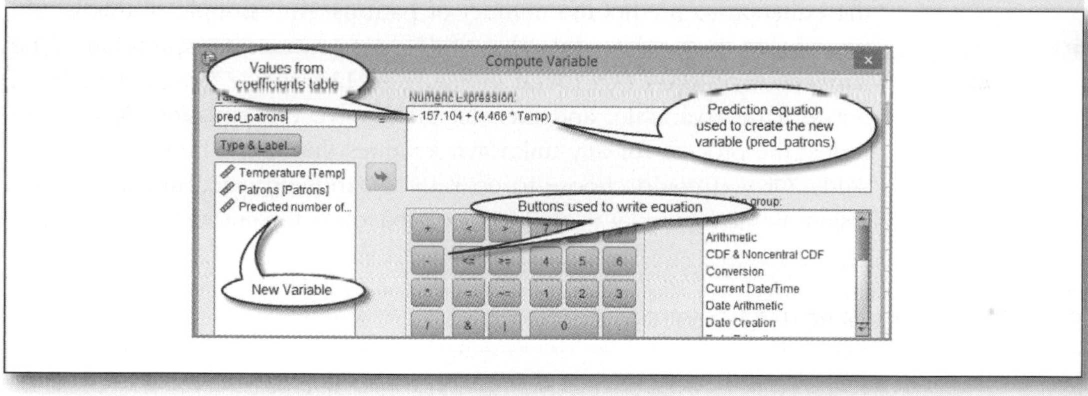

Once you click **OK**, SPSS creates a variable called "pred_patrons" and automatically inserts the new variable into your *swimmers* database. Your Output Viewer will open up, showing that a new variable was created and inserted into your original database. This message is shown in Figure 20.13. You will have to open your database to view the results of the data transformation via the prediction equation that you wrote.

Reading these predicted values in conjunction with the observed daily temperatures informs us as to what the equation would predict for each daily temperature. An example would be Case 1, where the day's temperature was 99 degrees, 281 patrons were observed, and the equation predicted 285. You would be safe to use 285 as an estimate of the number of patrons expected on days having a temperature of 99 degrees. When using the equation, you must, of course, consider any intervening circumstances, such as thunderstorms and holidays. Remember that the prediction is based on average observations.

Figure 20.13 Output Viewer Showing the Data Transformation Completed

```
DATASET ACTIVATE DataSet1.
DATASET CLOSE DataSet0.                          Prediction Equation
COMPUTE pred_patrons= - 157.04 + (4.466 * Temp).
VARIABLE LABELS  pred_patrons 'Predicted number of patrons'.
EXECUTE.
```

There are also other uses for the prediction equation, such as inserting a value into the equation that was not directly observed and then using the equation to predict the number of patrons. An example of this would be to insert the unobserved value of 83 degrees into the expression in the *Numeric Expression* box found in Figure 20.12 (replace "temp" with "83"), create a new variable, and then let SPSS solve the equation. You could repeat the process for any unknown x values that might be useful in your study. Or you could choose to do it by a handheld calculator as follows. Using $\hat{y} = a + bx$ or $\hat{y} = -157.104 + (4.466 \times 83)$, the answer is $\hat{y} = 213.57$.

△ 20.9 Research Question Answered

The single linear regression was chosen to investigate whether the number of swimming pool patrons was influenced by the day's temperature. The alternative hypothesis was that there was a direct relationship because the number of patrons was affected by the daily temperature. The null hypothesis was the opposite, that daily temperature did not affect the number of swimming pool patrons. How well did our single regression method answer these questions? Was the null hypothesis rejected? Was it possible to generate any statistical support for our alternative hypothesis?

We were able to support the proposition that single linear regression was applicable to our research question since the data assumptions were met through our statistical analysis (see Figures 20.3, 20.7, and 20.8). The prediction equation (model) developed was discovered to reduce the error in predicting the number of pool patrons by 92% (see Figure 20.9 for the *R Square* of .920, or 92%). The significant *F* of .000 provided evidence that there was an extremely low probability (less than .0005) that the daily temperature explanation of the variation in the number of pool patrons was the result of chance (see Figure 20.10). Empirical evidence was also supportive when the prediction was used to calculate predicted values, which were then compared with the actual observations (see Figure 20.14).

Figure 20.14   SPSS Variable View Showing the New Variable

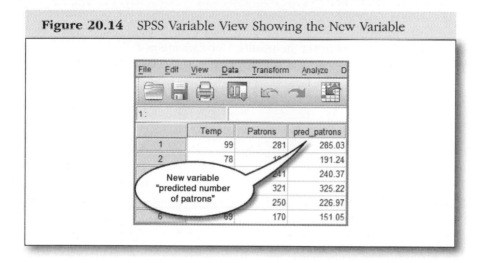

20.10 SUMMARY △

This chapter presented single linear regression analyses. With single linear regression, you have a single independent variable and one dependent variable. The object of the analysis is to develop a prediction equation that permits the estimation of the dependent variable based on the knowledge of the single independent variable. In the following chapter, we will extend the concept of one independent variable to multiple independent variables and call it multiple linear regression. Multiple linear regression results when you have two or more independent variables when attempting to predict the value of the dependent variable.

20.11 REVIEW EXERCISES △

20.1   Can you help the manager of a senior citizen center at the local library determine if there was any merit to her idea that the patron's age and the number of books checked out were related? Her thought was that as an individual got older, more books would be checked out. She would like to be able to predict the number of books that would be checked out by looking at a person's age. The manager is especially interested in the number of books checked out for those 65 years of age. She selected a random sample of 24 senior patrons and collected details of the age and the number of books checked out during a 4-week period. If you wish to help, select the correct statistical approach, write the null and alternative hypothesis, conduct the analysis, and interpret the results. Her data are as follows:

	age	books
1	62	2
2	67	6
3	65	4
4	70	10
5	66	7
6	63	4
7	67	6
8	65	4
9	63	2
10	68	9
11	65	5
12	69	7
13	68	7
14	62	3
15	64	4
16	70	9
17	63	2
18	68	6
19	64	3
20	65	5
21	69	8
22	64	3
23	62	2
24	68	7

20.2 An economist at a large university was studying the impact of gun crime on local economies. Part of his study sought information on the relationship between the number of gun control measures a lawyer/legislator introduced and his score awarded by the state bar on his knowledge of constitutional law. His idea was that low-scoring lawyers would introduce more gun control laws. He wished to quantify the strength and direction of any relationship and also see if the number of laws introduced could be predicted by knowing the legislator's constitutional law rating. One specific value he wished to predict was the number of laws introduced by the average score of 76, a value not directly observed in the data.

His research has thus far shown that gun control laws have a negative impact on local economies. The researcher selected a random sample of lawyers elected to office and then compiled public information on the two variables of interest ("gun control" and "state bar rating"). As a consulting statistician, your task is to select the correct statistical method, write the null and alternative hypotheses, do the analysis, and interpret the results. His data are as follows:

	const_score	gun_control
1	98	1
2	86	2
3	74	3
4	63	4
5	51	6
6	07	1
7	85	2
8	77	5
9	65	6
10	53	8
11	94	2
12	83	2
13	74	4
14	69	4
15	55	6
16	97	2
17	84	4
18	71	4
19	64	5
20	67	0
21	99	1
22	82	3
23	75	4
24	63	4

20.3 A deacon at St. Joseph the Worker Church had the theory that attendance at formal church services was a good indicator of the number of hours an individual volunteered. He randomly selected 12 individual volunteers and collected the required information. The deacon wanted to measure the strength and direction of any association. He also wanted a method whereby he might predict the number of hours volunteered by a person who attends church on average four times per month. Since you are an active volunteer and a student of statistics, he asked for your help. You have to select the appropriate statistical technique, write the null and alternative hypotheses, do the analysis, and interpret the results. The deacon's data are as follows:

	churchattend	hrsvolunteer
1	10	16
2	6	9
3	2	4
4	3	6
5	5	10
6	9	11
7	10	16
8	2	2
9	7	5
10	8	10
11	3	7
12	6	10

CHAPTER **21**

MULTIPLE LINEAR REGRESSION

△ 21.1 INTRODUCTION AND OBJECTIVES

In the previous chapter, we covered *single linear regression*. The single regression approach is used when you have one independent variable and one dependent variable. This chapter presents *multiple linear regression*, which is used when you have two or more independent variables and one dependent variable. The research question for those using multiple regression concerns how the multiple independent variables, either by themselves or together, influence changes in the dependent variable. You use the same basic concepts as with *single regression*, except that now you have multiple independent variables. The object of multiple linear regression is to develop a *prediction equation* that permits the estimation of the dependent variable based on the knowledge of the independent variables.

The data requirements for multiple linear regression are the same as for single linear regression. Sample size is always an issue with statistical methods, and the same is true for regression. We have kept our samples small to facilitate the visualization and input of data. One should keep in mind, though, that larger samples are usually better when performing most statistical analysis. The multiple regression example used in this chapter is as basic as possible—small sample size and only two independent variables, the minimum number required when using multiple regression. We ask the reader to be cognizant of the fact that more independent variables (and a larger sample size) could be analyzed using the same techniques described in this chapter.

OBJECTIVES

After completing this chapter, you will be able to

Describe the purpose of multiple linear regression

Input variable information and data for multiple linear regression

Describe the data assumptions required for multiple linear regression

Use SPSS to conduct multiple linear regression analysis

Interpret scatterplots concerning the data assumptions for regression

Interpret probability plots concerning the data assumptions for regression

Describe and interpret the SPSS output from multiple linear regression

Interpret ANOVA analysis as it relates to multiple linear regression

Interpret the coefficients table for multiple linear regression

Write the prediction equation and use it to make predictions

Write the prediction equation and use a calculator to make predictions

21.2 RESEARCH SCENARIO AND TEST SELECTION △

The researcher wants to understand how certain physical factors may affect an individual's weight. The research scenario centers on the belief that an individual's "height" and "age" (independent variables) are related to the individual's "weight" (dependent variable). Another way of stating the scenario is that age and height *influence* the weight of an individual. When attempting to select the analytic approach, an important consideration is the level of measurement. As with single regression, the dependent variable must be measured at the *scale* level (interval or ratio). The independent variables are almost always continuous, although there are methods to accommodate discrete variables. In the example presented above, all data are measured at the scale level. What type of statistical analysis would you suggest to investigate the relationship of height and age to a person's weight?

Regression analysis comes to mind since we are attempting to estimate (predict) the value of one variable based on the knowledge of the others, which can be done with a prediction equation. *Single regression* can be ruled out since we have two independent variables and one dependent variable. Let's consider *multiple linear regression* as a possible analytic approach.

We must check to see if our variables are approximately normally distributed. Furthermore, it is required that the relationship between the variables be approximately linear. And we will also have to check for homoscedasticity, which means that the variances in the dependent variable are the same for each level of the independent variables. Here's an example of homoscedasticity. A distribution of individuals who are 61 inches tall and aged 41 years would have the same variability in weight as those who are 72 inches tall and aged 31 years. In the sections that follow, some of these required data characteristics will be examined immediately, others when we get deeper into the analysis.

△   21.3 Research Question and Null Hypothesis

The basic research question (alternative hypothesis) is whether an individual's weight is related to that person's age and height. The null hypothesis is the opposite of the alternative hypothesis: An individual's weight is not related to his or her age and height.

Therefore, this research question involves two independent variables, "height" and "age," and one dependent variable, weight. The investigator wishes to determine how height and age, taken together or individually, might explain the variation in weight. Such information could assist someone attempting to estimate an individual's weight based on the knowledge of his or her height and age. Another way of stating the question uses the concept of prediction and error reduction. How successfully could we predict someone's weight given that we know his or her age and height? How much error could be reduced in making the prediction when age and height are known? One final question: Are the relationships between weight and each of the two independent variables statistically significant?

△   21.4 Data Input

In this section, you enter hypothetical data for 12 randomly selected individuals measured on *weight*, *height*, and *age*. You then use SPSS to analyze these data using multiple linear regression. As in recent chapters, detailed instructions on entering the variable information and the data are not given. The Variable View screen, shown in Figure 21.1, serves as a guide for the entry of variable information. Figure 21.1 contains the material needed to successfully enter all the variable information.

Figure 21.1 Variable View for Multiple Linear Regression for Three Variables

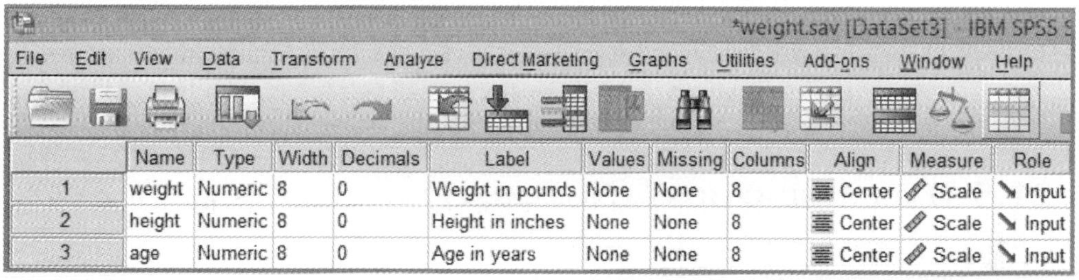

	Name	Type	Width	Decimals	Label	Values	Missing	Columns	Align	Measure	Role
1	weight	Numeric	8	0	Weight in pounds	None	None	8	Center	Scale	Input
2	height	Numeric	8	0	Height in inches	None	None	8	Center	Scale	Input
3	age	Numeric	8	0	Age in years	None	None	8	Center	Scale	Input

Figure 21.2 contains the data for the three variables on the 12 individuals. The table in Figure 21.2 is a copy of the Data View screen and therefore shows exactly what your data entry should look like.

Follow the bullet points below, and enter both the variable information and the data for the three variables; save the file as instructed.

Figure 21.2 Data View for Multiple Regression for Three Variables

	weight	height	age
1	115	62	41
2	140	62	21
3	125	62	31
4	125	64	21
5	145	64	31
6	135	64	41
7	165	72	41
8	190	72	31
9	175	72	21
10	150	66	31
11	155	66	31
12	140	64	21

- Start SPSS, and click **Cancel** in the *SPSS Statistics* opening window.
- Click **File**, select **New**, and click **Data**.
- Click **Variable View** (enter all the variable information as presented in Figure 21.1).
- Click **Data View** (carefully enter all the data for *weight*, *height*, and *age* given in Figure 21.2).
- Click **File**, then click **Save As**; type *weight* in the *File Name* box, and then click **OK**.

You have now entered and saved the data for an individual's weight, height, and age. In the next section, we check the distributions for normality.

△ 21.5 Data Assumptions (Normality)

As was done in the previous chapter, we first check the data distributions for normality.

- Click **Analyze**, select **Nonparametric Tests**, and then click **One-Sample** (the *One-Sample Nonparametric Tests* window opens).
- Click the Objective tab, and then click **Customize analysis**.
- Click the Fields tab (if your three variables are not in the *Test Fields* pane, then move them to it).
- Click the Settings tab, click **Customize tests**, and then click **Kolmogorov-Smirnov test**.
- Click **Options**, make sure **Normal** is checked, then click **OK**.
- Click **Run** (the Output Viewer opens showing Figure 21.3).

The results of the Kolmogorov-Smirnov (K-S) test indicate that two of the three variables are indeed normally distributed (see Figure 21.3). "Weight" and "Age" pass the test even when subjected to the Lilliefors correction factor, which is automatically applied to the K-S test by the SPSS program. We took a closer look at the variable "Height," which did not pass. We first did a *P-P plot*, which looked good. The appearance of the "Height" data in a histogram was not that good, but the mean (65.8) and standard deviation (3.95) were encouraging. We decided to conduct the K-S test without the Lilliefors correction. We were aware that this reduces the power of the K-S test to detect departures from normality, but we made the judgment to proceed. We next show the steps needed to conduct this procedure, which will improve the chances of our variable "Height" passing the K-S test.

Figure 21.3 The Kolmogorov-Smirnov Test for Normality—Three Variables

Hypothesis Test Summary

	Null Hypothesis	Test	Sig.	Decision
1	The distribution of Weight in pounds is normal with mean 146.667 and standard deviation 21.88.	One-Sample Kolmogorov-Smirnov Test	.200[1,2]	Retain the null hypothesis.
2	The distribution of Height in inches is normal with mean 65.833 and standard deviation 3.95.	One-Sample Kolmogorov-Smirnov Test	.200[1,2]	Reject the null hypothesis.
3	The distribution of Age in years is normal with mean 30.167 and standard deviation 7.93.	One-Sample Kolmogorov-Smirnov Test	.200[1,2]	Retain the null hypothesis.

Asymptotic significances are displayed. The significance level is .05.

[1] Lilliefors Corrected

[2] This is a lower bound of the true significance.

- Click **Analyze**, select **Nonparametric Tests**, and then click **One-Sample** (the *One-Sample Nonparametric Tests* window opens).
- Click the Objective tab, and then click **Customize analysis**.
- Click the Fields tab (make sure that *only* the variable "Height" is in the *Test Fields* pane).
- Click the Settings tab, click **Customize tests**, and then click **Kolmogorov-Smirnov Test**.
- Click the **Options** button just below **Kolmogorov-Smirnov Test**, and the *Kolmogorov-Smirnov Test Options* window opens (see Figure 21.4). Make sure **Normal** is clicked.
- In the *Distribution Parameters* pane, click **Custom**; in the *Mean* box, type *65.8*; and in the *Std. Dev.* box, type *3.95* (these values are easily obtained from the prior K-S test shown in Figure 21.3) (the window should now look like Figure 21.4). Finally, click **OK**.
- Click **Run** (the Output Viewer opens, showing Figure 21.5).

Using the K-S test without the Lilliefors correction is successful, and our "Height" variable passes the normality test. Based on this finding we make the decision to proceed with the regression analysis.

Figure 21.4   The *Kolmogorov-Smirnov Test Options* Window

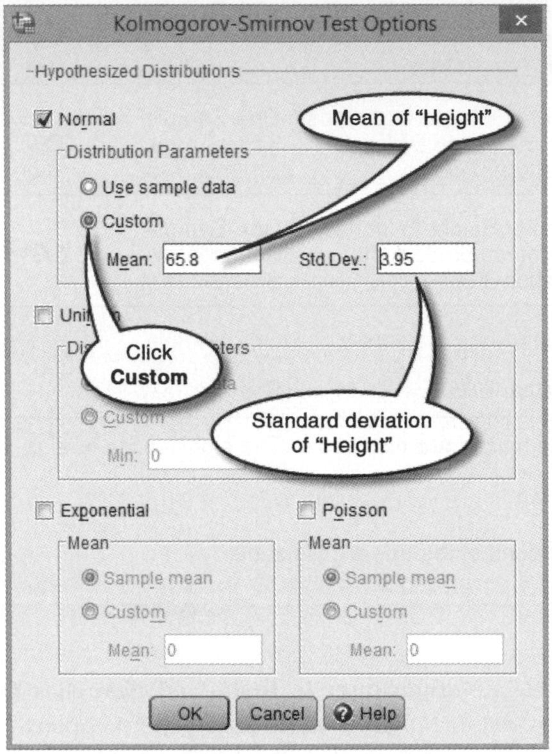

Figure 21.5   Kolmogorov-Smirnov Test for "Height" Without Lilliefors Correction

21.6 REGRESSION AND PREDICTION △

As we did with the single linear regression in Chapter 20, we now check the data for *linearity, equal variances,* and *normality of the error terms* (residuals). If it's not already running, open *weight.sav,* and follow the procedure presented next. Once you have completed all these analytic requests, you will see the output as presented in the following sections.

- Click **Analyze**, select **Regression**, and then click **Linear** (the *Linear Regression* window opens; see Figure 21.6 for its appearance after moving the variables).
- Click **Weight**, and then click the arrow next to the *Dependent:* box.
- Click **Height**, and then click the arrow next to the *Independent(s):* box.
- Click **Age**, and then click the arrow next to the *Independent(s):* box (at this point, your screen should look like Figure 21.6).

Figure 21.6 *Linear Regression* Window After Moving the Variables

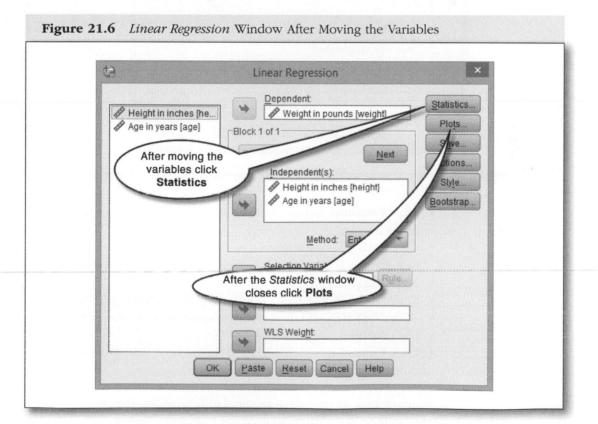

- Click **Statistics** (the *Linear Regression: Statistics* window opens; see Figure 21.7).
- Click **Estimates**, and then click **Model fit** (see Figure 21.7).

Figure 21.7   The *Linear Regression: Statistics* Window

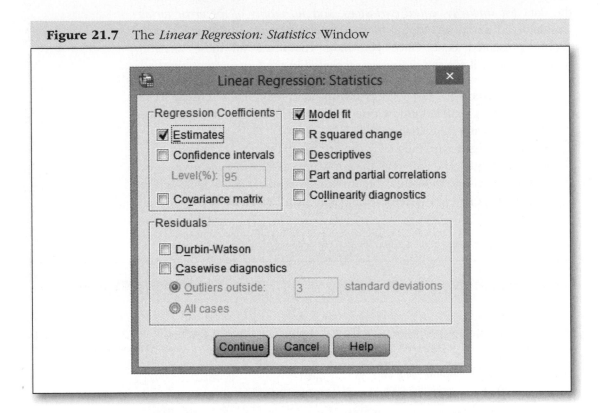

- Click **Continue** (returns to the *Linear Regression* window depicted in Figure 21.6).
- Click **Plots** (the *Linear Regression: Plots* window opens; see Figure 21.8) (actually, this is the same analytic request you made when doing single regression).
- Click ***ZPRED**, and then click the arrow beneath the *Y:* box.
- Click ***ZRESID**, and then click the arrow beneath the *X:* box.
- Click **Normal probability plot**.
- Click **Continue**, and then click **OK** (this final click produces all the output required to interpret our analysis).

You now have all the output required to finalize and interpret additional data assumptions and your multiple linear regression analysis.

Figure 21.8 The *Linear Regression: Plots* Window: Further Data Assumption Checks

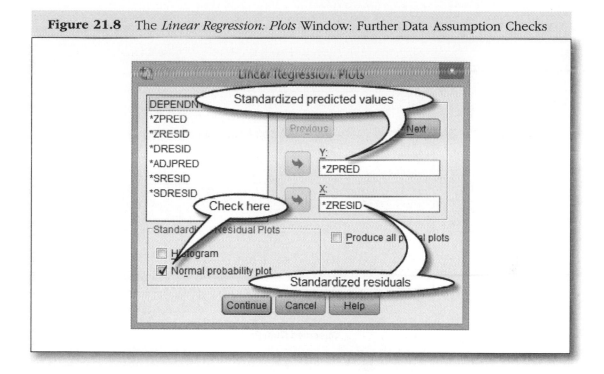

21.7 Interpretation of Output (Data Assumptions) △

Figure 21.8 shows that the *Normal probability plot* box was checked. The result is shown in Figure 21.9. The small circles are close to the diagonal line, which provides evidence that the *residuals (error terms)* are indeed normally distributed, which is a requirement of the linear regression procedure.

The final plot, Figure 21.10, results from the requests we made as shown in Figure 21.8. As with single variable regression, the scatterplot combines the standardized predicted values (*ZPRED) with the values for the standardized residuals (*ZRESID). Since the small circles follow no pattern—they are randomly dispersed in the scatteplot—we assume equality of variances. There are numerous appearances that Figure 21.10 may take on that would indicate unequal variances. One such appearance is referred to as the "bow tie" scatterplot. The "bow tie" scatterplot has the error terms bunched up along both verticals and tapering toward the middle, which is the zero point in Figure 21.10. There are many other shapes that indicate unequal variances.

Figure 21.9   Normal P-P Plot of Regression Standardized Residuals (Error Terms)

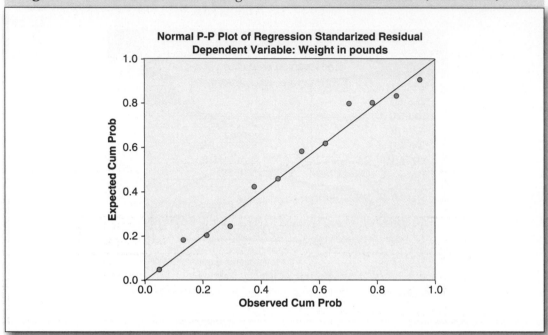

Figure 21.10   Scatterplot of Residuals: Lack of Pattern Indicates Equal Variances

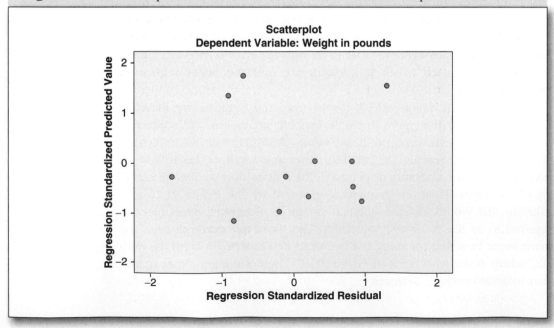

21.8 INTERPRETATION OF △
OUTPUT (REGRESSION AND PREDICTION)

The *Model Summary* shown in Figure 21.11 resulted from you clicking *Model Fit*, as depicted in Figure 21.7. The information provided in the *Model Summary* gives us information regarding the strength of the relationship between our variables.

Figure 21.11   Model Summary for Multiple Linear Regression

Model Summary[b]

Model	R	R Square	Adjusted R Square	Std. Error of the Estimate
1	.919[a]	.845	.811	9.515

a. Predictors: (Constant), Age in years, Height in inches

b. Dependent Variable: Weight in pounds

The value .919 shown in the *R* column of the table in Figure 21.11 shows a *strong* multiple correlation coefficient. It represents the correlation coefficient when both independent variables ("age" and "height") are taken together and compared with the dependent variable "weight." The *Model Summary* indicates that the amount of change in the dependent variable is determined by the two independent variables—not by one as in single regression. From an "interpretation" standpoint, the value in the next column, *R Square*, is extremely important. The *R Square* of .845 indicates that 84.5% (.845 × 100) of the variance in an individual's "weight" (dependent variable) can be explained by both the independent variables, "height" and "age." It is safe to say that we have a "good" predictor of weight if an individual's height and age are known. We next examine the ANOVA table shown in Figure 21.12.

The ANOVA table indicates that the mathematical model (the regression equation) can accurately explain variation in the dependent variable. The value of .000 (which is less than .05) provides evidence that there is a low probability that the variation explained by the model is due to chance. We conclude that changes in the dependent variable result from changes in the independent variables. In this example, changes in height and age resulted in significant changes in weight.

Figure 21.12   ANOVA Table Indicating a Significant Relationship Between the Variables

ANOVA[a]

Model		Sum of Squares	df	Mean Square	F	Sig.
1	Regression	4451.886	2	2225.943	24.588	.000[b]
	Residual	814.780	9	90.531		
	Total	5266.667	11			

a. Dependent Variable: Weight in pounds

b. Predictors: (Constant), Age in years, Height in inches

Prediction

As in the prior chapter, we come to that interesting point in regression analysis where we seek to discover the unknown. We accomplish such a "discovery" by using a prediction equation to make estimates based on our 12 original observations. Let's see how the *Coefficients* table in Figure 21.13 can assist us in making such a discovery possible.

Figure 21.13   Coefficient Values for Use in the Prediction Equation

Coefficients[a]

Model		Unstandardized Coefficients		Standardized Coefficients	t	Sig.
		B	Std. Error	Beta		
1	(Constant)	-175.175	48.615		-3.603	.006
	Height in inches	5.072	.727	.916	6.974	.000
	Age in years	-.399	.362	-.145	-1.103	.299

a. Dependent Variable: Weight in pounds

As with single linear regression, the *Coefficients* table shown in Figure 21.13 provides the essential values for the prediction equation. The prediction equation takes the following form:

$$\hat{y} = a + b_1 x_1 + b_2 x_2,$$

where $\hat{y}$ is the predicted value, a the intercept, b_1 the slope for "height," x_1 the independent variable "height," b_2 the slope for "age," and x_2 the independent variable "age."

You may recall that in the previous chapter on *single regression*, we defined the slope (*b*) and the intercept (*a*). Those definitions are the same for the multiple regression equation, except that we now have a slope for each of the independent variables.

The equation simply states that you multiply the individual slopes by the values of the independent variables and then add the products to the intercept—not too difficult. The slopes and intercepts can be found in the table shown in Figure 21.13. Look in the column labeled *B*. The intercept (the value for *a* in the above equation) is located in the (*Constant*) row and is 175.175. The value below this of 5.072 is the slope for "height," and below that is the value of -0.399, the slope for "age." The values for *x* are found in the *weight.sav* database. Substituting the regression coefficients, the slope and the intercept, into the equation, we find the following:

$$\hat{y} = 172.175 + (5.072 * \text{height}) + (-0.399 * \text{age}).$$

We are now ready to use the *Compute Variable* function of SPSS and make some predictions. The following bullet points assume that SPSS is running and the *weight* database is open.

- Click **Transform**, and then click **Compute Variable** (the *Compute Variable* window opens; see Figure 21.14, which shows the upper portion of the window, with the completed operations as described in the following bullet points).
- Click the *Target Variable* box, and then type *pred_weight*.

Figure 21.14 Completed *Compute Variable* Window—Prediction Equation

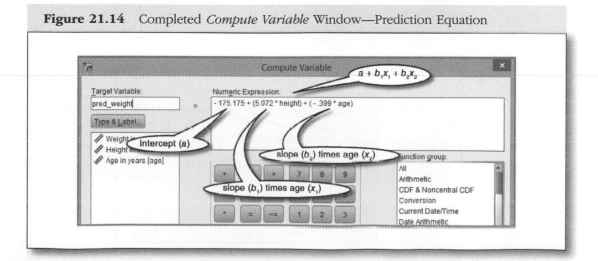

- Click the *Type & Label* box, and a window opens; then type *predicted weight* in the *Label* box.
- Click **Continue** (the *Compute Variable* window remains as shown in Figure 21.14).
- Click the *Numeric Expression* box, and type *-175.175 + (5.072 * height) + (-0.399 * age)* (if preferred, you could use the keypad in the *Compute Variable* window and the variable list to write this expression—try it, you may find it easier).
- Click **OK** at the bottom of this window (the bottom portion is not shown in Figure 21.14).

The new variable "pred weight" is automatically inserted into the *weight* database once you click **OK** in the above series of bullet points. If the database does not appear on your computer screen, then do the following:

- Click **weight.sav** at the bottom of your screen, which will open that database (you will see the new variable added to your database, as shown in Figure 21.15.)

Figure 21.15 shows the new Data View screen with the just created variable. Let's look at *Case 10* and interpret the newly created variable. *Case 10* shows that someone measured at 66 inches in height and aged 31 weighed 150 pounds. The prediction equation estimated that an individual possessing those two values (66 and 31) would weigh 147.21 pounds. If you read each of the cases for height and age, you can read the actual observed weight as well as the prediction.

Figure 21.15   Data View Showing a New Variable From Use of the Prediction Equation

	weight	height	age	pred_weight
1	115	62	41	122.93
2	140	62	21	130.91
3	125	62	31	126.92
4	125		21	141.05
5			31	137.06
6			41	133.07
7			41	173.65
8	190	72	31	177.64
9	175	72	21	181.63
10	150	66	31	147.21
11	155	66	31	147.21
12	140	64	21	141.05

New variable - results from the *Compute Variable* request

As we did for single regression, any values (for x_1 and x_2) that might be of special interest could be plugged into the equation and then solved for $\hat{y}$ (the predicted y value). One note of caution is that the values chosen should be within the range of the original observations to maintain accuracy. You can accomplish such predictions by substituting values into the *Numeric expression* panel in the *Compute Variable* window and creating a new variable as described in the previous chapter. The other method is to use a handheld calculator, as was done in the previous chapter.

21.9 RESEARCH QUESTION ANSWERED △

At the beginning of this chapter, we stated that the purpose of the research was to investigate whether a person's weight is influenced by his or her age and height. You might also recall that the null hypothesis was that age and height had no influence on a person's weight. How well did our multiple regression analysis answer these questions? And could we reject the null hypothesis and thereby provide evidence in support of our alternative hypothesis?

First, our questions concerning the required data assumptions for using multiple regression were answered in the affirmative. It was determined that multiple linear regression could be used (see Figures 21.3, 21.5, 21.9, and 21.10). Next, the prediction equation, which was developed from previous observations, was found to reduce the error in predicting weight by 84.5% (see Figure 21.11). Additional statistical evidence supporting the value of our prediction equation was provided with the finding of a significant F test. The significance was less than .05, indicating a low probability that the explanation of the variation in weight by using age and height was the result of chance (see Figure 21.12). Empirical evidence in support of the prediction equation was also observed. The *Compute Variable* function of SPSS was used to calculate predicted values that could directly be compared with our observations (see Figure 21.15).

21.10 SUMMARY △

In this chapter, we presented *multiple linear regression*. With multiple linear regression, you have two or more independent variables and one dependent variable. The object was to write a prediction equation that would permit the estimation of the value of the dependent variable based on the knowledge of two or more independent variables. In the next chapter, you learn about a third type of regression known as logistic regression. With logistic regression, you attempt to predict a binary dependent variable.

△   21.11 Review Exercises

21.1 This exercise is an extension of the senior center manager's problem in the previous chapter (Review Exercise 20.1). You may recall that the manager developed a prediction equation that estimated the number of books checked out at the library using the "patrons' age" as the *single independent* variable. For the current exercise, used to illustrate multiple linear regression, we add a second independent variable—"total years of education." Using the single variable, the model developed was able to account for 86% of the variance in the number of books checked out. Although the senior center manager was happy with that result, she wishes to add total years of education in the hope of improving her model. The manager wants to use a new equation (using two independent variables) to make predictions and then compare those predictions with the observed data to see how well it works. She also wishes to predict the number of books checked out by someone aged 63 with 16 years of education, which was not directly observed in her data. Use multiple linear regression, write the null and alternative hypotheses, conduct the analysis, write the prediction equations, make the predictions, and interpret the results. Her data are as follows:

	age	education	books
1	62	12	2
2	67	16	6
3	65	14	4
4	70	22	10
5	66	18	7
6	63	14	4
7	67	16	6
8	65	18	4
9	63	12	2
10	68	22	9
11	65	16	5
12	69	18	7
13	68	16	7
14	62	14	3
15	64	14	4
16	70	18	9
17	63	12	2
18	68	16	6

21.2 This problem is based on the single regression you did in the previous chapter. We just added another variable called "freedom index" to turn it into an example of multiple regression. You now have two independent variables ("constitutional law score" and "freedom index") and one dependent variable that counts the "number of measures introduced" by the legislator.

The political consultant wants to determine if the scores on knowledge of constitutional law and score of the freedom index are related to the number of gun control laws introduced. He also wishes to extend any findings into the realm of prediction by using it to estimate the number of measures introduced by a legislator rated average on both these independent variables. He also wishes to use the equation to predict for his data, which will permit him to examine the equation's performance when the predicted values are directly compared with the observed values. Use multiple linear regression, write the null and alternative hypotheses, conduct the analysis, write the prediction equations, make the predictions, and interpret the results. His data are as follows:

	const_score	gun_control	freeindex
1	98	1	28
2	86	2	23
3	74	3	26
4	63	4	24
5	51	6	14
6	97	1	25
7	85	2	21
8	77	5	23
9	65	6	18
10	53	8	12
11	94	2	21
12	83	2	26
13	74	4	15
14	69	4	19
15	55	6	10
16	97	2	21
17	84	4	21
18	71	4	18
19	64	6	17
20	57	8	18
21	99	1	30
22	82	3	23
23	75	4	14
24	63	4	28

21.3 As we have done in the previous two exercises, we bring forward from the previous chapter a single linear regression problem and add an additional variable. In that exercise, you had one independent variable, which was "the number of times an individual attended church during a month." For this current exercise, you will add another independent variable, which is "the number of times one prays in a day." The deacon of the church wants to see if the earlier prediction equation could be improved by adding this additional variable. As before, he wants to compare the performance of the new equation with the actual observed values. In addition, he wishes to predict the number of volunteer hours for those rated as average on the two independent variables. Use multiple linear regression, write the null and alternative hypotheses, do the analysis, and interpret the results. The deacon's new data are as follows:

	churchattend	pray	hrsvolunteer
1	10	6	16
2	6	5	9
3	2	1	4
4	3	2	6
5	5	5	10
6	9	6	11
7	10	7	16
8	2	2	2
9	7	3	5
10	8	6	10
11	3	2	7
12	6	4	10

CHAPTER 22

LOGISTIC REGRESSION

22.1 INTRODUCTION AND OBJECTIVES △

In the previous three chapters, we presented similar ways to statistically describe the relationships between individual variables. One chapter addressed *correlation*, another described *single linear regression*, and the third discussed *multiple linear regression*. Correlation seeks to measure the strength and direction of any identified relationship between two variables measured at the *scale* level. You may recall that the chapter "Pearson's Correlation and Spearman's Correlation" ended with an explanation of similar methods but for variables measured at the *ordinal* level. The next chapter, "Single Linear Regression," was actually an extension of the correlation chapter. However, the addition of regression presented the exciting possibility of being able to *predict* unknown events based on previously collected data. The example used resulted in the development of an equation to predict the number of patrons at a public swimming pool based on the daily temperature. The name *single* originates from the notion that we had one *independent* variable ("temperature" in our example). The third chapter, "Multiple Linear Regression," added the ability to have more than one independent variable when attempting to predict the value of a dependent variable. In our pool example, we might theorize that relative humidity and percentage of cloud cover might also influence the number of patrons using the public pool. Thus, if we chose to use the multiple regression technique, we would evaluate the influence of temperature, humidity, and degree of cloud cover on the number of people using the swimming pool. In the case of both single and multiple regression, we measured the dependent variable at the *scale* level (number of people).

We now come to the subject of this chapter, *logistic regression*. When using logistic regression, the dependent variable is measured at the *nominal* level. To be more specific, we refer to the dependent variable as *categorical* and having only *two* categories. This usually means that something does or does not happen. Examples might be that an individual passes or fails a class, a patient lives or dies as a result of an experimental treatment, or a climbing team reaches the summit or does not. The technical name, which is used by SPSS, for this type of regression is *binary logistic*. Just as with single and multiple regression, you may use logistic regression when you have single or multiple independent variables. In this chapter, we present examples where there are multiple independent variables measured at different levels, all of which are easily handled by the SPSS program.

Data assumptions are not as strict as when using single or multiple regression in that it is not necessary to have normally distributed data and equal variances. Sample size can be an issue when you have many independent variables and a small sample. Such a situation often results in empty cells in the omnibus chi-square test and can cause serious problems in your analysis. If you have a large number of independent variables, then you should consider either increasing the sample size or reducing the number of variables. Another concern would be with extreme values—it is best to eliminate outliers for the regression analysis. A third concern, perhaps the most important, is multicollinearity which results when the independent variables have a high correlation among one another. The ideal situation is to have weak correlation coefficients between the independent variables while the correlation between the independent and dependent variables is moderate to strong.

We should also mention that there are other logistic regression approaches that permit the use of multiple levels (categories) of the dependent variable. These procedures are referred to as *multinomial logistic regression*. Since we address the situation where there are two categories of our dependent variable, we use what is referred to as *binary logistic regression*.

OBJECTIVES

After completing this chapter, you will be able to

Determine when logistic regression should be used

Recode variables in a manner suitable for logistic analysis

Enter variable information and data in the format preparatory to SPSS analysis

Identify the data assumptions associated with the use of logistic regression

Conduct logistic regression analysis using SPSS

Interpret the results of the logistic regression

22.2 Research Scenario and Test Selection △

Sally, a successful real estate saleslady, was interested in developing a way to predict whether the first meeting with a potential home buyer would ultimately result in a sale. She consulted with her father, who happened to be a retired statistics professor. Her father initially thought that some type of regression equation might provide an answer. Sally and her father discussed some of the key variables that appeared to, at least by casual observation, provide some information as to whether a sale would be made or not. Information on these key variables was routinely obtained at the initial meeting with potential home buyers. Data were available for 30 such contacts on the following four variables:

1. Have the potential buyers been preapproved for a loan? (Coded as 0 = *no* and 1 = *yes*)

2. Do they ask questions—are they interested? (Coded as 0 = *no* and 1 = *yes*)

3. How much money do they have for a down payment? (In thousands of dollars)

4. Did they ultimately purchase a home from you? (Coded as 0 = *no* and 1 = *yes*)

The father–daughter team decided to use these questions as independent variables, in an attempt to predict a binary outcome of the dependent variable, sale or no sale (coded as 0 = *no* and 1 = *yes*).

The research would seek to develop an equation based on the independent variables measured at *nominal*, and *scale* level while the dependent variable was discrete (nominal). The logical choice for this analysis was logistic regression.

△   22.3 RESEARCH QUESTION AND NULL HYPOTHESIS

The aim of this research, based on logistic regression, was to accurately predict the outcome of a particular case based on the information provided by the variables selected for inclusion in the equation. Basically, there are two questions to study:

1. Could we predict whether a client would purchase a home if we knew whether he or she had been *preapproved for a loan*, whether he or she *showed an interest in buying* (by asking questions), and *the amount of money the client had for a down payment?*

2. If we could predict the outcome accurately, which variables were the most important? Which variables seemed to increase or decrease the outcome of buying a home?

The null hypothesis is the opposite of the alternative hypothesis (research questions) in that prior knowledge of the selected variables would not assist one predicting whether a client would or would not purchase a home.

△   22.4 DATA INPUT, ANALYSIS, AND INTERPRETATION OF OUTPUT

Let's begin the development of our prediction equation with the input of the data collected by Sally from randomly selected client contacts over a period of several weeks. We first set up the SPSS file by entering the variable information as shown in Figure 22.1.

- Start SPSS, click **File**, select **New**, and click **Data**.
- If not already clicked, click the Variable View tab.
- Enter all the variable information as shown in Figure 22.1 (be sure to enter all the values and value labels for your three categorical variables as indicated in the callouts).

Figure 22.1 Variable View for Logistic Regression—Real Estate Example

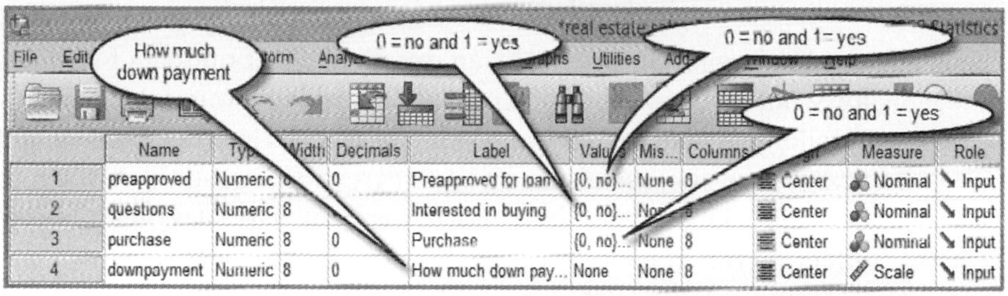

- Click the Data View tab, and enter all the data as shown in Figure 22.2. (Remember that the zeros (0s) represent an absence of the trait of interest while the ones (1s) represent the presence of the trait.)

Figure 22.2 Data View for Logistic Regression—Real Estate Example

	preapproved	questions	purchase	downpayment
1	0	1	0	10
2	0	0	0	10
3	0	1	1	80
4	0	0	0	10
5	0	1	0	80
6	1	0	0	10
7	1	1	1	90
8	0	0	0	80
9	1	0	1	10
10	0	0	0	20
11	1	1	1	70
12	1	0	0	10
13	1	1	1	90
14	1	0	0	90
15	1	0	0	10
16	0	0	0	20
17	1	1	1	80
18	0	0	0	20
19	0	0	0	90
20	0	0	0	20
21	1	0	0	80
22	0	0	0	10
23	0	1	0	50
24	1	1	1	70
25	0	1	0	90
26	0	0	0	70
27	1	1	0	20
28	0	0	0	70
29	0	1	0	20
30	1	0	0	80

- With all the data entered and checked for accuracy, click **File**, click **Save As** (the *Save As* window opens), in the *File name* box type *real estate logistic regression*, and then click **Save**.

We begin our analysis by an examination of our data for multicollinearity, where we look for high correlations between our independent variables. Smaller correlations between our predictors (independent variables) tend to increase the usefulness of our regression equation. A high correlation between our independent and dependent variables will have a positive effect on our equation. Follow the steps below:

- Click **Analyze**, click **Correlate,** and then click **Bivariate** (the *Bivariate* window opens).
- Move all four variables to the *Variables:* box.
- In the *Correlation Coefficients* panel, unclick **Pearson**, then check the box next to *Spearman* (your screen should now appear as shown in Figure 22.3).
- Click **OK** (the Output Viewer opens; see Figure 22.4).

Figure 22.3   The *Bivariate Correlations* Window—All Variables Selected

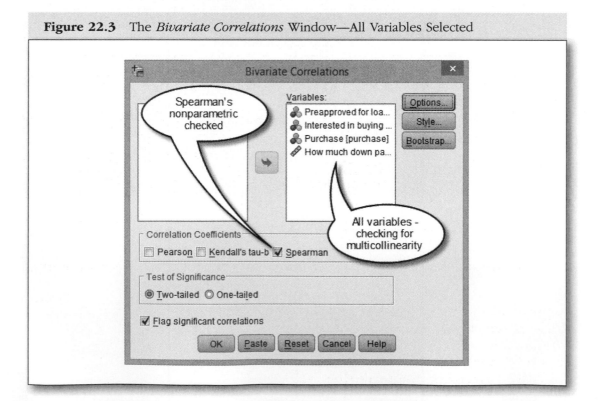

Once **OK** has been clicked, the Output Viewer opens with a table showing correlations for all the selected variables (see Figure 22.4). Looking at the table, we see that the values of the correlation coefficients are patterned correctly. The desired pattern is such that we want lower correlations among the independent variables and higher correlations for the independent variable–dependent variable relationships. As stated above, the observed pattern was good except for a correlation of .373 between the independent variables "interested in buying" and "how much down payment." We made the decision to leave the equation as is and proceed with the regression analysis as the effect should be minimal. Examine Figure 22.4 to see the pattern for these correlation coefficients—we have used callouts to bring attention to those values important to our work.

At this point, we are ready to conduct the actual logistic regression analysis of Sally's real estate data.

Figure 22.4 Spearman's Correlation Coefficients for Multicollinearity Check—Real Estate Example

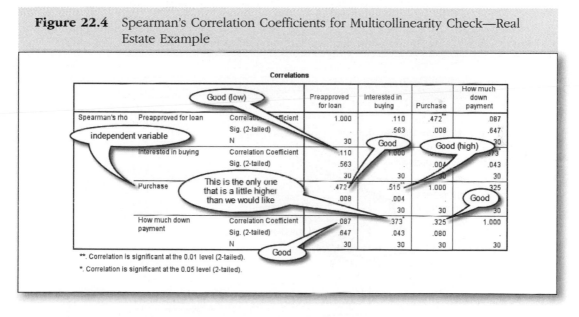

Assuming that SPSS is running and the real estate database is open, follow the steps below:

- Click **Analyze**, select **Regression**, and then click **Binary Logistic** (the *Logistic Regression* window opens).

- Click **purchase**, then click the arrow next to the *Dependent:* box.
- Click **preapproved**, then click the arrow next to the *Covariates:* box.
- Click **interested in buying**, then click the arrow next to the *Covariates:* box.
- Click **how much down payment**, then click the arrow next to the *Covariates:* box (at this point, your computer screen should look like Figure 22.5).
- Click the **Categorical** button (the *Logistic Regression: Define Categorical Variables* window opens).
- Click **preapproved**, then click the arrow that moves the variable to the *Categorical Covariates:* box.
- Click **interested in buying**, then click the arrow that moves it to the *Categorical Covariates:* box.
- Click **preapproved** once again (it's highlighted), then click **First** in the *Change Contrast* section, and then click **Change**.
- Click **question (interested in buying)**, then click **Change** (your screen should now appear as shown in Figure 22.6).

Figure 22.5 The *Logistic Regression* Window—Variables in Position for Analysis

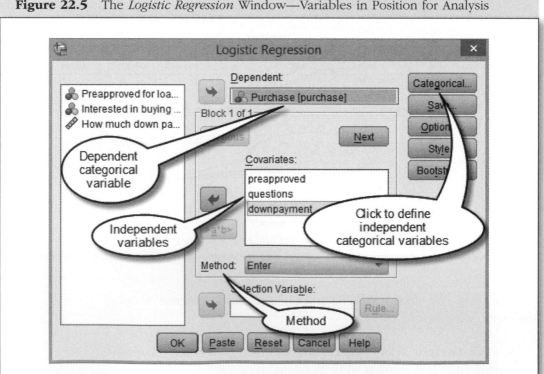

Figure 22.6 The *Logistic Regression: Define Categorical Variables* Window

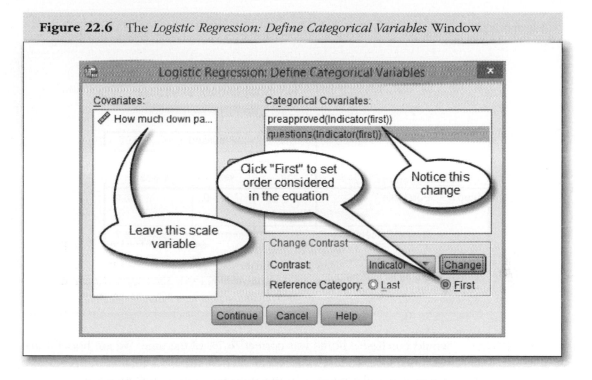

- Click **Continue** (which takes you back to the *Logistic Regression* window, Figure 22.5).
- Make sure **Enter** shows in the *Method* section, then click **OK** (the Output Viewer opens).

Clicking **OK** generates many tables that appear in the Output Viewer. Don't be overwhelmed as we will only look at those that directly address our effort to develop the prediction equation. At the beginning of the output, we skip over some tables that describe our sample and coding procedures. To begin our discussion, we look at two tables that come under the heading *Block 0: Beginning Block.*

Block 0 simply reports the findings when the SPSS program attempts to predict the outcome (purchase or not purchase) without using any of the independent variables. Basically, it only looks at the percentages of individuals in the sample who purchased or did not purchase a home. Look at Figure 22.7, titled *Classification Table*; the callout draws attention to 76.7% ((30 − 7)/30), which shows the difference between the SPSS original prediction that no one would purchase a home (see Row 1 in Figure 22.7) and the actual number of buyers who were observed. The program used the observed number of buyers (7) to determine that the prediction that no one

Figure 22.7 Classification Table With No Independent Variables in the Equation

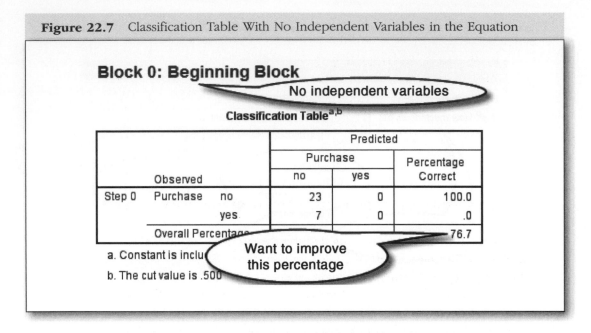

would purchase a home was correct 76.7% of the time. We are hopeful that when the independent variables are added into the equation our prediction capabilities will improve and this percentage value will increase.

The next table in the *Block 0* section is called *Variables in the Equation* and presents results of the *Wald* statistical test. Once again the results were obtained while using only the *Constant* in the equation (not one of the independent variables); therefore, it is of limited use. However, it can serve as a reference point. The results show a *Wald* statistic of 7.594, which is significant at .006 (see Figure 22.8).

Figure 22.8 Wald Test Results—With Only the Constant in the Equation

Variables in the Equation

		B	S.E.	Wald	df	Sig.	Exp(B)
Step 0	Constant	-1.190	.432	7.594	1	.006	.304

We next examine the more interesting portion of the SPSS output listed under the heading *Block 1: Method = Enter*. The following tables report the results when our independent variables are inserted into the equation. The first table presented (see Figure 22.9) is titled *Omnibus*

Figure 22.9 Chi-Square Omnibus Tests of Model Coefficients

Block 1: Method = Enter

Omnibus Tests of Model Coefficients

		Chi-square	df	Sig.
Step 1	Step	16.843	3	.001
	Block	16.843	3	.001
	Model	16.843	3	.001

Tests of Model Coefficients and tells us that the overall fit of our model is *good* as it surpasses the results where SPSS predicted (in *Block 0*) that no one would buy a home. This table reports a *goodness-of-fit* test showing a chi-square statistic of 16.843 with 3 degrees of freedom (*df*) and $N = 30$ with a significance level of $p < .001$. The p level of .001 informs us that the goodness-of-fit test can be taken seriously and provides evidence that we have a worthwhile model.

The *Model Summary* table presents additional information on the usefulness of our model following the insertion of the independent variables. This table is shown in Figure 22.10. The column titled *−2 Log likelihood* shows a value of 15.754, which is good. Smaller values are best, and such values can easily range into the hundreds; so we are satisfied that this test value of 15.754 adds credence to our model. There is some controversy over the value of the *Cox & Snell R Square* and the *Nagelkerke R Square* tests as some analysts say they do not qualify as true *R Square* values as in multiple regression. SPSS refers to them as *Pseudo R square* values. We report them

Figure 22.10 *Block 1 Model Summary*—Logistic Regression

Model Summary

Step	−2 Log likelihood	Cox & Snell R Square	Nagelkerke R Square
1	15.754[a]	.430	.648

a. Estimation terminated at iteration number 6 because parameter estimates changed by less than .001.

because we feel that they provide additional, albeit qualified, evidence that our model is valuable. The test values of .430 and .648 suggest that between 43% and 64.8% of the variability in the dependent variable is explained by our independent variables.

There are some statisticians who claim that the *Hosmer-Lemeshow Goodness of Fit Test* is the best test available to evaluate the fit of the logistic regression model. For this test to provide evidence of a good fit, we need to fail to reject the null hypothesis. Therefore, we want values greater than .05 in the *Sig.* column. Figure 22.11 shows a chi-square value of 5.654 at 8 *df* with a significance level of .686. Therefore, we have additional evidence that our model is reliable.

Figure 22.11   Model Fit Indicator Test (>.05 Desired)

Hosmer and Lemeshow Test

Step	Chi-square	df	Sig.
1	5.654	8	.686

The *Classification Table*, shown in Figure 22.12, serves the same purpose as the one in Figure 22.7. However, this time it shows the results when the independent variables are inserted into the equation. For this particular case, we wish to see how successful we are in predicting the purchase

Figure 22.12   *Classification Table* With All Independent Variables in the Equation

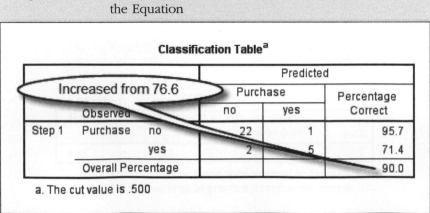

Classification Table[a]

			Predicted		
			Purchase		Percentage Correct
Observed			no	yes	
Step 1	Purchase	no	22	1	95.7
		yes	2	5	71.4
Overall Percentage					90.0

Increased from 76.6

a. The cut value is .500

category when the independent variables are used in the calculation. If we add the *yes* answers on the diagonal (2 + 1), subtract this from the total (30), and then divide by the total (30), we get a new overall prediction percent age of 90.0, as shown in Figure 22.12. The 90% value represents an increase of 13.4 (90 − 76.6) percentage points in predictive power, which provides additional evidence that our equation is useful.

The final regression output that we present is called *Variables in the Equation* and is shown in Figure 22.13. This table shows how each of the independent variables contributes to the equation. When looking at this table, we should pay special attention to the *Sig.* column. We see that the independent variables "preapproved" and "questions" ("interest in buying") have significance levels of .030 and .036, respectively. This informs us that both variables significantly contribute to the regression model. The variable "downpayment," with a signifinace level of .351, did *not* contribute to the equation. The *B* column specifies the weight of the variable's contribution, 3.288 for "preapproved" and 3.176 for "questions," and as expected, the weight of "downpayment" is almost nonexistent at .021.

Our original research question concerned whether we could pre-dict, based on three variables, whether an individual would purchase a home. The model that was developed used three independent variables: (1) "preapproved for a loan" (categorical/nominal), (2) "interested in buying" (categorical/nominal), and (3) "amount of money available for downpayment" (continuous/scale). The dependent variable was binary (categorical/nominal), where the outcome was whether they *purchased or did not purchase* a home.

The overall model, with all variables included, was found to be statisti-cally significant using a chi-square test with 3 *df*, *N* = 30, and a statistic of 16.843 at a significance level of .001. The strongest predictors of purchas-ing behavior were the variables "preapproved for loan" and "interest in

Figure 22.13 *Variables in the Equation* for Logistic Regression—Real Estate Example

Variables in the Equation

		B	S.E.	Wald	df	Sig.	Exp(B)
Step 1[a]	preapproved(1)	3.288	1.518	4.688	1	.030	26.777
	questions(1)	3.176	1.517	4.382	1	.036	23.946
	downpayment	.021	.022	.869	1	.351	1.021
	Constant	-6.143	2.244	7.492	1	.006	.002

a Variable(s) entered on step 1: preapproved, questions, downpayment.

buying." Individually, these two variables were found significant by the *Wald* statistic at .030 and .036. The variable "downpayment" was found *not* significant at .351.

You may recall that when we began our analysis of Sally's real estate data we calculated Spearman's correlation coefficients between all independent and dependent variables. The only correlation of concern was the value of .373 between "downpayment" and "interest in buying," as shown in Figure 22.4. We decided to leave both variables in the equation, and we discovered that it did not significantly affect the equation. We ran the logistic regression without the variable "downpayment," and the SPSS output changes were insignificant. With the two variables "interest in buying" and "preapproved," the omnibus test for the overall equation produced a chi-square test value of 15.911, 2 *df*, and a significance level of .000. The variable "downpayment" could thus be eliminated from the equation.

△   22.5 SUMMARY

This chapter completes our series on correlation and regression analysis. *Logistic regression* using multiple independent variables and a single binary dependent variable was illustrated. An equation was developed to assist a real estate salesperson in the identification of potential buyers based on the level of "interest in buying" and whether they had been "preapproved for a loan." A third variable, "amount available for the down payment," was also examined and found not to positively affect the equation. In the next chapter, we address principal component analysis as one of the major forms of factor analysis.

△   22.6 REVIEW EXERCISES

22.1   A major in the Air Force wanted to find a way to predict whether a particular airman would be promoted to sergeant within 4 years of enlisting in the military. He had data on many characteristics of individuals prior to enlistment. He chose three variables that he thought might be useful in determining whether they would get the early promotion. They are listed in the table showing the SPSS Variable View below. He selected a random sample of 30 individuals and compiled the required information. Your task is to develop a prediction equation that might assist the major in efforts to predict early promotion for his young airman. Write the research question(s) and the null and alternative hypotheses. The major's variable information and data are as follows:

	Name	Type	Width	De...	Label	Values	Missing		Align	Measure	Role
1	sports	Numeric	8	0	HS contact sports	{0, no}...	N	0=no 1=yes	Center	Nominal	Input
2	hunt	Numeric	8	0	hunting license	{0, no}...	None	0=no 1=yes		Nominal	Input
3	test	Numeric	8	0	induction test	None	None		Center	Scale	Input
4	sgt1	Numeric	8	0	4 years to sergeant	{0, no}...	None	8	Center	Scale	Input

	sports	hunt	test	sgt1
1	0	1	210	0
2	0	0	210	0
3	0	1	280	1
4	0	0	210	0
5	0	1	280	0
6	1	0	210	0
7	1	1	290	1
8	0	0	280	0
9	1	0	210	1
10	0	0	220	0
11	1	1	270	1
12	1	0	210	0
13	1	1	290	1
14	1	0	290	0
15	1	0	210	0
16	0	0	220	0
17	1	1	280	1
18	0	0	220	0
19	0	0	290	0
20	0	0	220	0
21	0	1	260	0
22	0	0	210	0
23	0	1	250	0
24	1	1	270	1
25	1	0	290	0
26	0	0	270	0
27	1	1	220	0
28	0	0	270	0
29	1	0	220	0
30	0	1	280	0

22.2 A social scientist wanted to develop an equation that would predict whether a male student would be successful in getting a date for the senior prom. The scientist had access to many student records and took a random sample of 40 students. She choose four characteristics that she felt would predict whether a male would get a date or not—a binary outcome. These variables are shown below in the SPSS Variable View. The *Label* column shows the description of the variable. Your job is to select the correct statistical approach and then assist the social scientist in developing the equation. Write the research question(s) and the null and alternative hypotheses. The variable information and data are as follows:

	Name	Type	Width	Decimals	Label	Values	Missing	Columns	Align	Measure	Role
1	work	Numeric	8	0	have personal income	{0, no}...	None	8	Center	Nominal	Input
2	height	Numeric	8	0	taller than 5'8"	{0, no}...	None	8	Center	Nominal	Input
3	grade	Numeric	8	0	GPA >3.5	{0, no}...	None	8	Center	Nominal	Input
4	activities	Numeric	8	0	3 or more	{0, no}...	None	8	Center	Nominal	Input
5	date	Numeric	8	0	date for prom	{0, no}...	None	8	Center	Nominal	Input

	work	height	grade	activities	date
1	0	1	0	0	0
2	0	0	0	1	0
3	0	1	1	0	1
4	0	0	0	0	0
5	0	1	0	0	0
6	1	0	0	0	0
7	1	1	0	0	1
8	0	0	1	0	0
9	1	0	0	1	1
10	0	0	0	0	0
11	1	1	1	0	1
12	0	0	0	0	0
13	1	0	0	1	1
14	0	0	1	0	0
15	1	0	0	0	0
16	0	0	0	0	0
17	1	1	1	0	1
18	0	0	0	0	0
19	0	0	0	1	0
20	0	0	0	0	0
21	1	0	1	0	0
22	0	0	0	0	0
23	0	1	0	0	0
24	1	1	1	0	1
25	0	1	0	0	0
26	0	0	1	0	0
27	0	1	0	1	0
28	0	0	0	0	0
29	0	1	0	0	0
30	1	0	0	0	0
31	0	1	0	0	0
32	0	1	0	0	0
33	1	0	0	0	0
34	0	0	0	0	0
35	1	1	1	0	0
36	0	1	1	0	1
37	1	0	0	1	0
38	0	0	1	0	0
39	1	1	0	0	0
40	0	0	0	1	0

22.3 For this review exercise, you will use the SPSS sample file titled *customer_dbase.sav*. You are a statistical consult with a contract to help a phone company executive develop a way to predict whether a customer would order the paging service. Based on prior experience, the executive feels that customers using voice mail ("voice"), caller ID ("callid") and electronic billing ("ebill") would also be inclined to utilize the paging service ("pager"). He is seeking statistical evidence and a written equation to support his intuitive feeling. He also wishes to utilize any equation that may result from the analysis to predict for future customers. Select the appropriate statistical method, open the database, select the variables, do the analysis, and then interpret the results.

CHAPTER 23

FACTOR ANALYSIS

△ **23.1 INTRODUCTION AND OBJECTIVES**

This chapter is different from prior chapters in that we don't test hypotheses that speculate about differences between populations. It is also dissimilar from our regression chapters in that we are not attempting to produce predictive equations with the intention of estimating unknown values based on past observations. The statistical technique discussed in this chapter presents a method whereby the researcher attempts to reduce the number of variables under study. The subject of this chapter, *factor analysis*, is a *data reduction* procedure—sounds pretty simple doesn't it? This data reduction is done by combining variables that are found to approximately measure the same thing. Part of the mathematical process involves the use of Pearson's correlation coefficient to determine the strength of the relationships between variables. You might think of two overlapping circles as separate variables with the intersection being a new *factor* that measures the same thing as the original two variables. Technically, this area of the intersection could be thought of as *shared variance*—more on this later. The *correlation coefficient* is simply a valid and reliable way to measure the strength of the relationship—it's the size of the intersection. For now, just think of *factor analysis* as a way to statistically simplify research by reducing the number of variables.

There are two basic types of factor analysis: (1) *descriptive*, often referred to as *exploratory* (we use both terms), and (2) *confirmatory*. We consider only the descriptive/exploratory approach in this book. There are several types of descriptive/exploratory factor analysis. The most common is known as *principal components*—this is the approach we use. (*Note:* When using *principal component factor analysis*, the term *component* is often substituted for *factor.*) If you require the testing of theories concerning the *latent* (hidden) structure underlying your variables, we recommend *confirmatory factor analysis* and a more advanced text. In this chapter, we do explore for and identify such latent variables, but we are not attempting to confirm or deny any hypothesis concerning the reality of their existence (beyond mathematical) and/or their usefulness.

This book strives to keep the technical/mathematical language that is used to describe factor analysis to a minimum. That being said, there are many terms that the beginning student/statistician might not be familiar with, and we will digress to define such terms when necessary. In lieu of providing the mathematical underpinnings, you are given step-by-step instruction on inputting the data and getting SPSS to accomplish the analysis of the data. Finally, we provide a careful explanation of each table and graph in the SPSS output. Our goal is to ensure that the reader has a complete understanding of the output, which will greatly enhance his or her ability to accurately interpret factor analysis.

Before getting started on actually doing factor analysis, we next introduce some of the specialized terms associated with the factor-analytic method. At the same time, we explain the process and the underlying logic of factor analysis.

First, you need to have a database composed of variables measured at the *scale* level (continuous data). The number of cases appropriate for successful factor analysis is not that easy to pin down. It is generally agreed that one consideration is the ratio of the number of cases to the number of variables. As a general rule, you should consider having 10 cases for each variable. Another requirement is that 150 cases should be a minimum number, with 300 to 1,000 much more desirable. Many researchers simply say, the more the better. That being said, another consideration, about sample size, is how your variables are actually related. For instance, if your initial data are such that *factor loadings* (more on this later) are high, then there is no reason why factor analysis could not be used with as few as 30 cases and three variables. However, such an application would more likely reside in the realm of confirmatory factor analysis. In addition to the above considerations, we will show how to request SPSS statistical tests that will examine your data and make a judgment as to their suitability for factor analysis.

Following the selection and input of your data, the first step in conducting the analysis is *factor extraction*. Remember that a *factor* is a group of correlated variables that represent a unique component of a larger group of variables (the intersection, as described above). Factor extraction is what SPSS does to find the fewest number of factors to represent the relationships among *all* the variables. The SPSS program *extracts factors* from your variables using a mathematical algorithm. Another term you will see is *factor loading*, which refers to the correlations between each variable and the newly created separate factors (components). Factor loadings are calculated using the *Pearson correlation coefficient*; therefore, they can range from +1 to −1. *Communalities* refers to the proportion of variability in the original variable that is accounted for by the high-loading factors.

Eigenvalue is simply a number that specifies the amount of variation in the original variables that can be attributed to a particular factor. These eigenvalues are used to help determine the number of factors to be selected. Part of the factor-analytic process is a procedure known as *factor rotation*. This is optional in SPSS, and you will be shown how to request this procedure. There are also many different methods of rotation designed for different types of factor loadings. An example of rotation would be when you may have correlated factors that require *oblique rotation* or uncorrelated factors calling for an *orthogonal* rotation. Whichever rotation method is chosen, the goal is to reorder the factor loadings to take full advantage of the difference in loadings. You want to maximize the degree to which each variable loads on a particular factor while minimizing variable loadings on all other factors.

Finally, you should be aware of the *scree plot*, which can be requested in the SPSS program. The scree plot uses a graph to map the eigenvalues of the newly produced factors and is of tremendous help in determining the number of factors that are appropriate.

OBJECTIVES

After completing this chapter, you will be able to

Understand the differences between the major factor-analytic methods

Determine when principal component factor analysis should be used

Select the correct variables and determine the suitability of data for factor analysis

Write research questions suitable for descriptive/exploratory factor analysis

Extract the correct number of factors from your variables

Rotate factors for maximum solution

Understand the logic behind principal component factor analysis

Interpret the SPSS output (results) from the factor analysis procedure

We hope that your mind has not "gone to the ski slopes or perhaps the beach" because of these last few paragraphs for we will try to breathe life into these definitions as we proceed through our example beginning in Section 23.2.

23.2 RESEARCH SCENARIO AND TEST SELECTION △

For this factor analysis demonstration, you will open the SPSS sample file titled *bankloan.sav*. This database is composed of fictitious data created to explore how the bank might reduce the number of loan defaults. The database consist of 850 past and prospective customers. We used all 850 cases to illustrate the *exploratory factor analysis principal component* approach to data reduction (*Note:* When the last 150 prospective customers were deleted, factor analysis detected no significant differences).

Since the *bankloan.sav* database also consisted of 10 variables measured at the scale level and a solid number of cases (850), we decided that factor analysis would be the appropriate statistical tool to describe and to explore the structure of the observable data (*manifest variables*). The major technique that will be used is principal component analysis since we were interested in discovering any underlying latent structures (factors or components) that were as yet unidentified. The idea was to reduce the 10 manifest variables into smaller clusters or components that would then serve to summarize the data.

23.3 RESEARCH QUESTION AND NULL HYPOTHESIS △

How many components (*latent* variables) might be identified among the 10 *manifest* variables that are being analyzed. If components are identified, how might they be interpreted? The null hypothesis states that there are no latent underlying structures and that all variables load equally.

23.4 DATA INPUT, ANALYSIS, AND INTERPRETATION OF OUTPUT △

We will begin the analysis by opening an SPSS sample database titled *bankloan.sav*. We would suggest that once it is opened, you save it in your documents file—it will then be easier to access in the future. Let's get started with the bullet points that will lead you through the input, analysis, and interpretation of the data.

- Start SPSS, Click **File**, select **Open**, and click **Data** (the *Open Data* window opens).
- In the C drive, locate and open *bankloan.sav* (see Section 3.5 of Chapter 3 if you need help).
- Click **Analyze**, select **Dimension Reduction**, and then click **Factor**.
- Click and move all the scale variables to the *Variable* box (see Figure 23.1).

Figure 23.1 *Factor Analysis* Window for Bank Loan

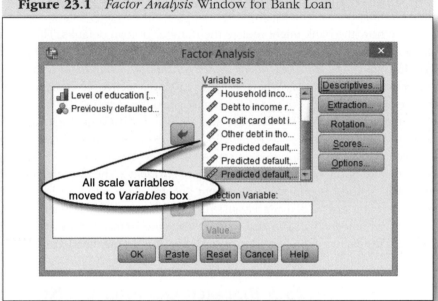

- Click **Descriptives** (the *Factor Analysis: Descriptives* window opens).
- Click **Variable descriptions**, and make sure **Initial Solutions** has been checked.
- In the *Correlation Matrix* panel, check **Coefficients**, **Significance levels**, and **KMO and Bartlett's test of sphericity**. (See Figure 23.2 for the window's appearance just prior to your next click.)
- Click **Continue** (you return to the *Factor Analysis* window).
- Click **Extraction**.
- Make sure *Principal components* shows in the *Method* section.
- Click **Correlation matrix** in the *Analyze* panel, and make sure both **Unrotated factor solution** and **Scree plot** are checked in the *Display* panel. In the *Extract* panel, make sure **Based on eigenvalue** is clicked and the value in the box next to *Eigenvalues greater than:* is 1 (the window should now look like Figure 23.3).

Figure 23.2 The *Factor Analysis: Descriptives* Window

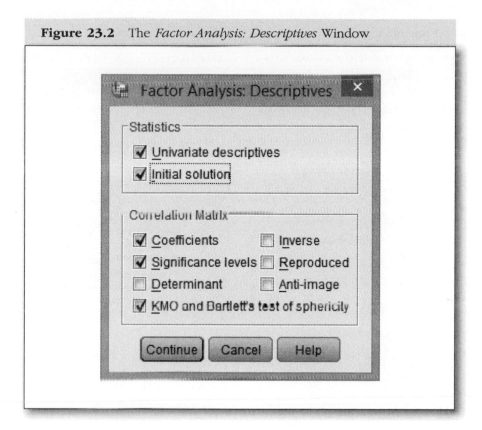

- Click **Continue**.
- Click **Rotation** (the *Factor Analysis: Rotation* window opens). In the *Methods* panel, click **Varimax**, and click **Continue**.
- Click **OK** (the Output Viewer opens with the results of our analysis).

Now that the tedious work surrounding the many calculations has been instantly accomplished by SPSS, let's see if we can understand *all* of the output.

The first two tables in the output we chose not to display. However, the first of these, the *Descriptives Statistics* table, should be examined to clarify that the intended variables were entered and the number of cases is correct. The second table, not shown here, is titled *Correlation Matrix*. This table should be examined more carefully, with the idea that *high* correlations are best when doing factor analysis. Thus, this table serves as a screening device for the data. The rule of thumb is that correlations exceeding .30 are desired (the more of them and the higher they are, the better). On reviewing the *bankloan.sav* correlation matrix, we felt that there were enough *moderate* to *strong* coefficients (>.30) to proceed with the analysis.

Figure 23.3   The *Factor Analysis: Extraction* Window

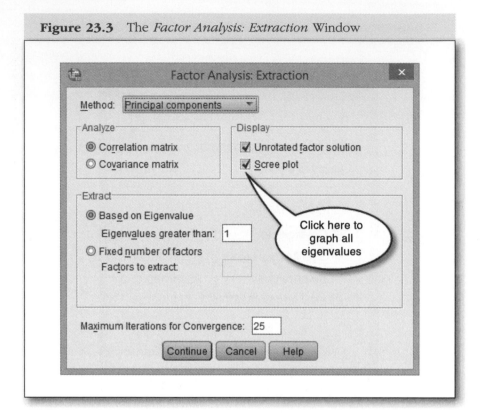

As we stated in this chapter's introduction, there are two tests available in the SPSS program that will further screen the data on two dimensions. The positive results of both tests are presented in Figure 23.3. The *Kaiser-Meyer-Olkin Measure of Sampling Adequacy* (KMO) reported a value of .717. Any value greater than .6 is considered an indication that the data are suitable for factor analysis. The next test result is for *Bartlett's Test of Sphericity*, which reported a chi-square of 10,175.535 at $df = 45$ and a significance level of .000. This also is a positive result, and we feel more confident that our final factor analysis will yield useful information (see Figure 23.4).

The next table shown in the output is titled *Communalities* and can be seen in Figure 23.5. You may recall that we briefly defined communalities in the Introduction as the proportion of variability in the original variable that is accounted for by the high-loading factors. To breathe some life into this definition, look at the value of .614 found in the first row under *Extraction* in the *Communalities* table. We can say that 61.4% (.614 × 100) of the variance in the variable "Age in years" can be explained by the *high-loading*

Figure 23.4 KMO and Bartlett's Test of Sphericity

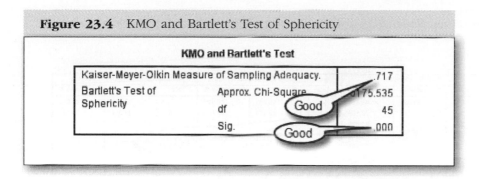

KMO and Bartlett's Test

Kaiser-Meyer-Olkin Measure of Sampling Adequacy.		.717
Bartlett's Test of Sphericity	Approx. Chi-Square	175.535
	df	45
	Sig.	.000

Figure 23.5 Communalities for Bank Loan

Communalities

	Initial	Extraction
Age in years	1.000	.614
Years with current employer	1.000	.740
Years at current address	.000	.418
Household income in thousands	1.000	.652
Debt to income ratio (x100)	1.000	.826
Credit card debt thousands	1.000	.732
		.756
		.96
2		.852
Predicted default, model 3	1.000	.824

This means that 61.4% of the variance in "Age in years" can be accounted for by the first 2 components

Extraction Method: Principal Component Analysis.

(*eigenvalues* >1) *components* (*factors*). These high-loading factors or components are identified in Figure 23.6 as Components 1 and 2. Another example is that 74% of the variability in "Years with current employer" can be explained by Components 1 and 2.

Figure 23.6   Total Variance Explained for Factor Analysis (Bank Loan Data)

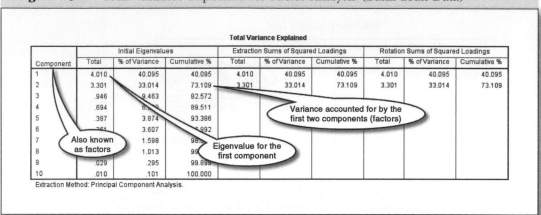

Total Variance Explained

Component	Initial Eigenvalues			Extraction Sums of Squared Loadings			Rotation Sums of Squared Loadings		
	Total	% of Variance	Cumulative %	Total	% of Variance	Cumulative %	Total	% of Variance	Cumulative %
1	4.010	40.095	40.095	4.010	40.095	40.095	4.010	40.095	40.095
2	3.301	33.014	73.109	3.301	33.014	73.109	3.301	33.014	73.109
3	.946	9.463	82.572						
4	.694	6.940	89.511						
5	.387	3.874	93.386						
6	.361	3.607	96.992						
7		1.598	98.590						
8		1.013	99.603						
9	.029	.295	99.899						
10	.010	.101	100.000						

Extraction Method: Principal Component Analysis.

In the table titled *Total Variance Explained* (see Figure 23.6), we actually see the *eigenvalues* for the 10 new components. Look at the column called *Initial Eigenvalues*, and notice the value of 4.010 for Component 1. This eigenvalue (4.010) is equivalent to 40.095% (4.010/10 × 100) of the total variance when all 10 variables are considered. The next row shows an *eigenvalue* of 3.301 for Component 2, which means that it accounts for 33.104% of the total variance for all variables. This percentage is *not* related to the variance of the first component; therefore, the two taken together (40.095 + 33.014) can be said to account for 73.109% of the variance for all variables (see the *Cumulative%* column in Figure 23.6).

The *scree plot* is shown in Figure 23.7, which graphs the *eigenvalues* on the *y*-axis and the 10 components on the *x*-axis. The scree plot is a widely accepted aid in selecting the appropriate number of components (factors) when interpreting your factor analysis. You simply select those components above the "elbow" portion—in this case, Components 1 and 2. As we saw in Figure 23.6, Components 1 and 2 account for 73.109% of the variance in all variables. We can say that our scree plot provides additional evidence in support of a two-component solution for our factor analysis problem.

Another table provided in the SPSS output is the *Component Matrix*, as seen in Figure 23.8. This table shows the *factor-loading* values for components with eigenvalues of 1.0 or more. This matrix presents loading values prior to rotation. The values in this table are interpreted in the same way as any correlation coefficient. By that we mean that zero indicates no loading, while minus values, such as −0.059 for Component 1, indicate that as the particular variable score increases, the component score decreases. Values

Figure 23.7 Scree Plot for Bank Loan Factor Analysis

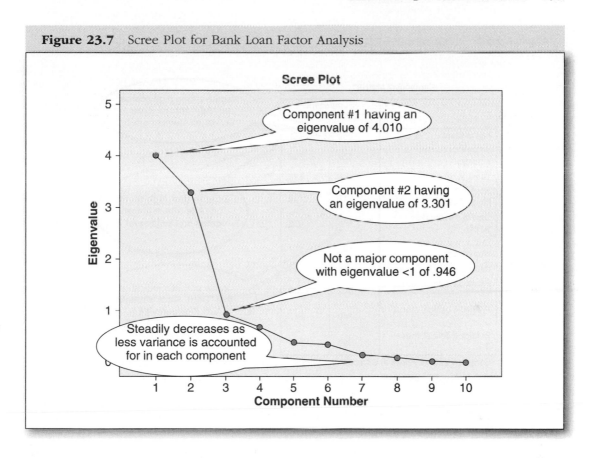

with a plus sign, such as 0.782 for Component 2, indicate that as the variable score increases, so does the component score. Remember that these values behave the same as correlation coefficients.

The information in this table can also assist in the naming of the newly created components. You look at the high-loading variables and what they measure, which can then suggest a name for a particular component. The purpose of the name is to describe what the component, the newly discovered latent variable, actually measures.

The final table that we present in our output is titled *Rotated Component Matrix* and is shown in Figure 23.9. This table was produced when we requested *Varimax* rotation as one of the last steps in setting up our analysis. The Varimax method of rotation is one of the most popular *orthogonal* approaches. *Orthogonal* means that we assume the components are uncorrelated. Whether this assumption is justified is another matter, but it is often done by those using the factor-analytic method. The other major type of

Figure 23.8 Component Matrix

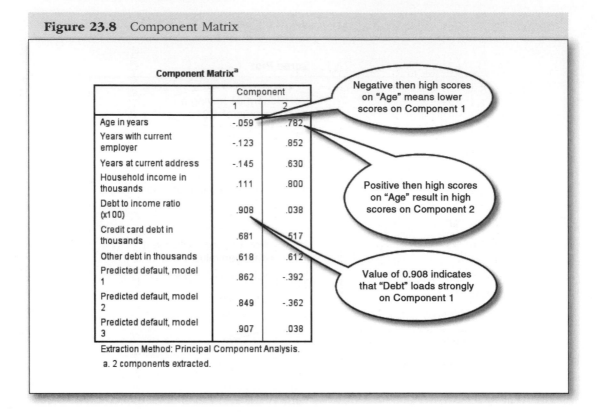

Component Matrix[a]

	Component 1	Component 2
Age in years	-.059	.782
Years with current employer	-.123	.852
Years at current address	-.145	.630
Household income in thousands	.111	.800
Debt to income ratio (x100)	.908	.038
Credit card debt in thousands	.681	.517
Other debt in thousands	.618	.612
Predicted default, model 1	.862	-.392
Predicted default, model 2	.849	-.362
Predicted default, model 3	.907	.038

Extraction Method: Principal Component Analysis.

a. 2 components extracted.

Negative then high scores on "Age" means lower scores on Component 1

Positive then high scores on "Age" result in high scores on Component 2

Value of 0.908 indicates that "Debt" loads strongly on Component 1

rotation (*oblique*) is generally used when you have some prior information that the factors may be correlated.

A look at Figure 23.9 shows that there were no improvements (changes) from the unrotated *Component Matrix* shown in Figure 23.8. The reader may wish to leave all settings the same in the factor analysis and try different rotation methods and look for changes.

Let's summarize our findings concerning the bank loan data and answer our original research questions. Recall that the research question was how many *components* (*latent* variables) might be identified among the 10 *manifest* variables that were being analyzed. If components were identified, how might they be interpreted? The null hypothesis stated that there were no latent underlying structures and that all variables loaded equally.

The data were first screened by creating a correlation matrix. The results revealed that there were many coefficients in the moderate-to-high range, which encouraged us to continue with the factor analysis. Next, we proceeded with the KMO test for sample adequacy, which resulted in a score of .717. Any value greater than .6 for the KMO test indicates that factor analysis

Figure 23.9 Rotated Component Matrix

Rotated Component Matrix[a]

	Component 1	Component 2
Age in years	-.060	.781
Years with current employer	-.124	.851
Years at current address	-.146	.630
Household income in thousands	.110	.800
Debt to income ratio (x100)	.908	.039
Credit card debt in thousands	.681	.518
Other debt in thousands	.617	.613
Predicted default, model 1	.862	-.391
Predicted default, model 2	.850	-.361
Predicted default, model 3	.907	.039

Extraction Method: Principal Component Analysis.
Rotation Method: Varimax with Kaiser Normalization.

a. Rotation converged in 2 iterations.

can be used. The next test, Bartlett's, also indicated that the data were suitable for factor analysis, with a chi-square test significant at .000. Following this screening of the manifest variable data, we continued with principle component factor analysis.

The analysis provided evidence of the underlying structure and the existence of two components (latent variables). The first component explained 40.095% of the variance, while the second explained 33.014%. Together they explained 73.109% of the variance in all our variables. Now comes the fun part, naming our newly discovered components. For help with this, we look at the *Component Matrix* and the *Rotated Component Matrix*. Remember that there was very little change when we did the *Varimax* rotation, so we can select either matrix. Let's look at the unrotated matrix for Component 1 and identify those variables with high loadings. All of the high loadings are associated with "debt," beginning with "Debt to income ratio" (.908) and ending with "Predicted default" (.907); therefore, we call this the "Debt to Default" component. For Component 2, high loadings were on variables that measured "responsibility" traits such as age, steady employment, continual domicile, and household income. It also loaded a little less on amount of debt but not on any of the default items. We will name Component 2 "Responsible Debtor."

It will take much more research to substantiate the initial identification of these components. However, we feel that the reader has received a good introduction to *principal component factor analysis* as a rather powerful research tool. Hopefully, the reader will have gotten a taste of potential rewards when discovering the unknown while studying a mass of numbers. Go get a database and try it—we think you will like it!

△ 23.5 SUMMARY

The major subject area covered in this chapter was the *factor-analytic* method when used as a *descriptive/exploratory* statistical tool. The other major use of *factor analysis* was described as *confirmatory*—where the user seeks to generate evidence in support of hypotheses concerning the underlying structure of observed data. The descriptive/exploratory factor analysis method known as *principal component analysis* was presented in detail. Confirmatory analysis was left for another time and place.

An example of factor analysis was given that began with a determination of the suitability of a particular database for the analysis and ended with a summarization and interpretation of the SPSS output. The next two chapters present the nonparametric statistic known as chi-square.

△ 23.6 REVIEW EXERCISES

23.1 For this exercise, you will use the SPSS sample file called *customer_dbase.sav*. This database is composed of 5,000 cases and 132 variables. You will select the first 10 *scale* variables and search for underlying *latent* variables (*factors*) within these variables. The idea is that you must explore the data in an attempt to reduce the 10 variables into smaller *clusters*, referred to as *components* in *principle component factor analysis*. Write the null and alternative hypotheses, open the database, select the correct statistical approach, and search for any underlying latent factors in the first 10 scale variables.

23.2 This review exercise uses the same SPSS sample file database as in the previous example (*customer_dbase.sav*). However, this time you will examine eight scale variables dealing with pet ownership (#29 through #36). You are to look for underlying latent variables that would permit the reduction of these eight variables. Write the null and alternative hypotheses, select the correct procedures, examine the initial statistics, and interpret the results.

23.3 For this review exercise, you will use the SPSS sample file titled *telco. sav*. This database has 1,000 cases and 22 variables measured at the *scale* level. You will select the first 16 of the scale variables (up to *wireten/wireless over tenure*, #25) and attempt to identify any underlying factor(s) that would permit data reduction. State the null and alternative hypotheses, select the statistical method, and proceed with the analysis.

CHAPTER 24

CHI-SQUARE
GOODNESS OF FIT

△ 24.1 INTRODUCTION AND OBJECTIVES

Chi-square, designated by the symbol χ^2, is a very popular nonparametric statistical test that is appropriate when data are in the form of frequency counts or percentages, or proportions that can be converted to frequencies. Chi-square is appropriate for *nominal* or *ordinal* data and can be used to compare frequencies occurring in categories. Basically, there are two major types of chi-square problems: (1) *goodness of fit* and (2) *test of independence*. The first type, goodness of fit, is the subject of this chapter; therefore, we reserve an explanation of the second type (independence) for the following Chapter 25. We use the goodness-of-fit approach when we wish to determine if the observed counts, percentages, or proportions match some expected (theoretical) values. For example, if you roll a fair die 900 times, you would expect, in terms of probability, 150 ones, 150 twos, 150 threes, 150 fours, 150 fives, and 150 sixes. This expectation is based on the idea that a fair die should result in one sixth of the 900 rolls landing on each of the six die faces an equal number of times. Since 900/6 = 150, we expect the die to stop rolling on each of the six die faces about 150 times. The question is how much divergence from these expected frequencies you are willing to accept before you decide that the die is somehow rigged and not fair. Chi-square will provide objective evidence to help you answer that question.

There are situations where *equality* of frequencies or proportions in various categories (as in the above example) is *not* the question for the goodness-of-fit test. An example might be that we wish to determine if our data's distribution approximates the normal curve. One way to answer this question of normality is to use the chi-square test to determine if the data diverge from the normal curve to a greater extent than would be expected by chance. Another example would be that if we randomly select samples of cats from a feline population in a certain city, we would probably not expect the same frequencies for each breed of cat. The chi-square goodness-of-fit test allows you to specify (theorize) frequencies and then see whether the observed category of frequencies differ from your theoretical frequencies. When using the chi-square test, you choose a level of significance such as .05, apply the test, and interpret the outcome in terms of the size of the chi-square value. The mathematics involved in computing χ^2 involves subtracting the expected value from the observed value, squaring the result, dividing by the expected value, and then adding all the cases. Clearly, the closer the expected values are to the observed values, the smaller the value of χ^2. Larger chi-square values means larger differences in the observed versus expected values, which may then lead to rejection of the null hypothesis. The null hypothesis states that there is no difference in the observed and expected values. The chi-square goodness-of-fit test allows you to determine whether the observed group frequencies for a single variable differ from what is expected by chance.

The reader will note that this chapter deviates slightly from previous ones in that we will provide two examples of the goodness-of-fit test. We do this to clearly illustrate different SPSS approaches that depend on how the data were collected. The *Nonparametric Tests/Legacy Dialogs* method is the first to be illustrated and is used when you already have the number of observed values in each category counted. The second method, *Nonparametric Tests/One Sample*, is used when you have a large number of cases for a *nominal* or *ordinal* variable and simply have the SPSS program review the data file and count the number of observations in each category. SPSS then calculates the chi-square value, interprets the results, and either retains or rejects the null hypothesis.

302 U<small>SING</small> IBM® SPSS® S<small>TATISTICS</small>

OBJECTIVES

After completing this chapter, you will be able to

Describe the purpose of the chi-square goodness-of-fit statistical test

Explain the computation of χ^2

Use chi-square to test hypotheses regarding frequencies

Interpret the results of an SPSS chi-square goodness-of-fit statistical test

Interpret the research question and null hypothesis

Use the *Nonparametric Tests/Legacy Dialogs* method to calculate χ^2

Use the *Nonparametric Statistics/One Sample* approach to calculate χ^2

△ 24.2 RESEARCH SCENARIO AND TEST SELECTION: LEGACY DIALOGS

We and especially Hilda, our trusted research assistant, are particularly fond of DOTS® candy made by the Tootsie Roll® Company. She enjoys the different flavors and was interested in determining if the five flavors (orange, strawberry, cherry, lime, and lemon) were equally distributed in a number of boxes. In short, did the boxes contain the same number of each flavor? On first glance, the *chi-square goodness-of-fit* test appeared to be the way to investigate this question because we are dealing with nominal data and frequencies of occurrence of each flavor of candy (categories). Assumptions related to the chi-square test are as follows: (a) observations must be independent and (b) sample size should be relatively large, such that the expected frequencies for each category are at least 1 and the expected frequencies are at least 5 for 80% or more of the categories. The data met the assumptions, so we decided to proceed with the chi-square test.

△ 24.3 RESEARCH QUESTION AND NULL HYPOTHESIS: LEGACY DIALOGS

A quick inspection of several boxes of the candy led us to question the belief that different flavors of candies would usually be distributed equally in the boxes if there were not a compelling reason to do otherwise. Perhaps the company did a marketing study and found that most people prefer a particular flavor.

We state the null and alternate hypotheses:

H_0: There is no difference in the frequency of each flavor of candy (i.e., the numbers of candies of each flavor are equal).

H_A: There is a difference in the frequency of each flavor of candy (i.e., the numbers of candies of each flavor are *not* equal).

24.4 DATA INPUT, ANALYSIS, AND △ INTERPRETATION OF OUTPUT: LEGACY DIALOGS

We purchased 13 boxes of DOTS from various stores in the area (and yes, Hilda ate most of the candies). We counted the total number of pieces in each box and tallied the number of flavors. A summary of our results is shown in Figure 24.1.

Figure 24.1   Frequency of the Five Flavors

Orange	Strawberry	Cherry	Lime	Lemon
131	117	263	147	123

What is the expected number for each flavor of candy? Since the total number of pieces of candy is 781, and we assume an equal number for the five flavors, the expected number for each flavor is $781/5 = 156.2$.

We ask you, the reader, to perform a chi-square test using the summarized data in Figure 24.1 and the *Legacy Dialogs* approach. Use SPSS's default setting of .05 level of significance to determine if the flavors are equally distributed. Is the number of each flavor the same? We will take you through the process step-by-step. Let's get started.

- Start SPSS, click **File**, select **New**, and click **Data**.
- Click the Variable View tab, and type *Flavor* as the first variable. Set type of variable to *Numeric*, decimals to 0, *Align* to *Center*, and *Measure* to *Nominal*. Click the cell below *Values*, and in the *Value Labels* window type the following: *1 = Orange, 2 = Strawberry, 3 = Cherry, 4 = Lime,* and *5 = Lemon*. Click **OK**.
- Type *Frequencies* for the second variable, set type of variable to *Numeric*, decimals to 0, *Align* to *Center*, and *Measure* to *Nominal*.
- Click the Data View tab, and enter data as shown in Figure 24.2.

Figure 24.2   Data View Screen for the *Legacy Dialogs* Chi-Square Test

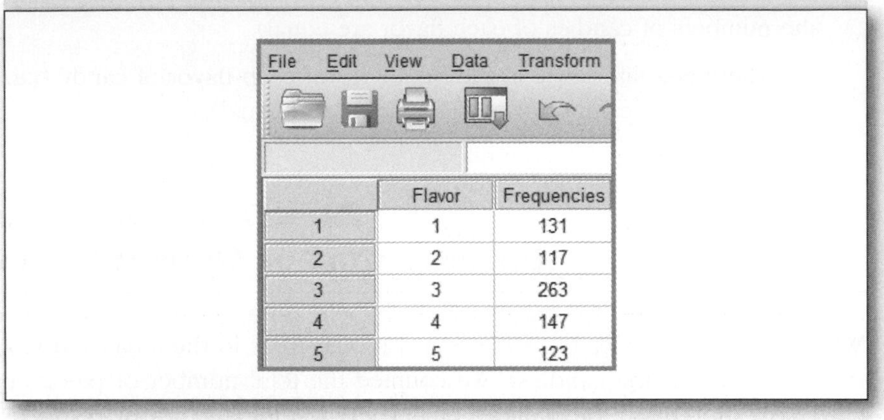

	Flavor	Frequencies
1	1	131
2	2	117
3	3	263
4	4	147
5	5	123

- Click **File**, click **Save As** (the *Save As* window opens), in the *File Name* box type *Flavor*, then click **Save**.
- Click **Data** on the Main Menu, and then click **Weight Cases** (the *Weight Cases* window opens).
- Click **Weight cases by**, and then click **Frequencies** in the left panel; click the right arrow to move it to the *Frequency Variable* box. (*Note:* After you click *Weight cases*, later when you attempt to close the database, SPSS will ask you if you want to save the changes; if so, click **Yes**.)
- Click **OK**.
- Click **Analyze**, select **Nonparametric Tests**, select **Legacy Dialogs**, and then click **Chi-Square** (a window titled *Chi-Square Test* will open).
- Click **Flavo**r in the left panel, and click the right arrow to place it in the *Test Variable List* box.
- Click **All categories equal**.
- Click **OK**.

Figure 24.3 shows the observed and expected frequencies of each flavor and the differences (residuals) between the observed and expected frequencies.

Although the SPSS table lists the significance level as .000 in Figure 24.4, it does not mean that there is no level of significance. SPSS rounds values, so the actual value is less than .0005, which SPSS rounds to .000.

The rather large computed value of χ^2 is 94.53, which is shown in Figure 24.4, and the level of significance is .000. Since .000 is less than .05, we reject the null hypothesis. By rejecting the null hypothesis (that the flavors are equally distributed), we now have evidence in support of our (Hilda's) theory that the flavors are *not* equally distributed.

Figure 24.3 Observed and Expected Frequencies for Each Flavor

Flavor

	Observed N	Expected N	Residual
Orange	131	156.2	-25.2
Strawberry	117	156.2	-39.2
Cherry	263	156.2	106.8
Lime	147	156.2	-9.2
Lemon	123	156.2	-33.2
Total	781		

Figure 24.4 Chi-Square Tests for Flavor Frequencies

Test Statistics

	Flavor
Chi-Square	94.525[a]
df	4
Asymp. Sig.	.000

a. 0 cells (0.0%) have expected frequencies less than 5. The minimum expected cell frequency is 156.2.

24.5 RESEARCH SCENARIO AND TEST SELECTION: ONE SAMPLE

For this demonstration, you will use the SPSS sample database titled *work-prog.sav*. You have used this database before—it consists of eight variables and 1,000 cases. Once the database is opened, you will conduct a chi-square test on the categorical variable called "marital" and labeled *Marital status*. The *Marital status* variable ("marital") for these 1,000 cases is categorized as 0 = *Unmarried* and 1 = *Married*.

The investigator wished to study the sample data in order to learn more about the marital status as it may relate to the other seven variables. Some of the theories she was developing required that the categories of the unmarried and married be equal. An initial review of the data indicated that this might not be true. She wished to test her theory using a statistical test. The chi-square *goodness-of-fit* test was selected because the data were nominal (categorical) and she wished to test for the equality or nonequality of marital status in the sample.

Δ 24.6 Research Question and Null Hypothesis: One Sample

The researcher suspects that the unmarried and married categories in the sample are not equal. The question is whether there is a significant difference in the marital status of the members of the sample studied. The null and alternative hypotheses are stated below:

H_0: There is no significant difference in the numbers of unmarried and married individuals in the sample (i.e., the categories are equal).

H_A: There is a significant difference in the numbers of unmarried and married individuals in the sample (i.e., the categories are unequal).

Δ 24.7 Data Input, Analysis, and Interpretation of Output: One Sample

- Start SPSS, click **File**, select **Open**, and click **Data**.
- In the *Open Data* window, find *workprog.sav* in the SPSS sample files. Click **workprog.sav**, then click **Open** (if you need help in finding and opening the *workprog.sav* file, go back to Section 3.5).
- Click **Analyze**, select **Nonparametric Tests**, then click **One Sample** (the *One-Sample Nonparametric Tests* window opens)**.**
- Click the Objective tab, then click **Customize analysis**.
- Click the Fields tab, and use the arrow to make sure that "Marital status" is the only variable in the *Tests Fields* panel.
- Click the Settings tab, click **Customize tests**, and check the box next to *Compare observed probabilities to hypothesized (Chi-Square test)* (see Figure 24.5).
- Click **Options** (in the *One-Sample Nonparametric Tests* window).
- In the *Chi-Square Test Options* window (see Figure 24.6), in the *Choose Test Options* panel, click **All categories have equal probability**, then click **OK**.

Figure 24.5 One-Sample Nonparametric Tests for Marital Status

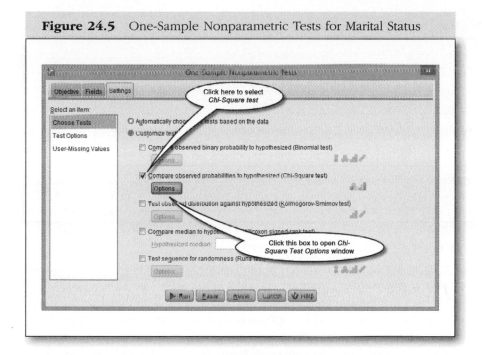

Figure 24.6 *Chi-Square Test Options* Window for Marital Status

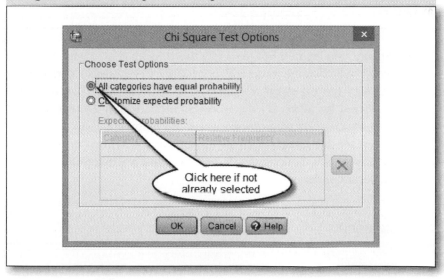

- Click **Run** (the Output Viewer opens with the *Hypothesis Test Summary*, as shown in Figure 24.7).

Figure 24.7 Hypothesis Test Summary for Marital Status

Hypothesis Test Summary

	Null Hypothesis	Test	Sig.	Decision
1	The categories of Marital status occur with equal probabilities.	One-Sample Chi-Square Test	.230	Retain the null hypothesis.

Asymptotic significances are displayed. The significance level is .05.

The *Nonparametric Tests/One Sample* procedure results in a single output titled *Hypothesis Test Summary*. This test result states the null hypothesis, in this case that the categories of *Marital status* occur with equal probabilities, with a significance level of .230, and that we *retain* the null hypothesis. Therefore, we can say that the researcher now has objective evidence that the groups are equal and she may confidently proceed with additional testing.

△ 24.8 SUMMARY

In this chapter, you learned how to apply the chi-square goodness-of-fit test to determine whether the observed occurrences of a single variable are consistent with the expected frequencies. You used two data-processing methods offered by SPSS: (1) the *Nonparametric/Legacy Dialogs* and (2) *Nonparametric/One Sample* approaches. In the following chapter, you will apply chi-square to a situation involving *two* variables. The *chi-square test of independence* is a nonparametric test designed to determine whether two variables are independent or related.

△ 24.9 REVIEW EXERCISES

24.1 A medical biologist was studying three bacteria that were known to be equally present in samples taken from the healthy human digestive system. We shall label them as A, B, and C. The transformed numbers of bacteria observed by the scientist were A = 3,256, B = 2,996, and

C = 3,179. The question is whether the difference from the expected shown in these values qualifies as being statistically significant. Write the null and alternative hypotheses, input the data, run the analysis, and interpret the results.

24.2 For this problem, you will open the SPSS *sample file telco.sav* and use the first variable, named "region" (labeled *Geographic indicator*). The variable "region" represents five different zones in which the 1,000 cases reside. The researcher believes that the cases are not equally distributed among the five zones. You are to write the null and alternative hypotheses and then study the variable "region" in an attempt to develop evidence in support of the researchers' hypothesis.

24.3 A high school principal was concerned that several of her teachers were awarding "A" grades at vastly different rates—she had heard many complaints from students and parents. She compiled the data on the teachers and decided to see if there was in fact a statistically significant difference. Her first look at the grades gave her the idea that there was indeed a difference. Can you write the null and alternative hypotheses, select the appropriate statistical test, conduct the analysis, and then interpret the results? The data are as follows—the teacher's name and the number of "A" grades are given in parentheses: Thomas (6), Maryann (10), Marta (12), Berta (8), and Alex (10).

CHAPTER 25

CHI-SQUARE TEST OF INDEPENDENCE

Δ 25.1 INTRODUCTION AND OBJECTIVES

The *chi-square test of independence* is a nonparametric test designed to determine whether two categorical variables are *independent* or *related*. The term *independence* derives from the fact that when using this chi-square test, the null hypothesis states that the variables are independent (not related). This test is designed to be used with data that are expressed as frequencies. The *discrete* data to be analyzed (the observations) are placed in a *contingency* table, and the *expected* values are calculated. As in the chi-square goodness-of-fit test, the *observed* values are then compared with *expected values*.

A *contingency table* is an arrangement in which a set of objects is classified according to two criteria of classification, one criterion entered in rows and the other in columns. A contingency table may have more than two columns and more than two rows. For example, we may have a column variable "education" having three categories: (1) no high school degree, (2) high school degree, and (3) associate degree. The row variable "gender" might have the two categories of male and female. In this scenario, we could use the chi-square test to seek evidence of a relationship between gender and level of education.

As in the previous chapter, we present two examples in our demonstration of the chi-square test of independence. The first example uses the data already summarized, so you will input in the same manner as with the first example in Chapter 24. The second example will utilize a database from the SPSS sample files.

OBJECTIVES

After completing this chapter, you will be able to

Describe the purpose of the chi-square test of independence

Write the null and alternative hypotheses for the chi-square test of independence

Interpret the contingency table for the chi-square test of independence

Use SPSS to conduct chi-square tests of independence

Interpret the results of an SPSS test for chi-square

25.2 RESEARCH SCENARIO △
AND TEST SELECTION: SUMMARIZED DATA

A researcher had heard that color blindness is related to gender in certain populations. He wished to determine if this were true for a group of individuals for whom he had collected data. Because the sample involves data measured at the nominal level and is concerned only with frequencies, the researcher decided that a *chi-square test of independence* would be appropriate. Assumptions related to the test are as follows: (1) observations must be independent and (2) sample size should be relatively large, such that the expected frequencies for each category are at least 1 and the expected frequencies are at least 5 for 80% or more of the categories.

25.3 RESEARCH QUESTION AND △
NULL HYPOTHESIS: SUMMARIZED DATA

The researcher is aware that color blindness is a genetic trait that may be related to gender. He selects a representative sample of males and females to investigate the question. The basic research question is whether the proportion of males who are color blind is the same as the proportion of females.

We state the null and alternative hypotheses as follows:

H_0: Color blindness is independent of gender (i.e., color blindness and gender are not related).

H_A: Color blindness is not independent of gender (i.e., color blindness and gender are related).

Δ 25.4 DATA INPUT, ANALYSIS, AND INTERPRETATION OF OUTPUT: SUMMARIZED DATA

As we have done in recent chapters we next show the researcher's summarized data as entered into the Variable View and Data View screens of the SPSS program. The summarized data represent 1,000 individuals, of which 480 are males and 520 are females. Figure 25.1 shows the Variable View in SPSS and serves as a guide for the reader to enter the variable information. At this time, enter all the variable information, as shown in Figure 25.1.

The Data View screen, shown in Figure 25.2, provides the data and shows how the data are to be entered for this demonstration. At this time, enter all the data, as shown in Figure 25.2.

Figure 25.1   Variable View for Chi-Square Test of Independence for the Color-Blind Study

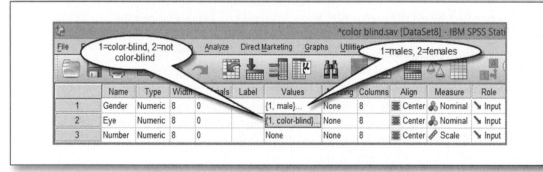

Figure 25.2   Data View Screen for Chi-Square Test of Independence

	Gender	Eye	Number
1	1	1	38
2	1	2	442
3	2	1	6
4	2	2	514

This may seem a little out of place, but for this type of analysis you must weigh the cases prior to saving the database. Follow the next series of bullets to accomplish this necessary step.

- Click **Data** on the Main Menu, and then click **Weight Cases** at the bottom of the drop-down menu (the *Weight Cases* window opens).
- Click **Weigh cases by**, click **Number**, and then click the arrow to move it to the *Frequency Variable* box.
- Click **OK**.

Once you have entered the variable information and the data as shown in Figures 25.1 and 25.2 and have weighed the cases, you should save the file as described below. Assuming you have now entered the variable information, data, and weights, follow the bullets given next.

- Click **File**, click **Save As** (the *Save As* window opens), and in the *File name* box type *color blind*.
- Click **Save**.
- Click **Analyze**, select **Descriptive Statistics**, and then click **Crosstabs**.
- Click **Gender**, then click the arrow to place it in the Row(s): box.
- Click **Eye**, then click the arrow to place it in the Column(s): box (see Figure 25.3).
- Click the **Statistics** button (a window called *Crosstabs: Statistics* opens).
- In this window, click **Chi-square**, and then click **Continue**.
- Click **Cells** (a window called *Crosstabs: Cell Display* opens).
- In this window, click **Observed** and **Expected** (found in the *Counts* panel), and then click **Continue**.
- Click **OK** (the Output Viewer opens with Figures 25.4 and 25.5).

The first table produced by SPSS is the *Case Processing Summary*, which can be useful in finding the problems with the database—we don't show this table here. The next table is shown in Figure 25.4 and is labeled by SPSS as *Gender * Eye Crosstabulation*. This is the *contingency table* discussed at the beginning of this chapter. We requested that it display a minimal amount of information, such as *observed* and *expected* values as well as row and column totals. We did this to reduce clutter and encourage thinking of chi-square as the difference between the *observed* and *expected* values. In our next example (Section 25.5), we show how to generate additional useful information that will be shown in the contingency table output.

The third table produced by SPSS is called *Chi-Square Tests* and is shown in Figure 25.5. This is where we look to see if the null hypothesis of independence is rejected or retained. This important information is

Figure 25.3   Test of Independence *Crosstabs* Window

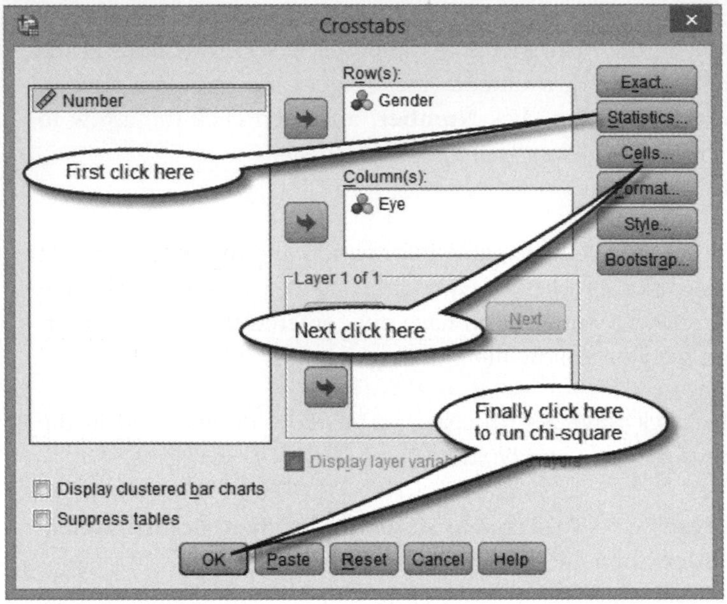

Figure 25.4   Contingency Table for the Gender/Color-Blind Study

Gender * Eye Crosstabulation

			Eye		
			color-blind	not color-blind	Total
Gender	male	Count	38	442	480
		Expected Count	21.1	458.9	480.0
	female	Count	6	514	520
		Expected Count	22.9	497.1	520.0
Total		Count	44	956	1000
		Expected Count	44.0	956.0	1000.0

found in the row titled *Pearson Chi-Square* and below the column titled *Asymp. Sig. (2-sided)*. We see the value of .000, which means that the null

Figure 25.5   Chi-Square Test of Independence for the Gender/Color-Blind Study

Chi-Square Tests

	Value	df	Asymp. Sig. (2-sided)	Exact Sig. (2-sided)	Exact Sig. (1-sided)
Pearson Chi-Square	27.139[a]	1	.000		
Continuity Correction[b]	25.555	1	.000		
Likelihood Ratio	29.773	1	.000		
Fisher's Exact Test				.000	.000
Linear-by-Linear Association	27.112	1	.000		
N of Valid Cases	1000				

a. 0 cells (0.0%) have expected count less than 5. The minimum expected count is 21.12.

b. Computed only for a 2x2 table

hypothesis of independence is rejected. Thus, we now have statistical evidence in support of the researcher's idea that there is a relationship between gender and color blindness for his sample. Depending on his sampling technique, he may or may not extend his hypothesis to a larger population.

25.5 RESEARCH SCENARIO AND TEST SELECTION: RAW DATA △

This example will illustrate the chi-square test of independence using raw data from the SPSS sample files. The particular file you will use is titled *customer_dbase.sav*. It is a hypothetical database that contains 134 variables on 5,000 customers. It was created to simulate the use of a company's existing database to make special offers to only those individuals most likely to reply. As is often done in this type of research, we will use the data for another purpose: to explore the relationship between customers' gender and their job satisfaction. There are two variables that we have selected to address this question: (1) "gender," labeled as *Gender*, where 0 = *male* and 1 = *female*, and (2) "jobsat," labeled as *Job satisfaction*, where 1 = *highly dissatisfied*, 2 = *somewhat dissatisfied*, 3 = *neutral*, 4 = *somewhat satisfied*, and 5 = *highly satisfied*.

The variable "gender" is classified as a *nominal* variable, while "jobsat" is classified as *ordinal*; therefore, chi-square would be the appropriate hypothesis-testing tool. Furthermore, since we are interested in studying the data in an effort to determine if the two variables are related, the *chi-square test of independence* is the test of choice.

△ 25.6 Research Question and Null Hypothesis: Raw Data

What are the percentages for each of the job satisfaction categories for males and females? Are the proportions of males for the five job satisfaction categories significantly different from the proportions for females?

We next state the null and alternative hypotheses:

H_0: Responses to the level of job satisfaction question are independent of gender (i.e., job satisfaction answers are the same for male and female customers).

H_A: Responses to the level of job satisfaction question are related to gender (i.e., job satisfaction answers are different for males and females).

△ 25.7 Data Input, Analysis, and Interpretation of Output: Raw Data

Let's get started by opening the selected database from the SPSS sample files. Once again, we will assume that you know how to navigate the program files to locate and open the *customer_dbase.sav*, while reminding you that help can be found in Section 3.5 (Chapter 3) of this book.

- Start SPSS, click **File**, select **Open**, then click **Data**.
- In the *Open Data* window, go to your C drive, and locate and click **customer_dbase.sav**, which moves it to the *File name* box, then click **Open** (see Section 3.5 for help in locating and opening this sample file).
- Click **Analyze**, select **Descriptive Statistics**, and then click **Crosstabs** (the *Crosstabs* window opens; see Figure 25.6).
- Click **Gender**, then click the arrow to move it to the *Row(s):* box.
- Scroll to and click **Job satisfaction**, then click the arrow to move it to the *Column(s):* box.
- Click **Statistics** (the *Crosstabs: Statistics* window opens), then click **Chi-square** (see Figure 25.7).
- Click **Continue** (to return to the *Crosstabs* window).
- Click **Cells** (the *Crosstabs: Cell Display* window opens) in the *Counts* panel; make sure both **Observed** and **Expected** are clicked, and then in the *Percentages* panel click **Row**, **Column**, and **Total** (see Figure 25.8).
- Click **Continue**, and then click **OK**, which initiates the analysis and opens the Output Viewer (see Figures 25.9 and 25.10).

Figure 25.6   Crosstabs for Chi-Square Test of Independence ("Gender" and "Job Satisfaction")

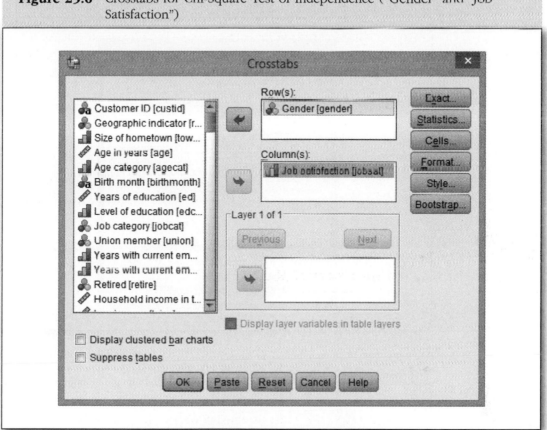

Next, we will take you through an explanation of the output while also answering the research questions. First, we look at the contingency table (crosstabulation) that is shown in Figure 25.9 to answer the first question regarding the percentages for each of the job satisfaction categories for males and females.

Let's look at the first cell for *Highly dissatisfied* and *Male* and interpret the information—understand one cell, and you will understand them all. The first *row*, called *Count*, reports the actual frequency for male customers who reported being highly dissatisfied with their job (475). The *Expected Count* row has the value calculated by the chi-square analysis (473.6). The third *row*, *% within Gender*, tells us that 19.4% of the males reported being highly dissatisfied with their jobs. You can confirm this by adding together all the values for males in each of the job satisfaction categories (19.4 + 21.8 + 20.8 + 20.3 + 17.7 = 100). Since we are interested in comparing males and females, a quick look at the female category shows 19.3%, or no practical

Figure 25.7   The *Crosstabs: Statistics* Window ("Gender" and "Job Satisfaction")

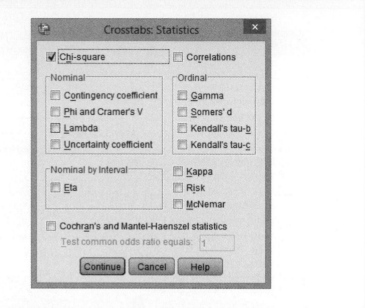

Figure 25.8   The *Crosstabs: Cell Display* Window

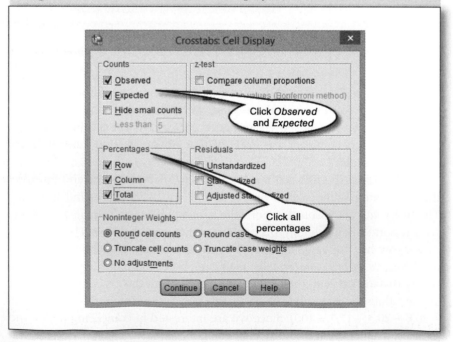

Figure 25.9 Chi-Square Test of Independence: "Gender" and "Job Satisfaction" Crosstabulation

Gender * Job satisfaction Crosstabulation

			Job satisfaction					Total
			Highly dissatisfied	Somewhat dissatisfied	Neutral	Somewhat satisfied	Highly satisfied	
Gender	Male	Count	475	535	510	496	433	2449
		Expected Count	473.6	509.9	534.9	496.7	434.0	2449.0
		% within Gender	19.4%	21.8%	20.8%	20.3%	17.7%	100.0%
		% within Job satisfaction	49.1%	51.4%	46.7%	48.9%	48.9%	49.0%
		% of Total	9.5%	10.7%	10.2%	9.9%	8.7%	49.0%
	Female	Count	492	506	582	518	453	2551
		Expected Count	493.4	531.1	557.1	517.3	452.0	2551.0
		% within Gender	19.3%	19.8%	22.8%	20.3%	17.8%	100.0%
		% within Job satisfaction	50.9%	48.6%	53.3%	51.1%	51.1%	51.0%
		% of Total	9.8%	10.1%	11.6%	10.4%	9.1%	51.0%
Total		Count	967	1041	1092	1014	886	5000
		Expected Count	967.0	1041.0	1092.0	1014.0	886.0	5000.0
		% within Gender	19.3%	20.8%	21.8%	20.3%	17.7%	100.0%
		% within Job satisfaction	100.0%	100.0%	100.0%	100.0%	100.0%	100.0%
		% of Total	19.3%	20.8%	21.8%	20.3%	17.7%	100.0%

Figure 25.10 Chi-Square Test of Independence for "Gender" and "Job Satisfaction"

Chi-Square Tests

	Value	df	Asymp. Sig. (2-sided)
Pearson Chi-Square	4.704[a]	4	.319
Likelihood Ratio	4.705	4	.319
Linear-by-Linear Association	.392	1	.531
N of Valid Cases	5000		

a. 0 cells (0.0%) have expected count less than 5. The minimum expected count is 433.96.

difference. The next row, *% within Job Satisfaction*, shows 49.1%, which means that of all the customers who said that they were highly dissatisfied with their jobs, 49.1% were males. Looking again at the *Female* category, we find that 50.9% of the highly dissatisfied were females—that is, no real difference. The final row, *% of Total*, means that 9.5% of *all* the customers could be classified as male and highly dissatisfied with their job. You may also find the *row* and *column totals* helpful in your quest to make these data more understandable. Of course, we could continue making comparisons using the remaining 17 cells, but we will leave that to the reader. In the next table, we decide whether to retain or reject the null hypothesis.

Next, we look at Figure 25.10 and, as in the previous example, we look at the *Pearson Chi-Square* row and the column titled *Asymp. Sig. (2-sided)*. The value shown of .319 (>.05) tells us that we must retain the null hypothesis. The null hypothesis was stated as follows: *Responses to the level of job satisfaction question are independent of gender (i.e., job satisfaction answers are the same for male and female customers)*. Our analysis provided statistical evidence that there is no statistically significant difference for males and females on the job satisfaction question.

△ 25.8 Summary

In this chapter, you learned how to use SPSS to analyze data in a chi-square contingency table, how to interpret the output, and how to decide whether to reject or fail to reject the null hypothesis based on the results.

This is the final chapter in this book. At this time, your arsenal and understanding of descriptive and inferential statistics are substantial, as is your understanding of data analysis and which statistical tests are appropriate for your data. Most important, you have a solid command and understanding of SPSS and the various analyses and procedures available to help you progress in your research or simply to give you the tools to understand the research of others.

△ 25.9 Review Exercises

25.1 A community activist believed that there was a relationship between membership in the police SWAT Team and prior military experience. He collected data from several police departments in an effort to support his belief. He found that there were 57 members of the SWAT team with prior military experience and 13 members with no prior military service. There were also 358 police personnel who had military experience but were not members of SWAT and another 413 with no military experience and not members of SWAT. You must write the null and alternative hypotheses, select the correct statistical method, do the analysis, and interpret the results.

25.2 For this exercise, you will open the SPSS sample file *bankloan.sav* and determine if there is a relationship between gender and the size of their hometown for these 5,000 bank customers. The bank official conducting the research believes that "size of hometown" is definitely related to "gender." Your task is to assist the bank official in uncovering evidence in support of his belief. Write the null and alternative hypotheses, select the appropriate statistical method, conduct the analysis, and interpret the results.

25.3 A nutritionist was developing a healthy-eating educational program and was seeking evidence to support her belief that males and females do not consume the same amount of vegetables. She conducted a survey that categorized people by gender and whether they consumed low, medium, or high amounts of vegetables. The numbers for males were low = *29*, medium = *21*, and high = *16*. The numbers for females were low = *21*, medium = *25*, and high = *33*. Write the null and alternative hypotheses, select the correct test, do the analysis (include percentages for all categories), and interpret the results.

APPENDIX A

CLASS SURVEY DATABASE
(ENTERED IN CHAPTER 5)

Table A.1 Variables and Attributes for *class_survey1.sav*

Variables and Attributes (Properties)										
Name	Type	Width	Decimals	Label	Values	Missing	Columns	Align	Measure
class	Numeric	8	0	Morning or Afternoon Class	1 = *Morning* 2 = *Afternoon*	None	8	Left	Nominal
exam1_pts	Numeric	8	0	Points on Exam One	None	None	8	Left	Scale
exam2_pts	Numeric	8	0	Points on Exam Two	None	None	8	Left	Scale
predict_grde	Numeric	8	0	Student's Predicted Final Grade	1 = *A* 2 = *B* 3 = *C* 4 = *D* 5 = *F*	None	8	Left	Nominal
gender	Numeric	8	0	Gender	1 = *Male* 2 = *Female*	None	8	Left	Nominal
anxiety	Numeric	8	0	Self-rated Anxiety Level	1 = *Much anxiety* 2 = *Some anxiety* 3 = *Little anxiety* 4 = *No anxiety*	None	8	Left	Ordinal
rate_inst	Numeric	8	0	Instructor Rating	1 = *Excellent* 2 = *Very good* 3 = *Average* 4 = *Below average* 5 = *Poor*	None	8	Left	Ordinal

Table A.2 Data for *class_survey1.sav*

student	class	exam1_pts	exam2_pts	predict_grde	gender	anxiety	rate_inst
1	1	100	83	1	2	4	2
2	1	50	68	3	2	2	1
3	1	78	68	3	2	2	1
4	1	50	78	3	1	2	1
5	1	97	74	2	2	3	2
6	1	41	71	3	2	2	1
7	1	30	72	3	1	1	2
8	1	31	83	2	1	1	1
9	1	71	63	2	2	2	1
10	1	85	89	1	2	3	1
11	1	86	93	2	2	2	1
12	1	67	64	2	2	1	1
13	1	52	100	2	1	2	1
14	1	88	83	1	2	4	1
15	1	25	23	1	1	1	1
16	1	100	100	1	2	2	2
17	1	14	71	3	2	2	2
18	1	60	75	3	2	1	2
19	2	93	84	1	2	2	1
20	2	94	93	1	1	4	1
21	2	90	89	1	1	2	2
22	2	78	80	2	1	2	1
23	2	50	84	3	2	1	1
24	2	74	50	2	2	3	1
25	2	62	93	2	1	4	1
26	2	80	81	2	2	2	1

(Continued)

(Continued)

student	class	exam1_pts	exam2_pts	predict_grde	gender	anxiety	rate_inst
27	2	87	97	1	1	2	1
28	2	25	61	3	2	1	1
29	2	50	82	2	1	2	1
30	2	99	93	2	2	3	1
31	2	50	64	3	2	1	1
32	2	100	100	1	2	2	1
33	2	66	62	3	2	2	1
34	2	50	100	3	2	1	1
35	2	100	94	1	2	4	1
36	2	26	53	4	2	1	3
37	2	41	72	4	2	1	1

APPENDIX B

BASIC INFERENTIAL STATISTICS

B.1 INTRODUCTION △

The purpose of this book is to provide information that enhances one's ability to use the power of SPSS to conduct *descriptive* analysis and various *inferential* statistical tests. The purpose of this appendix is to provide the reader with some basic concepts concerning inferential statistics. Some of the concepts presented here are not necessarily needed to complete the SPSS operations but are certainly useful in understanding the process used to generate the output. If you require information on basic descriptive statistics, turn to Chapter 4.

As you peruse this appendix, you may see information that is also presented in the text as part of the description of various inferential statistical methods. In the following sections, you will see most of the major areas of inferential techniques presented—many that were only briefly mentioned in the text. We hope this appendix provides a "one-stop" place where you can obtain basic information on many aspects of inferential statistics not covered in the text.

As the term *inferential* implies, one uses inferential statistics to make inferences concerning a defined population of interest. As opposed to descriptive statistics, which describes "what is," inferential statistics predicts what "might be," based on the results of selecting representative or random samples from a defined population of interest.

A researcher may be interested in selecting a sample from the population of males attending a certain university. Defining and describing this population would be relatively straightforward by simply consulting the university records. Another researcher may be interested in investigating the number and breeds of cats living in a certain city. Defining and describing this population of cats would be nearly impossible. The bottom line is that it may be easy to "talk" about a population but unless you are able to define and describe that population adequately, you may have difficulty selecting a representative sample. However, there are methods of selecting samples, although not random, that closely represent a given population. Let's begin our discussion of inferential statistics with a brief discussion of populations and samples.

△ B.2 Populations and Samples

Drawing a random sample from a defined population of interest is the only method to ensure that the sample best represents the actual population. The term *random* does not mean "willy-nilly," as the term might suggest. On the contrary, if a sample is truly random, it means that every member of the population has an equal and independent chance of being selected. However, the procedure for selecting such a sample is often quite difficult, or perhaps impossible, given the size and nature of the population. Fortunately, there are other methods of selecting samples that may obviate this problem.

One of the key things to remember is that the *population* is defined by the researcher. If you, the researcher, plan to study high school students, it is perfectly legitimate to define the population as all students attending Boys Town High School. The point is that the population can be large or small—it's up to the person conducting the research. It could be all high school students in the country or perhaps a single school, as mentioned in the example above.

△ B.3 Sampling Procedures

Only through random sampling of every element in a population can one assume that the sample is representative of that population. Since such sampling is not always possible, other methods of sampling have been developed, including stratified, cluster, and systematic.

Simple random sampling: This involves selecting a sample such that all individuals in the defined population have an equal and independent chance of inclusion in the sample. A random procedure such as a random number table or a computer program is used to select the sample.

Stratified sampling: This divides a population into subgroups and obtains random samples from each subgroup to ensure representation of each subgroup in the sample.

Cluster sampling: In this method, intact groups, not individuals, are randomly selected.

Systematic sampling: Every *i*th individual is selected from a list of individuals in a population.

Stratified, cluster, and systematic sampling are not random samples because each member of the defined population does not have an equal and independent chance of being selected. However, these are used by researchers when no other alternatives are viable. Researchers using these methods of sampling often employ inferential statistical procedures to analyze the results. This process is generally accepted as long as the extent of generalization is clearly explained. For instance, if the researcher conducting the study of Boys Town High School students limits the findings to that high school, then everything is fine. If the researcher attempts to extend any findings to, perhaps, all high schools in the city, then there is a generalization problem. Such results are open to question because the samples are not actually representative of the total population (i.e., all high schools in the city).

B.4 Hypothesis Testing   Δ

In hypothesis testing, two hypotheses are created, the null hypothesis (H_0) and the alternative hypothesis (H_A), only one of which can be true. The null hypothesis states that there is no *significant* difference in what we observed. Hypothesis testing is the process of determining whether to reject or retain the null hypothesis. If the null is rejected, it means that differences cannot be attributed to chance movement of the data, referred to as random error. The alternative hypothesis states that there is a significant difference. This is the researcher's idea or reason for doing the research.

As an example of the above scenario, we posit that there is a significant difference between two methods of teaching reading. We set the null hypothesis to say that there is *no* difference in teaching methods and that any differences can be attributed to chance movements in the data. In statistics, we say that any difference we observe is the result of random error. The alternative hypothesis is the opposite (and the researcher's idea) that the two teaching methods yield significantly different results. If we reject the null hypothesis, we now have evidence that the alternate hypothesis is true, meaning that a difference we observe is caused by a systematic difference between groups. For example, a group taught by reading Method A scored significantly higher on the reading test than a group taught by Method B. In this example, the systematic difference is in the teaching method.

In the vast majority of research designs, we hope to reject the null. However, there are times when we wish to retain it. An example is when we use the Kolmogorov-Smirnov one-sample test, which has the null that a series of continuous values are distributed the same as a normal curve. In most cases, we hope it is normal; thus, we hope to retain (not reject) the null in this particular case.

△   B.5 Parametric Statistical Tests

SPSS offers two types of hypothesis tests—*parametric* and *nonparametric*. The appropriate type of test for your data depends not only on the level of measurement of the data (*nominal*, *ordinal*, or *scale*) but also on the assumptions regarding populations and measures of dispersion (variance). Parametric tests are used to analyze interval (scale) data, whereas nonparametric tests are used to analyze ordinal (ranked) data and nominal (categorized) data.

Parametric tests are those that assume a certain distribution of the data (usually the normal distribution), an interval (scale) level of measurement, and equality (homogeneity) of variances. Probably, the most critical assumption regarding parametric tests is that the population from which a random sample is selected has a normal distribution. For example, if we select a sample from the population of all men living in a certain city and compute the mean height on that sample, the assumption is that "height" is a variable that is normally distributed in that population of men.

Many parametric tests, such as the *t* test, are quite robust, meaning that even though the assumptions may not be met, the results of the test are still viable and open to statistical interpretation. In other words, moderate violations of parametric assumptions have little or no effect on substantive conclusions in most instances. We describe some SPSS procedures you can use to make determinations concerning the normality of populations.

Parametric tests should not be used unless the following have been determined: (a) the samples are random, (b) the values are independent, (c) the data are normally distributed, and (d) the samples have equal or at least similar variances. However, researchers often take some latitude in applying these constraints. For example, if using a Likert-type scale in which each response is assigned a point value, say from 1 to 5, researchers often assume an interval level of measurement among the items, when in actuality the Likert-type scale should be considered as ordinal data, because one cannot show that the distance between a choice of 1 and 2 is equal to the distance between 2 and 3, and so on. When such decisions are made, one must critically consider the outcomes of research using such scales.

Also, in practice, levels of measurement are often downgraded from scale to ordinal or nominal, mainly for convenience and for use of a more familiar test such as chi-square. This is not really good practice, because one is casting out information: There is more information contained in interval data than in ordinal data.

The *power* of a statistical test is defined as its ability to detect a significant difference in a sample of data. The power of a parametric test is greater than that of a nonparametric test. Consequently, one should use a parametric test whenever assumptions such as normality of distributions and equivalences of variance can be met.

In many chapters in this book, we describe various parametric tests, including assumptions related to these tests. We follow some of these parametric tests with a description of a nonparametric test that may be used in their place in case assumptions for the parametric tests cannot be met.

B.6 Nonparametric Statistical Tests   △

When assumptions cannot be met with your data, you should investigate the use of *nonparametric* statistics to analyze your data.

Nonparametric tests are designed for databases that may include counts, classifications (nominal), and ratings. Such tests may be easier for a layperson to understand and interpret. Assumptions are usually quite minimal when using nonparametric tests.

Statistical analyses that do not depend on the knowledge of the distribution and parameters of the population are called nonparametric or *distribution-free* methods. Nonparametric tests use *rank* and *frequency* information to assess differences between populations. Nonparametric tests are used when a corresponding parametric test may be inappropriate. Nonparametric tests are useful when variables are measured at the ordinal or nominal level. And nonparametric tests may be applied to interval data if the assumptions regarding normality are not met.

In many of the chapters of this book, we present the nonparametric alternative to the parametric test.

△ B.7 DATA TRANSFORMATION

Alternatives exist when applying statistical tests to your data. If appropriate, you may wish to transform data via a mathematical transformation to produce a variable more closely resembling a normal distribution. Or you may prefer to apply appropriate nonparametric statistical tests. Section 6.4 in Chapter 6 of this book provides step-by-step instructions on transformation of data using SPSS.

△ B.8 TYPE I AND TYPE II ERRORS

When a researcher states a null hypothesis, he or she is attempting to draw conclusions (make decisions) about a population based on samples drawn from that population. If the researcher rejects a null hypothesis that is true, a Type I error has occurred. If he or she fails to reject a null hypothesis that is false, a Type II error has occurred. What may be the consequences of Type I and Type II errors?

A school district has decided to try a new method of teaching reading compared with the standard method. Suppose the null hypothesis, which states that there is no difference between these methods of teaching reading as based on students' reading scores, is rejected when it is, indeed, true that there is no difference, indicating a Type I error. The district may then introduce the new method of teaching reading at considerable cost to the district when the efficacy of both methods is the same. Suppose the contrary occurs, and we fail to reject the null hypothesis when it is, indeed, false. Then, the district will simply continue to use the customary method of teaching at no additional cost. However, the district would be withholding the use of a method of teaching reading that is superior to the customary method.

Most researchers would consider a Type I error the more costly, so they would wish to minimize the chance (probability) of making such an error.

△ B.9 TESTS OF SIGNIFICANCE

A test of significance is not indicative of *importance* but refers to a statistical level of probability. Significance tests determine the probability of making a Type I error (a preselected probability level we are willing to take if the decision we make is wrong). Typical levels for significance tests are .05 (5 out of 100) and .01 (1 out of 100). A preselected probability level, known as *level of significance* (denoted as *alpha*, or α), serves as a criterion to determine whether to reject or fail to reject the null hypothesis. If after performing a statistical analysis, a researcher obtains a probability of .05 or less, he or she will, by convention, reject the null hypothesis, meaning that the .05 level or less is the maximum Type I error rate the researcher is willing to accept.

Each statistical test in SPSS lists the exact level of significance in an output table. If this level is less than the alpha value (e.g., .01 or .05) you have chosen, then you would reject the null hypothesis. If the significance value is larger than, let's say, .01, you would fail to reject the null hypothesis.

B.10 Practical Significance Versus Statistical Significance △

The results of a test of significance may be statistically significant, but the actual difference may be minimal and of little importance. For example, a researcher may find a statistically significant difference in two methods of teaching reading as determined by students' scores on a reading test, but the actual difference in these scores may be so minimal that it would not benefit a school district to change the method of teaching reading.

An effect size is a measure permitting a judgment of the relative importance of a difference or relationship by indicating the size of the difference. In statistics, an effect size is a measure of the strength of the relationship between two variables in a statistical population, or a sample-based estimate of that quantity. An effect size calculated from data is a descriptive statistic that conveys the estimated magnitude of a relationship without making any statement about whether the apparent relationship in the data reflects a true relationship in the population. In that way, effect sizes complement inferential statistics such as *p* values.

Four methods of computing effect size are Pearson's *r* correlation, Cohen's *d*, eta-squared, and Cramer's *V*. Although SPSS does not provide direct calculations of these effect sizes, you can use the output provided by SPSS to compute these values.

B.11 One- and Two-Tailed Tests △

If the researcher has evidence that a difference would only occur in one direction, then a one-tailed test of significance would be appropriate. An example might be an owner of a coffee vending service who suspects that her machines are dispensing too much coffee. She might obtain a random sample of her machines and determine the amount they are dispensing and then test whether the observed amounts are greater than some desired amount. However, if the researcher has no evidence of direction, then a two-tailed test of significance is appropriate. If we apply the concept of a two-sided test to the coffee machine scenario, we would say that she is concerned whether the machines are dispensing too little or too much of the liquid. Each statistical test in SPSS lists results for a two-tailed test. If the test is actually one-tailed, it is easier to obtain a significant difference at a given level of significance. For a significance level of .05, the rejection area for a directional test is all on one side of the curve, making it easier to reject the null. For a two-sided (nondirectional) test, the .05 area is divided by 2; thus, you have a rejection area of .025 on each side of the curve.

△ B.12 Degrees of Freedom

The term *degrees of freedom* is defined as the number of observations free to vary around a parameter. In short, the number of degrees of freedom is the maximum number of variates that can freely be assigned before the rest of the variates are completely determined.

Let's consider a concrete example to help clarify the concept of degrees of freedom. Consider the following: $2 + 4 + 6 + ? = 20$. You can determine the missing number by subtracting the sum of the first three numbers from 20, yielding $20 - 12 = 8$. There are no degrees of freedom. Now, consider the following: $2 + 4 + ? + ? = 20$. There are an unlimited combination of numbers, the sum of which is 14, including $7 + 7$, $6 + 8$, $9 + 5$, $3.3 + 10.7$, and so on. However, if you choose, for example, the number 11 as one of the missing numbers, then the other number is determined, and in this case, it is 3. Consequently, in this example, there is 1 degree of freedom.

Each statistical test of significance in SPSS has a particular formula for determining the degrees of freedom. You will see this indicated in output tables as *df*.

APPENDIX C

ANSWERS TO REVIEW EXERCISES

CHAPTER 1: FIRST ENCOUNTERS △

1.1 You have classified the size of several fish that were caught in a "catch and release" fishing contest for children as small, medium, and large. The number of fish caught by the children are 32 small, 21 medium, and 11 large. *Note.* When inputting these data and information, you are *not* required to enter the names for the categories of the fish (small, medium, large). SPSS calls these categories *Labels* and *Label Values.* You will learn to input this information in a later chapter. Input the variable information and data, and build a frequency table and a bar graph. Name and save the database in the *Documents* section of your computer.

Answer:

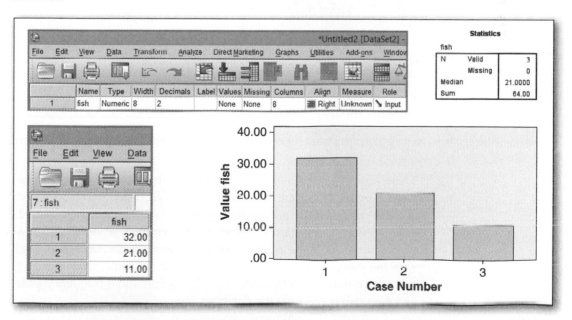

1.2 One day you are sitting in your professor's office getting help on regression analysis. His phone rings; he apologizes but says that he must take the call. As you wait for him to end his phone call, you scan his bookshelves and make mental notes of the titles. You arrive at the following: 15 books on introductory statistical analysis, 12 on advanced statistics, 3 on factor analysis, 8 on various regression topics, 13 on research methods, and 2 on mathematical statistics. You think to yourself, "Wow! This guy must have an exciting life!" As in the previous problem, don't concern yourself with the category labels for the textbooks. For now, just input the data and variable information, build a bar chart, generate a descriptive table, and name and save the database.

Answer:

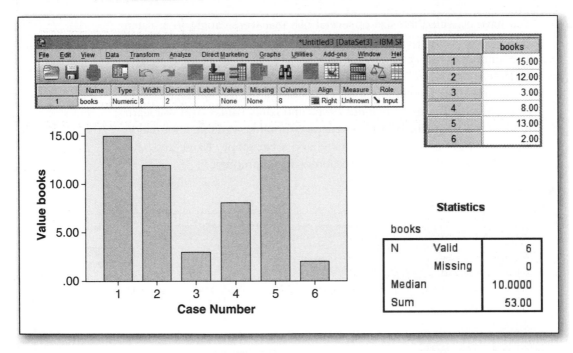

1.3 There was a quarter-mile drag race held at the abandoned airport last week. The makes of the winning cars were recorded by an interested fan. The results of her observations were as follows: Chevrolets won 23 races, Fords won 19 times, Toyota won 3, Hondas won 18, and KIAs won 8 races. As in the previous two problems, don't concern yourself with the categories' labels for the makes of the cars. Your task is to enter these data into SPSS, generate a bar graph and a frequency table, and then name and save the database.

Answer:

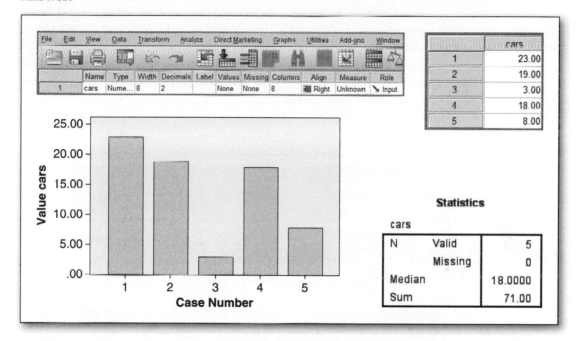

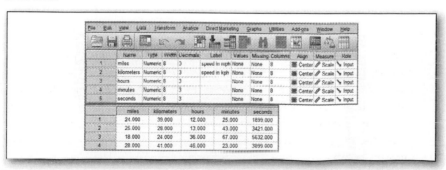

CHAPTER 2: NAVIGATING IN SPSS Δ

2.1 You have designed a data-collecting instrument that has the following five variables measured at the *scale* level (*labels* are given in parentheses; *decimals* are set to 3 and *align* to *center*): (1) miles (speed in miles per hour), (2) kilometers (speed in kilometers per hour) (3) hours, (4) minutes, and (5) seconds. Input this information into the Variable View screen, and then enter four cases of fabricated data in the Data View screen.

Answer:

2.2 You must set up the SPSS Data Editor to analyze the three variables listed below on 30,000 individuals. The variables are (1) age (label is *age in years*, no decimals, *center-aligned* and *scale* data), (2) education (label is *years beyond H.S.*, no decimals, *center-aligned* and *scale* data), and (3) family (label is *number of siblings*, no decimals, *center-aligned* and *scale* data). Make up and enter data for three cases—now you only have 29,997 more to enter!

Answer:

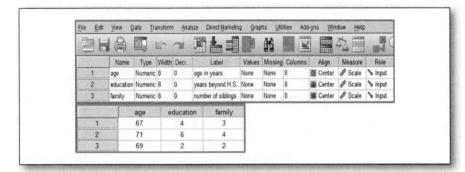

2.3 You are the range safety officer at a long-distance firearms training facility. You have collected the ballistic information on four rifles—data are given below. You would like to set up a data file in SPSS to collect many hundreds of similar cases in the future. The variables are (1) caliber (2 decimals, *center-aligned* and *scale* data), (2) five hundred (2 decimals, label is *500-yard drop in feet*, *center-aligned* and *scale* data), (3) one thousand (2 decimals, label is *1,000-yard drop in feet*, *center-aligned* and *scale* data), and (4) weight (no decimals, label is *bullet weight in grains*, *center-aligned* and *scale* data). Set up the SPSS Variable View page for this officer. There is no need to enter data on this exercise—unless you're a military veteran!

Answer:

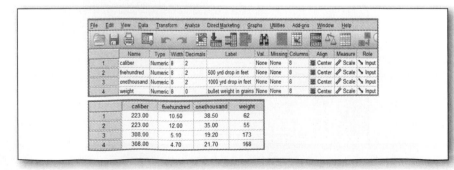

CHAPTER 3: GETTING DATA IN AND OUT OF SPSS △

3.1 With this review exercise, you will open an SPSS sample file *workprog. sav*. Once it is opened, you will save it in your document files, making it easy to access as you will need this database frequently as you progress through this book.

Answer:

	Name	Type	Width	Decimals	Label	Values	Mi...	Col...	Align	Measure	Role
1	age	Nu...	4	0	Age in years	None	None	6	Right	Scale	Input
2	marital	Nu...	4	0	Marital status	{0, Unmarrie...	None	7	Right	Nomi...	Input
3	incbef	Nu...	8	2	Income before the program	None	None	10	Right	Scale	Input
4	incaft	Nu...	8	2	Income after the program	None	None	10	Right	Scale	Input
5	ed	Nu...	4	0	Level of education	{1, Did not c...	None	6	Right	Ordinal	Input
6	gender	String	1	0	Gender	{f, Female}...	None	6	Left	Nomi...	Input
7	reside	Nu...	4	0	Number of people in household	None	None	6	Right	Scale	Input
8	prog	Nu...	4	0	Program status	None	None	6	Right	Nomi...	Input

3.2 In this exercise, you must import an Excel file from your computer and show the appearance of the *Open Excel Data Source* window and the first six rows of the Variable View screen. There should be an Excel file used as a demonstration for the Excel data program within your system files. Its name is *demo [AB6401]*, and it will be opened as an SPSS spreadsheet; examine the file, and observe that you can analyze the data as in any other SPSS spreadsheet.

Answer:

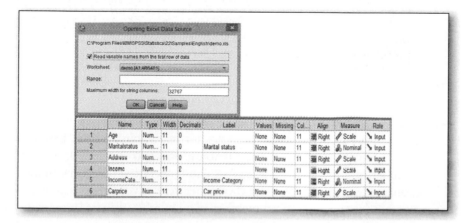

3.3 Open another one of SPSS's sample files called *customer_dbase.sav* (it has 132 variables and 5,000 cases), and save it in your document files. It's another database that you will use several times throughout this book.

Answer:

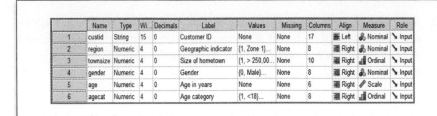

	Name	Type	Wi..	Decimals	Label	Values	Missing	Columns	Align	Measure	Role
1	custid	String	15	0	Customer ID	None	None	17	Left	Nominal	Input
2	region	Numeric	4	0	Geographic indicator	{1, Zone 1}...	None	8	Right	Nominal	Input
3	townsize	Numeric	4	0	Size of hometown	{1, > 250,00...	None	10	Right	Ordinal	Input
4	gender	Numeric	4	0	Gender	{0, Male}...	None	8	Right	Nominal	Input
5	age	Numeric	4	0	Age in years	None	None	6	Right	Scale	Input
6	agecat	Numeric	4	0	Age category	{1, <18}...	None	8	Right	Ordinal	Input

△ CHAPTER 4: LEVELS OF MEASUREMENT

4.1 The following is a list of variables that an investigator wishes to use to measure the health and survival of a particular species of earthworm: age, length, weight, moisture content, breed, environmental factors, and acid content of the soil. Your job is to assist this researcher in specifying the correct levels of measurement for these key variables and set up Variable View in SPSS.

Answer:

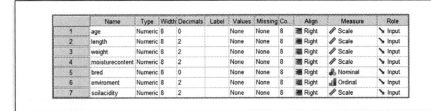

	Name	Type	Width	Decimals	Label	Values	Missing	Co..	Align	Measure	Role
1	age	Numeric	8	0		None	None	8	Right	Scale	Input
2	length	Numeric	8	2		None	None	8	Right	Scale	Input
3	weight	Numeric	8	2		None	None	8	Right	Scale	Input
4	moisturecontent	Numeric	8	2		None	None	8	Right	Scale	Input
5	bred	Numeric	8	0		None	None	8	Right	Nominal	Input
6	enviroment	Numeric	8	2		None	None	8	Right	Ordinal	Input
7	soilacidity	Numeric	8	2		None	None	8	Right	Scale	Input

4.2 A social researcher is interested in measuring the level of religiosity of a sample of senior citizens. Help her in establishing the levels of measurement for the following variables: pray (do you pray?), services (number of times you attend formal church services per year), money (donated to church), volunteer (hours per year of volunteer assistance), member (are you an official member of a church?), discuss (how many times each week do you discuss religious doctrine?), and times pray (how many times per week do you pray?). Input these data into the SPSS Variable View spreadsheet.

Answer:

	Name	Type	Width	Decimals	Label	Values	Missing	Columns	Align	Measure	Role
1	pray	Numeric	8	0		None	None	8	Right	Nominal	Input
2	services	Numeric	8	0		None	None	8	Right	Scale	Input
3	money	Numeric	8	2		None	None	8	Right	Scale	Input
4	volunteer	Numeric	8	0		None	None	8	Right	Scale	Input
5	member	Numeric	8	0		None	None	8	Right	Nominal	Input
6	discuss	Numeric	8	0		None	None	8	Right	Scale	Input
7	timespray	Numeric	8	0		None	None	8	Right	Scale	Input

4.3 A political consultant wished to measure the level of politicization of the candidates for a job at the White House. He decided that the following variables would provide at least some of the evidence required to assess the extent of their interest in politics: vote (did you vote in the previous election?), letter (the most recent letter sent to a politician), meeting (the most recent political meeting attended), donate (how much money did you donate to a politician in the past year?), candidate (have you run for office?), and party (are you a member of a political party?). Your job is to input these variables and their labels into SPSS and specify their levels of measurement.

Answer:

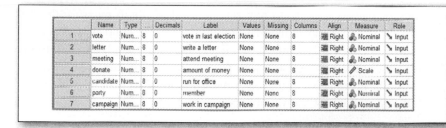

	Name	Type	Decimals	Label	Values	Missing	Columns	Align	Measure	Role
1	vote	Num... 8	0	vote in last election	None	None	8	Right	Nominal	Input
2	letter	Num... 8	0	write a letter	None	None	8	Right	Nominal	Input
3	meeting	Num... 8	0	attend meeting	None	None	8	Right	Nominal	Input
4	donate	Num... 8	0	amount of money	None	None	8	Right	Scale	Input
5	candidate	Num... 8	0	run for office	None	None	8	Right	Nominal	Input
6	party	Num... 8	0	member	None	None	8	Right	Nominal	Input
7	campaign	Num... 8	0	work in campaign	None	None	8	Right	Nominal	Input

CHAPTER 5: ENTERING VARIABLES AND DATA AND VALIDATING DATA

5.1 The highway patrol officer wants to set up an SPSS file to record traffic violations. She wishes to record data at the *nominal*, *ordinal*, and *scale* levels of measurement. The first item of interest (the largest source of income for the highway patrol) is speeding. Input three variables that could record speed at each level of measurement. The next item of interest is vehicle violations—in the same database set up a variable at the correct level of measurement and with three categories if necessary. Impaired driving is another important violation. How would you measure and record information for this violation? Show all this in the same database.

Answer:

If you were actually entering data for the first variable ("speed"), then the observed speeds would be recorded in miles per hour (mph). The following are the suggested values for the above categories, but you can dream up your own—(1) speedcat: 1 = *<10 mph*, 2 = *>10 and <20 mph*, 3 = *>20 mph*; (2) speedyesno: 1 = *yes* and 2 = *no*; (3) vehicle: 1 = *serious danger*, 2 = *minor danger*, and 3 = *both*; and (4) impaired: 1 = *intoxicated*, 2 = *medical reason*, and 3 = *tired driver*.

5.2 A child psychologist is investigating the behavior of children in the play area of Balboa Park. Help him set up an SPSS file to measure the following variables on individual children: time the child was observed, pieces of play equipment used, other children in the play area, interaction with others, interaction with parent, and child's general demeanor. Input these variables into SPSS at measurement levels of your choosing but appropriate for the variable being analyzed.

Answer:

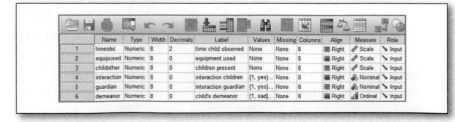

The following are the suggested values for the categorized variables—(1) interaction: 1 = *yes* and 2 = *no*, (2) *guardian*: 1 = *yes* and 2 = *no*, and (3) demeanor: 1 = *sad*, 2 = *happy*, and 3 = *very happy*.

5.3 The following sample data were collected by the owner of a private 640-acre forest reserve. He did a sample of 10 acres as a trial survey for the entire reserve. He needs to set up and test a computer file system using SPSS's Data Editor. The 10-acre sample was subdivided into 2.5 parcels, with each yielding the following data: hardwood trees, softwood trees, new-tree growth, stage of decay for fallen trees, soil moisture content, and crowded conditions. Your database will have four cases ($4 \times 2.5 = 10$) and seven variables. Enter some fictitious data for the newly created variables on the four 2.5-acre plots.

Answer:

	Name	Type	Width	Decimals	Label	Values	Mis...	Columns	Align	Measure	Role
1	hardwood	Numeric	8	0		None	None	8	Right	Scale	Input
2	softwood	Numeric	8	2		None	None	8	Right	Scale	Input
3	newgrowth	Numeric	8	0	new tree growth	{1, low}...	None	8	Right	Ordinal	Input
4	decay	Numeric	8	0	stage of decay	{1, low}...	None	8	Right	Ordinal	Input
5	soil	Numeric	8	2	moisture content	None	None	8	Right	Scale	Input
6	crowding	Numeric	8	0	new growth crowded	{1, yes}...	None	8	Right	Nominal	Input

The following are the suggested coding for the categorized variables— (1) newgrowth: 1 = *low*, 2 = *moderate*, and 3 = *high*; (2) decay: 1 = *low*, 2 = *moderate*, and 3 = *high*; and (3) crowding; 1= *yes* and 2 = *no*.

	hardwood	softwood	newgrowth	decay	soil	crowding
1	10	45	1	1	53.25	1
2	23	48	2	1	43.98	2
3	13	64	1	1	24.67	1
4	16	47	3	2	34.56	1

CHAPTER 6: WORKING WITH DATA AND VARIABLES △

6.1 An urban planner was tasked with recording the walking speeds (in miles per hour) of people at a downtown government center. He recorded walking speeds of the same individuals over a period of 5 days. A sample of the data is provided below along with the variable information. There are three variables. You are to set up a database in SPSS and then use the *Compute Variable* feature to create a fourth variable called "avgspeed." There are five cases: Variable 1: *speed1* has the label of *1-walking speed* with speeds of 3.20, 3.18, 1.40, 3.26, and 2.57. Variable 2: *speed2* has the label of *2-walking speed* with speeds of 3.34,

3.61, 2.10, 3.12, and 2.82. Variable 3: *speed3* has the label of *3-walking speed* with speeds of 3.25, 3.24, 1.97, 3.41, and 2.98. Save the database, which has the new variable ("avgspeed") as speed—you may use it in future review exercises.

Answer:

	Name	Type	Width	Decimals	Label	Values	Missing	Columns	Align	Measure	Role
1	speed1	Numeric	8	2	1-walking speed	None	None	8	Center	Scale	Input
2	speed2	Numeric	8	2	2-walking speed	None	None	8	Center	Scale	Input
3	speed3	Numeric	8	2	3-walking speed	None	None	8	Center	Scale	Input

	speed1	speed2	speed3
1	3.20	3.34	3.25
2	3.18	3.61	3.24
3	1.40	2.10	1.97
4	3.26	3.12	3.41
5	2.57	2.82	2.98

	speed1	speed2	speed3	avgspeed
1	3.20	3.34	3.25	3.26
2	3.18	3.61	3.24	3.34
3	1.40	2.10	1.97	1.82
4	3.26	3.12	3.41	3.26
5	2.57	2.82	2.98	2.79

6.2 Use the data from the variable ("avgspeed") created in the previous exercise to recode the values into a *nominal* (string) variable. Your task is to use SPSS's *Recode into Different Variables* feature and form two categories for the average walking speeds of the individuals. The two categories are based on the average speed of 2.9 miles per hour for all walkers. All speeds above the mean are to be classified as *Fast*; those below the mean are classified as *Slow*. You will create a new nominal or string variable called "catspeed."

Answer:

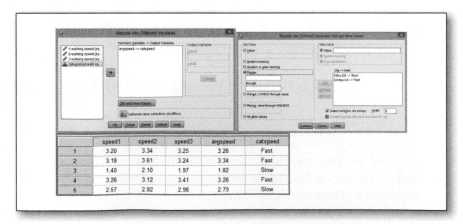

	speed1	speed2	speed3	avgspeed	catspeed
1	3.20	3.34	3.25	3.26	Fast
2	3.18	3.61	3.24	3.34	Fast
3	1.40	2.10	1.97	1.82	Slow
4	3.26	3.12	3.41	3.26	Fast
5	2.57	2.82	2.98	2.79	Slow

6.3 You have been given the following scale data (test scores): 100, 109, 114, 118, 125, 135, 135, 138, 139, and 140. You must set up a database in SPSS and then transform the data using the *Compute Variable* and *arithmetic* functions to calculate new variables giving the log and square root of the original test scores. You must end up with a database consisting of 10 cases and three variables named "test," "logtest," and "sqrttest."

Answer:

	test	logtest	sqrttest
1	100	4.61	10.00
2	109	4.69	10.44
3	114	4.74	10.68
4	118	4.77	10.86
5	125	4.83	11.18
6	135	4.91	11.62
7	135	4.91	11.62
8	138	4.93	11.75
9	139	4.93	11.79
10	140	4.94	11.83

CHAPTER 7: USING THE SPSS HELP MENU △

7.1 How can you use the SPSS *Help* function to get information on transforming data in an attempt to get a more normal distribution of values?

Answer:

Use the **Help** button on the Main Menu and *Topics* to obtain this information.

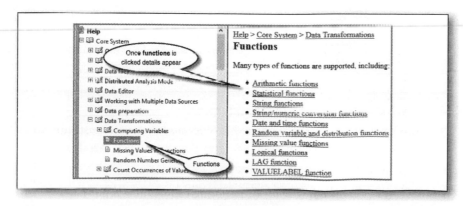

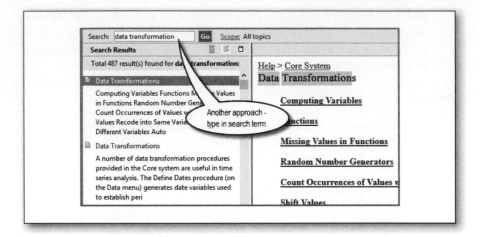

7.2 You need help to find out how to summarize categorical (*nominal*) data for a PowerPoint presentation, and you want to use a graph made from data stored in SPSS. Hint: Use the SPSS Help on the Main Menu and the *Tutorial*.

Answer:

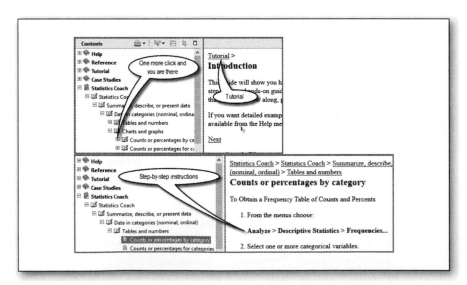

7.3 You have a large database opened in SPSS, and now you must summarize and describe the data it contains. You decide to click **Analyze** on the

Main Menu, then **Frequencies**; the *Frequencies* window opens, and you realize that you need help—what do you do next?

Answer:

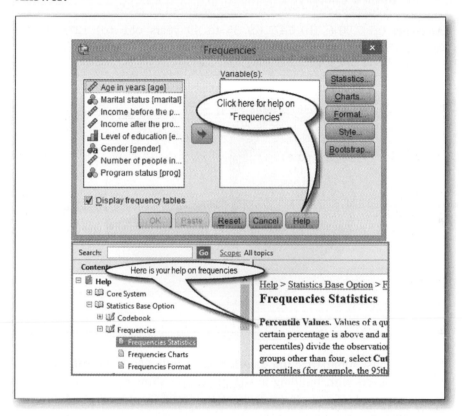

CHAPTER 8: CREATING GRAPHS △
FOR NOMINAL AND/OR ORDINAL DATA

8.1 You are given the task of building a pie chart that summarizes the five age categories of 582 individuals recorded in an SPSS sample file titled *satisf.sav*. The variable is named "agecat" and labeled *Age category*. Build a 3-D pie graph that displays the names of the categories, the numbers of observations, and the percentages of the total for each slice. Also, answer the following questions: (a) What percentage of people are 18 to 24 years old? (b) How many individuals are 50 to 64 years old? (c) What is the largest age category? (d) What is the quantity, and its respective percentage, for the individuals who are 25 to 34 years old? (e) What is the quantity, and its respective percentage, for the smallest category?

Answer: (a) 7.90%; (b) 147; (c) 35 to 49 years old; (d) 127, 21.82%; (e) 32, 5.50%

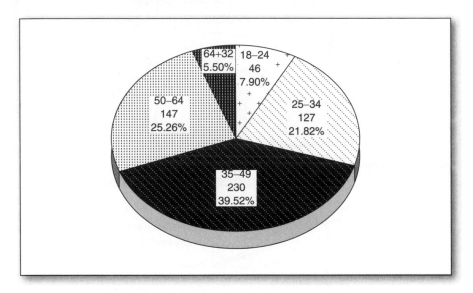

8.2 You must summarize the data for 8,083 individuals. Specifically, you are tasked with building a 3-D pie graph showing the number and percentage of people in each of four regions. Open *dmdata3.sav*, select the discrete variable named "Region," and build a pie graph. Embellish the graph to display the names of the categories, the numbers of observations, and the percentages of the total for the slices to answer the following questions: (a) What is the largest region, and what percentage of the total respondents does it represent? (b) Which region is the smallest, and how many individuals does it have? (c) What is the number of customers, and its respective percentage, for the *North*? (d) What is the number of customers, and its respective percentage, for the *West*? (e) Rank the regions from the smallest to the biggest.

Answer: (a) East, 25.79%; (b) South, 1,965; (c) 1,967, 24.34%; (d) 2,066, 25.56%; (e) South, North, West, East

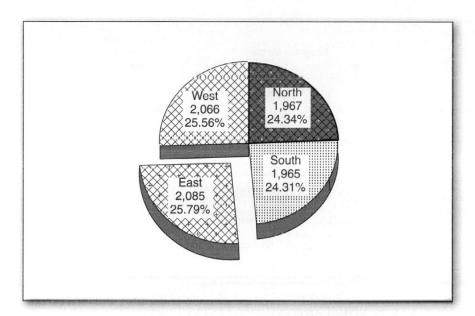

8.3 You must visually display the relationship between two categorical variables using the population pyramid graphing method. The variables you use are found in the SPSS sample file called *workprog.sav*. The variable "prog" (program status) is the split variable, and "ed" (level of education) is the distribution variable. Build the population pyramid to compare these discrete variables split into two groups of program status, 0 and 1. It has been determined that the categories of 0 and 1 are approximately equal; therefore, these distributions can be directly compared. Also, look at the finished graph, and answer the following questions: (a) By looking at the graph, does it appear that program status and level of education are related? (b) Which of the six categories contains the most observations? (c) Which category has the least number of observations?

Answer: (a) Looking at it should strongly suggest that there are no differences between level of education and program status. This should be confirmed with a chi-square test. (b) Most of the observations are contained in the "did not complete high school" category, with a program status of "0"—244 individuals. (c) The least number of observations are found in the "some college" category, with a program status of "0"—92 individuals.

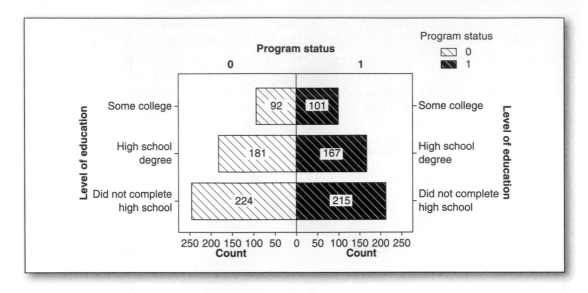

△   CHAPTER 9: GRAPHS FOR CONTINUOUS DATA

9.1   For this exercise, you will build a simple histogram using the SPSS sample file known as *autoaccidents.sav*. Open the file, select the variable named "accident" and labeled *Number of accidents in the past 5 years*, and build the simple histogram. Use the graph you build to answer the following questions: (a) Are the data skewed to the right, skewed to the left, or normal? (b) What is the average number of accidents for this group? (c) How many of this group reported one accident? (d) How many people had four accidents? (e) Of the 500 people in this group, how many had no accidents? (f) How many people had three accidents?

Answer: (a) positively skewed, (b) 1.72, (c) 140, (d) 40, (e) 122, (f) 63

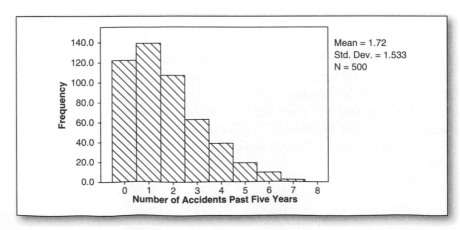

9.2 Use the same database as in the previous exercise (*autoaccidents.sav*), and select the variable named "Age" having the label *Age of insured.* Build the boxplot, and answer the following questions: (a) What are the minimum and maximum ages, excluding outliers and extremes? (b) What is the age of the outlier? (c) What are the limits of the interquartile range? (d) What is the interquartile range for this age distribution? (e) What is the median value? (f) Does the graph depict normally distributed data or perhaps a negative or positive skew?

Answer: (a) 22 and 64, (b) 68 years of age, (c) 35 and 48, (d) 48 - 35 = 13, (d) 41, (e) not normal and has a slight positive skew (must be confirmed with the Kolmogorov-Smirnov test)

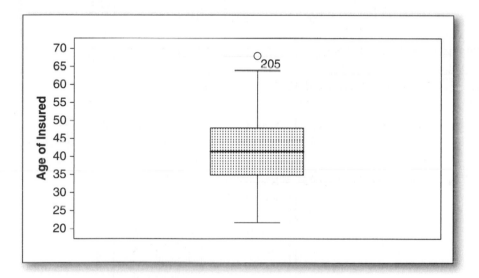

9.3 For this exercise, you will open the SPSS sample file called *workprog. sav* and select *two* discrete variables and *one* continuous variable to build a histogram paneled on both columns and rows. For the *x*-axis use the variable labeled as *Age in years* and named "Age." The discrete variable named "Marital" and labeled as *Marital status* will be the *column-paneled* variable. The *row-paneled* variable is named "Ed" and labeled as *Level of education.* Look at the finished graph, and answer the following questions: (a) Which of the row panels (*Level of Education*) contains the most people? (b) Which of the six groups had the most participants in the age interval 17.5 to 18.5 years, and how many are in that group? (c) Which group had a distribution of ages that most closely resembled a normal distribution? (d) What is the shape of the distribution for unmarried individuals with a high school degree?

Answer: (a) Most of the individuals in the work program did not complete high school. (b) The group of married people with a high school education was the most populated, with 70 individuals. (c) Of all the groups, the group of married individuals without a high school diploma most closely resembles the normal distribution. (d) Positive skew.

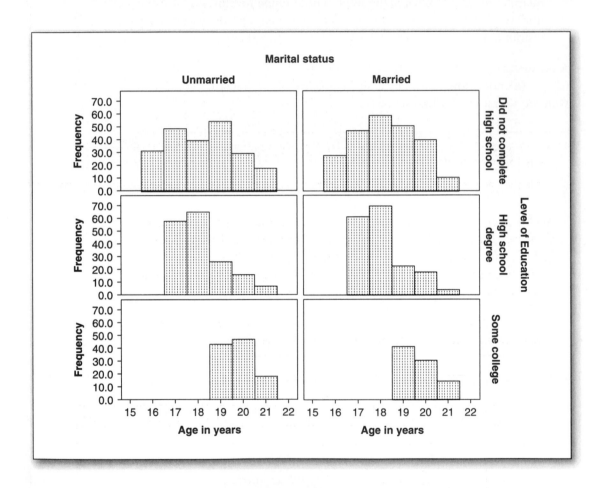

Δ CHAPTER 10: PRINTING DATA VIEW, VARIABLE VIEW, AND OUTPUT VIEWER SCREENS

10.1 Open the SPSS sample file called *workprog.sav*, and print the data for the first six variables of the first 10 cases.

Answer:

	age	marital	incbef	incaft	ed	gender
1	16	0	8.00	12.00	1	m
2	17	0	8.00	10.00	2	f
3	17	0	8.00	11.00	1	f
4	19	1	9.00	18.00	1	f
5	18	1	7.00	12.00	1	f
6	17	1	8.00	15.00	1	f
7	17	1	8.00	13.00	2	f
8	21	0	9.00	22.00	1	f
9	18	0	7.00	18.00	1	m
10	16	0	7.00	9.00	1	m

10.2 You have a meeting with your research group, and you wish to discuss some of the variables you have selected for a project. The database is stored in an SPSS sample file called *workprog.save*. What is the quickest and easiest way to print out all the information about your study variables?

Answer: Use the *File/Display Data File Information* method to generate the *Variable Information* table in the Output View as follow:

Variable Information

Variable	Position	Label	Measurement Level	Role	Column Width	Alignment	Print Format	Write Format
age	1	Age in years	Scale	Input	6	Right	F4	F4
marital	2	Marital status	Nominal	Input	7	Right	F4	F4
incbef	3	Income before the program	Scale	Input	10	Right	F8.2	F8.2
incaft	4	Income after the program	Scale	Input	10	Right	F8.2	F8.2
ed	5	Level of education	Ordinal	Input	6	Right	F4	F4
gender	6	Gender	Nominal	Input	6	Left	A1	A1
reside	7	Number of people in household	Scale	Input	6	Right	F4	F4
prog	8	Program status	Nominal	Input	6	Right	F4	F4

Variables in the working file

10.3 You must analyze several variables of the *workprog.sav* database (SPSS sample file) and then prepare a clean handout for your colleagues. You need basic descriptive statistics on all your scale data and a frequency table for level of education for the categorical variables. Generate the output, give it the title "Work Program Study," and print the "cleaned" output.

Answer:

Work Program Study

Descriptive Statistics

	N	Minimum	Maximum	Mean	Std. Deviation
Age in years	1000	16	21	18.48	1.353
Income before the program	1000	6.00	14.00	8.9540	1.63663
Income after the program	1000	7.00	36.00	16.5930	4.67067
Number of people in household	1000	1	7	1.86	1.079
Valid N (listwise)	1000				

Level of education

		Frequency	Percent	Valid Percent	Cumulative Percent
Valid	Did not complete high school	459	45.9	45.9	45.9
	High school degree	348	34.8	34.8	80.7
	Some college	193	19.3	19.3	100.0
	Total	1000	100.0	100.0	

▵   CHAPTER 11: BASIC DESCRIPTIVE STATISTICS

11.1 Open the SPSS sample database called *bankloan.sav*, and calculate the following statistics for all variables measured at the *scale* level: *mean, std. deviation, variance, range, minimum, maximum, S.E., mean, kurtosis,* and *skewness*. Print the SPSS descriptive statistics output produced in answer to your request.

Answer:

Descriptive Statistics

	N	Range	Minimum	Maximum	Mean		Std. Deviation	Variance	Skewness		Kurtosis	
	Statistic	Statistic	Statistic	Statistic	Statistic	Std. Error	Statistic	Statistic	Statistic	Std. Error	Statistic	Std. Error
Age in years	850	36	20	56	35.03	.276	8.041	64.665	.335	.084	-.658	.168
Years with current employer	850	33	0	33	8.57	.232	6.778	45.940	.863	.084	.379	.168
Years at current address	050	34	0	34	8.37	.236	6.895	47.541	.924	.084	.257	.168
Household income in thousands	850	433.00	13.00	446.00	46.6753	1.32202	38.54305	1485.567	3.701	.084	22.486	.168
Debt to income ratio (x100)	850	41.20	.10	41.30	10.1716	.23047	6.71944	45.151	1.125	.084	1.398	.168
Credit card debt in thousands	850	20.55	.01	20.56	1.5768	.07292	2.12584	4.519	3.702	.084	19.500	.168
Other debt in thousands	850	35.15	.05	35.20	3.0788	.11658	3.39880	11.552	3.206	.084	16.635	.168
Predicted default, model 1	850	.99928	.00012	.99940	.2585762	.00880636	.25674727	.066	1.071	.084	.247	.168
Predicted default, model 2	850	.99942	.00004	.99946	.2578660	.00917303	.26743757	.072	1.055	.084	.120	.168
Predicted default, model 3	850	.87346	.07464	.94810	.2585786	.00582671	.16987638	.029	1.525	.084	2.078	.168
Valid N (listwise)	850											

11.2 Open the SPSS sample database called *bankloan.sav*, and build frequency tables for all variables measured at the *nominal* or *ordinal* level.

Answer:

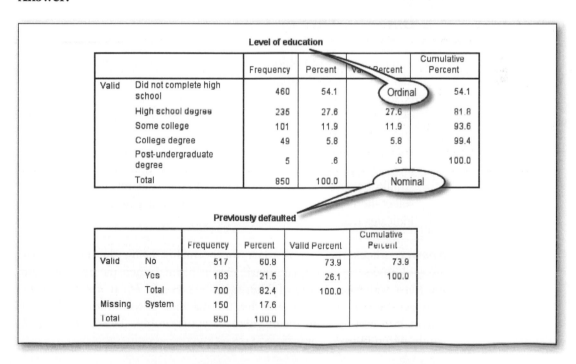

Level of education

		Frequency	Percent	Valid Percent	Cumulative Percent
Valid	Did not complete high school	460	54.1	*Ordinal*	54.1
	High school degree	235	27.6	27.6	81.8
	Some college	101	11.9	11.9	93.6
	College degree	49	5.8	5.8	99.4
	Post-undergraduate degree	5	.6	.6	100.0
	Total	850	100.0	*Nominal*	

Previously defaulted

		Frequency	Percent	Valid Percent	Cumulative Percent
Valid	No	517	60.8	73.9	73.9
	Yes	183	21.5	26.1	100.0
	Total	700	82.4	100.0	
Missing	System	150	17.6		
Total		850	100.0		

11.3 Open the SPSS sample database called *bankloan.sav*, and determine if the variables "age" and "household income in thousands" are normally distributed.

Answer:

Hypothesis Test Summary

	Null Hypothesis	Test	Sig.	Decision
1	The distribution of Age in years is normal with mean 35.029 and standard deviation 8.04.	One-Sample Kolmogorov-Smirnov Test	.000[1]	Reject the null hypothesis.
2	The distribution of Household income in thousands is normal with mean 46.675 and standard deviation 38.54.	One-Sample Kolmogorov-Smirnov Test	.000[1]	Reject the null hypothesis.

Asymptotic significances are displayed. The significance level is .05.

[1] Lilliefors Corrected

◬ Chapter 12: One-Sample *t* Test and a Binomial Test of Equality

12.1 You are a seller of heirloom garden seeds. You have several machines that automatically load the seeds into packages. One of your machines is *suspected* of sometimes being inaccurate and of underloading or overloading the packages. You take a random sample of 20 packages and record their weights as follows: 3.09, 2.74, 2.49, 2.99, 3.22, 2.51, 2.28, 3.54, 2.52, 3.20, 3.09, 2.56, 3.43, 3.25, 2.69, 2.49, 3.30, 2.69, 2.89, and 3.57. In the past, it has been determined that the average weight should be 2.88 ounces. Write the null and research hypotheses and select the correct statistical test to determine if your sample evidence indicates that the machine is malfunctioning.

Answer: Check the sample data for normality using the Kolmogorov-Smirnov one-sample nonparametric test. If normal, then use the one-sample *t* test with a test value of 2.88. H_0: $\mu = 2.88$ and H_A: $\mu \neq 2.88$. You fail to reject the null and determine that the machine is within tolerance—it is not malfunctioning. The mean value of your sample, 2.92, is not significantly different from 2.88.

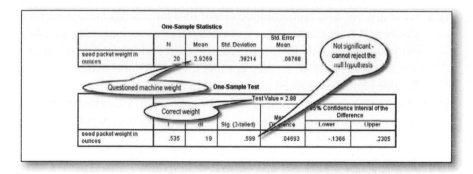

12.2 There is an annual race to reach the top of Kendall Mountain in the small town of Silverton, Colorado. A health scientist believes that the time to reach the top has significantly changed over the past 10 years. The average time for the first 12 runners to reach the summit in the 2005 race was 2.15 hours. He records the times for the top 12 runners in 2015 as follows: 1.43, 1.67, 2.13, 2.24, 2.45, 2.50, 2.69, 2.86, 2.92, 2.99, 3.35, and 3.36 hours, respectively. You must now write the null and alternative hypotheses in an attempt to produce evidence in support of your research hypothesis.

Answer: H_0: $\mu = 2.15$ and H_A: $\mu \neq 2.15$. The significant t test of .044 allows you to reject the null and, therefore, provides evidence in support of the alternative. There is now statistical evidence that the times are significantly different. The direction indicates runners in the 2015 race having taken significantly more time to reach the summit. Let's blame it on global warming!

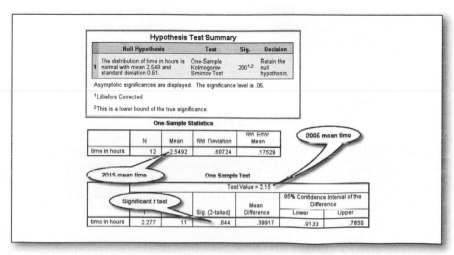

12.3　An instructor in a community college auto mechanics class asked his students if they thought that gas consumption, as measured in miles per gallon (mpg), had improved for the Ford Focus in the past 5 years. It was a team project, so they got together and found that on average the Ford Focus was rated at 24.3 mpg in 2010. No data for the current year (2015) were available, so they worked out a random sampling plan and collected the following current consumption data for 12 Ford Focus vehicles: 26.4, 26.5, 26.9, 26.9, 26.4, 26.4, 26.8, 26.9, 26.4, 26.9, 26.8 and 26.4 mpg. The plan was to somehow compare their sample data with the 5-year-old data—can you help these students?

Answer: First of all, we note that we have *scale* data, and we will be comparing *mean* values for miles per gallon. A *t* test comes to mind, but do we have a normal distribution? Let's check for normality with the nonparametric one-sample Kolmogorov-Smirnov test. The test indicates that the distribution is *not* normal, so we decide on the nonparametric one-sample binomial test. In the *Settings* window, we use the binomial test and the *Binomial Options* window with a *Hypothesized proportion* of 0.5 and a *Custom cut point* of 24.3 (the 2010 average miles per gallon). The results indicate that we must reject the null hypothesis. There is *not* an equal distribution of categories above and below the mean of 26.642. The students must conclude that there is a significant change in the miles per gallon over the past 5 years. Furthermore, we can conclude that gas mileage has increased for the Ford Focus.

Hypothesis Test Summary

	Null Hypothesis	Test	Sig.	Decision
1	The distribution of Miles per Gallon is normal with mean 26.642 and standard deviation 0.24.	One-Sample Kolmogorov-Smirnov Test	.024[1]	Reject the null hypothesis.

Asymptotic significances are displayed. The significance level is .05.

[1] Lilliefors Corrected

Hypothesis Test Summary

	Null Hypothesis	Test	Sig.	Decision
1	The categories defined by Miles per Gallon <=24.300 and >24.300 occur with probabilities 0.5 and 0.5.	One-Sample Binomial Test	.000[1]	Reject the null hypothesis.

Asymptotic significances are displayed. The significance level is .05.

[1] Exact significance is displayed for this test.

CHAPTER 13: INDEPENDENT-SAMPLES t TEST △
AND MANN-WHITNEY U TEST

13.1 Two 12-man teams of Marines were randomly selected from Marine Corps Air Stations Miramar and Yuma to be compared on their Combat Fitness Test. Their scores ranged from a low of 263 to a perfect score of 300. Miramar scores: 267, 278, 295, 280, 268, 286, 300, 276, 278, 297, 298, and 279. Yuma scores: 263, 272, 286, 276, 267, 284, 293, 270, 272, 296, 279, and 274. The Yuma team leader and researcher had the idea that the scores were unequal. Can you help the Yuma team leader write the null and alternative hypotheses and select the appropriate test(s) to see if there is evidence in support his idea?

Answer: The data were measured at the scale level, and you are looking at the differences between means. Furthermore, the samples are independent. If the distributions for both teams can be shown to approximate the normal curve with equal variances, we would select the t test for independent samples. The Kolmogorov-Smirnov test indicates that both teams had scores that are normally distributed, so we proceed with the t test. The t test indicates that the variances are equal. Levene's test fails to reject the null of equal variances. The null hypothesis is stated as H_0: $\mu_1 - \mu_2 = 0$, while the team leader's idea is stated as the alternative hypothesis, H_A: $\mu_1 - \mu_2 \neq 0$. We fail to reject the null; therefore, there is no statistical evidence to support the team leader's idea that the Combat Fitness Test scores are unequal. The Marines are equally prepared for combat.

Hypothesis Test Summary

	Null Hypothesis	Test	Sig.	Decision
1	The distribution of MCAS Miramar is normal with mean 283.500 and standard deviation 11.54.	One-Sample Kolmogorov-Smirnov Test	.188[1]	Retain the null hypothesis.
2	The distribution of MCAS Yuma is normal with mean 277.667 and standard deviation 10.23.	One-Sample Kolmogorov-Smirnov Test	.200[1,2]	Retain the null hypothesis.

Asymptotic significances are displayed. The significance level is .05.

[1] Lilliefors Corrected

[2] This is a lower bound of the true significance.

Group Statistics

	Airstation	N	Mean	Std. Deviation	Std. Error Mean
Combat Fitness Test	MCAS Miramar	12	283.50	11.540	3.331
	MCAS Yuma	12	277.67	10.228	2.952

Independent Samples Test										
		Levene's Test for Equality of Variances		t-test for Equality of Means						
		No significant difference in mean scores for Miramar and Yuma					Mean Difference	Std. Error Difference	95% Confidence Interval of the Difference	
		F		df	Sig. (2-tailed)				Lower	Upper
Combat Fitness Test	Equal variances assumed	.438	.515	1.310		.204	5.833	4.451	-3.398	15.065
	Equal variances not assumed			1.310	21.687	.204	5.833	4.451	-3.406	15.073

13.2 The local bank president had the idea that the money held in individual savings accounts would be significantly different for males and females. A random sample of the dollars in male and female savings accounts was recorded as follows. Males: 5,600, 5,468, 5,980, 7,890, 8,391, 9,350, 10,570, 12,600, 8,200, 7,680, 6,000, and 8,900. Females: 4,900, 5,200, 5,000, 7,000, 8,000, 9,050, 9,900, 12,000, 8,000, 7,500, 5,900, and 8,500. Write the null and alternative hypotheses, and select the correct test to seek evidence in support of the bank president's contention that male and female saving habits are significantly different.

Answer: You have scale data, and the task is to determine if there is a significant difference between the mean savings amounts for males and females. Since you have two independent groups (males and females) and scale data, we first think of a t test. Check to see if the male and female savings distributions are normal by using the Kolmogorov-Smirnov test. The test provides evidence that they are normal, so we proceed with the t test for independent samples. Let's state the null hypothesis as H_0: $\mu_1 - \mu_2 = 0$ and the alternative (the bank president's idea) as H_A: $\mu_1 - \mu_2 \neq 0$.

The t test shows equal variances (see Levene's test) and a *Sig.* of .595, so we fail to reject the null hypothesis. The t test does not detect a significant difference in the savings habits of males and females. There is *no* statistical evidence in support of the bank president's contention.

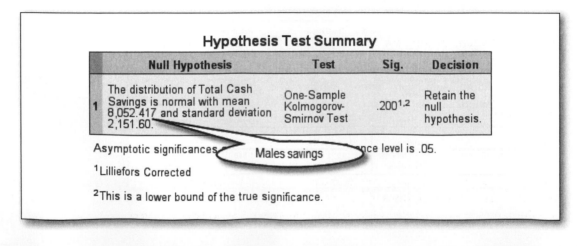

Hypothesis Test Summary

	Null Hypothesis	Test	Sig.	Decision
1	The distribution of Total Cash Savings is normal with mean 8,052.417 and standard deviation 2,151.60.	One-Sample Kolmogorov-Smirnov Test	.200[1,2]	Retain the null hypothesis.

Asymptotic significances   Males savings   nce level is .05.

[1] Lilliefors Corrected

[2] This is a lower bound of the true significance.

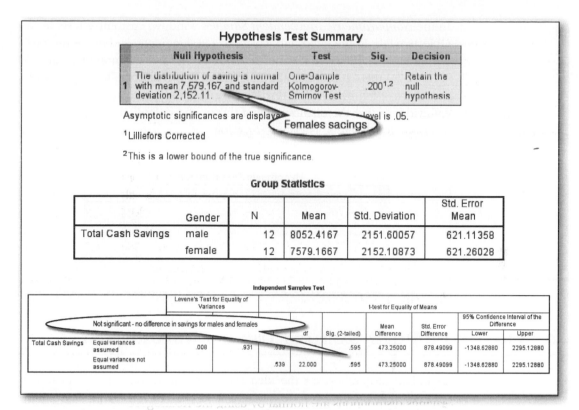

Hypothesis Test Summary

	Null Hypothesis	Test	Sig.	Decision
1	The distribution of saving is normal with mean 7,579.167 and standard deviation 2,152.11.	One-Sample Kolmogorov-Smirnov Test	.200[1,2]	Retain the null hypothesis

Asymptotic significances are displayed. ~~level is .05.~~

Females sacings

[1]Lilliefors Corrected

[2]This is a lower bound of the true significance.

Group Statistics

	Gender	N	Mean	Std. Deviation	Std. Error Mean
Total Cash Savings	male	12	8052.4167	2151.60057	621.11358
	female	12	7579.1667	2152.10873	621.26028

Independent Samples Test

		Levene's Test for Equality of Variances		t-test for Equality of Means					95% Confidence Interval of the Difference	
		F	Sig.	t	df	Sig. (2-tailed)	Mean Difference	Std. Error Difference	Lower	Upper
		Not significant - no difference in savings for males and females								
Total Cash Savings	Equal variances assumed	.008	.931	.539		.595	473.25000	878.49099	-1348.62880	2295.12880
	Equal variances not assumed			.539	22.000	.595	473.25000	878.49099	-1348.62880	2295.12880

13.3 For this review exercise, you will select and open the SPSS sample file called *bankloan.sav*. You will test for significant differences in the categories of education ("ed") and whether they have previously defaulted on a loan ("default"). There are five educational categories and two for the "default" variable. Write the alternative hypothesis and null hypothesis, and use the appropriate statistical test to see if the distribution of levels of education is the same for the categories that had previously defaulted.

Answer: You are looking for evidence of significant differences between five categories of education and whether the individual defaulted on a loan. Since you have categorical data for both variables, you have to rule out a *t* test. You do have independent groups (those who defaulted and those who did not), but you need a test that will compare the ranks rather than the means. The Mann-Whitney *U* Test is designed to look for significant differences between independent groups measured at the categorical level.

The alternative hypothesis is that the distributions of level of education are *not the same* for the categories of those who have previously defaulted on a loan. The null hypothesis is that the distributions of level of education are the same across the categories of those who have previously defaulted.

In the *Nonparametric Independent Samples* window, customize tests in the Objective tab; in the Fields tab, customize assignments, and move the ordinal variable "education" to the *Test Field* box and the nominal variable "default" to the *Group's* box. In the Settings tab, customize tests, and select the *Mann-Whitney* test. The test results indicate that you now have statistical evidence that the distributions of those who defaulted on loans are not the same for different levels of education.

Hypothesis Test Summary

	Null Hypothesis	Test	Sig.	Decision
1	The distribution of Level of education is the same across categories of Previously defaulted.	Independent-Samples Mann-Whitney U Test	.001	Reject the null hypothesis.

Asymptotic significances are displayed. The significance level is .05.

△   CHAPTER 14: PAIRED-SAMPLES *T* TEST AND WILCOXON TEST

14.1 The researcher has the idea that listening to hard rock music can directly influence one's perception of the world. The professor randomly selects a group of 15 individuals from his "Introduction to Psychology" lecture hall class of 300 freshman students. He wishes to test his theory on his sample of 15 students by giving them the World Perception Test (WPT), having them listen to loud hard rock music, and then regiving the WPT. The test results are as follows. The WPT premusic scores are 16.34, 16.67, 17.18, 16.73, 16.37, 16.91, 17.32, 16.61, 16.67, 17.23, 17.26, 16.70, 16.43, 17.32, and 16.52. The WPT postmusic scores are 16.62, 16.49, 16.91, 16.61, 16.34, 16.85, 17.12, 16.43, 16.49 17.20, 16.70, 16.55, 16.25, 17.06, and 16.37. Write the null and alternative hypotheses, and see if you can produce statistical evidence in support of the professor's idea that listening to hard rock music can actually change your perception of the world.

Answer: The data were measured at the scale level, so you can look for differences in means, which suggests some type of a *t* test. Since you are measuring the same individuals in a pretest–posttest design, you should use the paired-samples *t* test to look for differences between the means. The null hypothesis is written as H_0: $\mu_1 - \mu_2 = 0$, while the alternative hypothesis (the professor's idea) is H_A: $\mu_1 - \mu_2 \neq 0$.

There is a requirement that both pre- and posttest scores be distributed normally, so we first check for normality using the Kolmogorov-Smirnov test. The outcome of that test is positive, so we proceed with the pair-samples t test. The results of the t test show a *Sig.* of .000, indicating that we can reject the null hypothesis of equality. We can say that we now have statistical evidence in support of the professor's contention that listening to hard rock music changes one's perception of the world.

Hypothesis Test Summary

	Null Hypothesis	Test	Sig.	Decision
1	The distribution of Pre Music WPT Score is normal with mean 16.819 and standard deviation 0.36.	One-Sample Kolmogorov-Smirnov Test	.129[1]	Retain the null hypothesis.
2	The distribution of Post Music WPT Score is normal with mean 16.640 and standard deviation 0.32.	One-Sample Kolmogorov-Smirnov Test	.200[1,2]	Retain the null hypothesis.

Asymptotic significances are displayed. The significance level is .05.

[1]Lilliefors Corrected

[2]This is a lower bound of the true significance.

Paired Samples Statistics

		Mean	N	Std. Deviation	Std. Error Mean
Pair 1	Pre Music WPT Score	16.8191	15	.35669	.09210
	Post Music WPT Score	16.6404	15	.32012	.08266

Paired Samples Test

		Paired Differences							
		Mean	Std. Deviation	Std. Error Mean	Lower	Upper	t		Sig. (2-tailed)
Pair 1	Pre Music WPT Score - Post Music WPT Score	.17863	.12718	.03284	.10820	.24906	5.440	14	.000

Reject the null hypothesis of equal means

14.2 Data were collected from 10 major oil well drilling operators that recorded the number of hours lost per week due to work-related accidents. A rigorous safety program was instituted, and the number of lost hours was once again recorded following the introduction of the safety program. The presafety program values are 41, 55, 63, 79, 45, 120, 30, 15, 24, and 24. The postsafety program values are 32, 49, 50, 73, 43, 115, 32, 9, 22, and 19.

A research consultant was hired to examine the data and determine if the safety program significantly changed the weekly hours lost from on-job injuries. Write the null and alternative hypotheses, and select and conduct the appropriate test to seek evidence that the program was successful.

Answer: You have data measured at the scale level; therefore, you can look for significant differences between means—this suggests a *t* test. You are measuring the same group twice in a pretest–posttest design, therefore the paired-samples *t* test would be the correct test. You should first check for normality, then conduct the *t* test.

The Kolmogorov-Smirnov test indicates normality for both distributions. The null hypothesis is written as H_0: $\mu_1 - \mu_2 = 0$, while the alternative hypothesis is H_A: $\mu_1 - \mu_2 \neq 0$. The *t* test results in a *Sig.* value of .003, informing us that the null can be rejected; we now have evidence that there is a significant difference in the mean number of weekly hours lost due to on-the-job injury. Next, we can look at the *Paired Samples Statistics* table and see that the mean hours lost due to injury was reduced from 49.6 to 44.4. The consultant can now say that the safety program was a success.

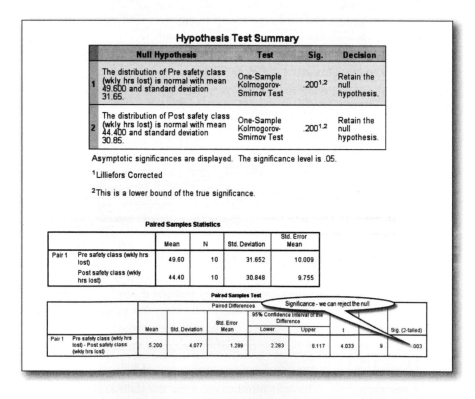

Hypothesis Test Summary

	Null Hypothesis	Test	Sig.	Decision
1	The distribution of Pre safety class (wkly hrs lost) is normal with mean 49.600 and standard deviation 31.65.	One-Sample Kolmogorov-Smirnov Test	.200[1,2]	Retain the null hypothesis.
2	The distribution of Post safety class (wkly hrs lost) is normal with mean 44.400 and standard deviation 30.85.	One-Sample Kolmogorov-Smirnov Test	.200[1,2]	Retain the null hypothesis.

Asymptotic significances are displayed. The significance level is .05.

[1]Lilliefors Corrected

[2]This is a lower bound of the true significance.

Paired Samples Statistics

		Mean	N	Std. Deviation	Std. Error Mean
Pair 1	Pre safety class (wkly hrs lost)	49.60	10	31.652	10.009
	Post safety class (wkly hrs lost)	44.40	10	30.848	9.755

Paired Samples Test

		Paired Differences					t		Sig. (2-tailed)
		Mean	Std. Deviation	Std. Error Mean	95% Confidence Interval of the Difference Lower	Upper			
Pair 1	Pre safety class (wkly hrs lost) - Post safety class (wkly hrs lost)	5.200	4.077	1.289	2.283	8.117	4.033	9	.003

Significance - we can reject the null

14.3 The chemical engineer added a chemical to a fast-burning compound that changed the oxygen consumption once the reaction started. He did a pretest to measure the oxygen consumption index; then he added the chemical and recorded the posttest oxygen index. The pretest values are 120, 139, 122, 120, 124, 120, 120, 125, 122, 123, 126, and

138. The posttest values are 121, 140, 123, 121, 125, 122, 121, 126, 123, 124, 127, and 139. Write the null and alternative hypotheses, select the correct test, and look for differences between the pretest and posttest oxygen index values.

Answer: Just by looking at the values in both distributions, you might conclude that they are not normal. To be sure, conduct the Kolmogorov-Smirnov test, which agrees with your initial assessment of nonnormality. Since you have nonnormal data, you select the related-samples sign test and the related-samples Wilcoxon Signed Rank Test. Select the *Analyze, Nonparametric, Related Sample* approach, and click **Customize analysis** in the Objective tab; move both variables to the *Test Fields* box, then click **Customize tests** in the Settings tab, and select the *sign* and *Wilcoxon* tests.

The null hypothesis is written as H_0: median$_{pretest}$ = median$_{posttest}$, while the alternative hypothesis is H_A: median$_{pretest}$ ≠ median$_{posttest}$. Both the sign rank and the Wilcoxon tests find statistical evidence in support of the alternative hypothesis, since the null of equality is rejected. The addition of the chemical additive significantly changed the oxygen index for this compound.

Hypothesis Test Summary

	Null Hypothesis	Test	Sig.	Decision
1	The distribution of Oxygen Index is normal with mean 124.917 and standard deviation 6.67.	One-Sample Kolmogorov-Smirnov Test	.017[1]	Reject the null hypothesis.
2	The distribution of Oxygen Index with change is normal with mean 126.000 and standard deviation 6.61.	One-Sample Kolmogorov-Smirnov Test	.014[1]	Reject the null hypothesis.

Asymptotic significances are displayed. The significance level is .05.

[1] Lilliefors Corrected

Hypothesis Test Summary

	Null Hypothesis	Test	Sig.	Decision
1	The median of differences between Oxygen Index and Oxygen Index with change equals 0.	Related-Samples Sign Test	.000[1]	Reject the null hypothesis.
2	The median of differences between Oxygen Index and Oxygen Index with change equals 0.	Related-Samples Wilcoxon Signed Rank Test	.001	Reject the null hypothesis.

Asymptotic significances are displayed. The significance level is .05.

[1] Exact significance is displayed for this test.

△　Chapter 15: One-Way ANOVA and Kruskal-Wallis Test

15.1 An El Salvadorian pig farmer, Jose, had the idea to add a by-product from the production of cane sugar to his pig feed. The idea was that the pigs would eat more, gain weight, and be worth more at market time. He had 24 weaner pigs weighing from 20 to 40 pounds. He randomly divided the pigs into three groups of eight. He concocted three different feed types, each containing different levels of the cane sugar by-product (*low* sugar, *medium,* and *high*). The farmer decided to record the pounds of feed consumed by each pig for 1 week. The pigs fed the low-sugar feed consumed 8.5, 8.0, 13.2, 6.8, 6.45, 6.0, 9.12, and 9.75 pounds. The pigs fed the medium-sugar feed consumed 10.99, 10.5, 9.67, 8.61, 10.92, 12.8, 9.03, and 9.45 pounds. The pigs fed high-sugar feed consumed 10.39, 9.97, 13.78, 12.69, 12.8, 9.67, 9.98, and 10.67 pounds.

Your task is to seek evidence that there is a significant difference in consumption for the three different feed types. Write the null and alternative hypotheses. If statistical significance is determined, then identify which groups contribute to overall significance. Once you complete the analysis, answer Jose's question about whether he should add the cane sugar by-product to his pig feed and which of the three feeds is the best.

Answer: You have scale data; therefore, you can look for differences between the means. There are three means for the pounds of feed consumed; therefore t tests are not appropriate—you must use the ANOVA procedure, which provides for three or more means. You do the Kolmogorov-Smirnov test to find if all three distributions are normal—they are. You next write the null hypothesis as H_0: $\mu_1 = \mu_2 = \mu_3$ and the alternative hypothesis as H_A: One or more of the three feed types result in unequal feed consumption. If the null hypothesis (H_{01}) is rejected, then the following additional null hypotheses should be tested: H_{02}: $\mu_1 = \mu_2$, H_{03}: $\mu_1 = \mu_3$, H_{04}: $\mu_2 = \mu_3$.

The ANOVA test finds a significant difference (reject the null) between the means, so we look to the post hoc analysis *Scheffe's Multiple Comparisons* to specify the means that contributed to the significant F value (12.499). We find the only significant difference in feed consumption to be between the high- and low-sugar feed. We recommend that Jose continue to use the high-sugar feed based on the evidence that the pigs did eat more.

Hypothesis Test Summary

	Null Hypothesis	Test	Sig.	Decision
1	The distribution of weaner pigs weekly consumption (lbs) is normal with mean 8.478 and standard deviation 2.32.	One-Sample Kolmogorov-Smirnov Test	.200[1,2]	Retain the null hypothesis.

Asymptotic significances are displayed. The significance level is .05.

Hypothesis Test Summary

	Null Hypothesis	Test	Sig.	Decision
1	The distribution of weaner pigs weekly consumption (lbs) is normal with mean 10.246 and standard deviation 1.35.	One-Sample Kolmogorov-Smirnov Test	.200[1,2]	Retain the null hypothesis.

Asymptotic significances are displayed. The significance level is .05.

Hypothesis Test Summary

	Null Hypothesis	Test	Sig.	Decision
1	The distribution of weaner pigs weekly consumption (lbs) is normal with mean 11.244 and standard deviation 1.60.	One-Sample Kolmogorov-Smirnov Test	.177[1]	Retain the null hypothesis

Asymptotic significances are displayed. The significance level is .05.

Descriptives

weaner pigs weekly consumption (lbs)   Highest feed consumption and highest sugar content

	N	Mean	Std. Deviation	Std. Error	95% Confidence Interval for Mean Lower Bound	Upper Bound	Minimum	Maximum
low	8	8.4775	2.31842	.81968	6.5393	10.4157	6.00	13.20
medium	8	10.2463	1.34779	.47652	9.1195	11.3730	8.61	12.80
high	8	11.2438	1.59646	.56443	9.9091	12.5784	9.67	13.78
Total	24	9.9892	2.08080	.42474	9.1105	10.8678	6.00	13.78

ANOVA

Miles per Gallon (1)   Significant difference

	Sum of Squares	df	Mean Square	F	Sig.
Between Groups	26.896	2	13.448	12.499	.000
Within Groups	45.189	42	1.076		
Total	72.085	44			

Multiple Comparisons

Dependent Variable: weaner pigs we...   significant difference between low and high sugar

Scheffe

(I) sugarlevel	(J) sugarlevel	Mean Difference (I-J)	Std. Error	Sig.	95% Confidence Interval Lower Bound	Upper Bound
low	medium	-1.76875	.90094	.170	-4.1411	.6036
	high	-2.76625*	.90094	.020	-5.1386	-.3939
medium	low	1.76875	.90094	.170	same as above	
	high	-.99750	.90094	.551	-3.3698	1.3748
high	low	2.76625*	.90094	.020	.3939	5.1386
	medium	.99750	.90094	.551	-1.3748	3.3698

*. The mean difference is significant at the 0.05 level.

15.2 A chemical engineer had three different formulas for a gasoline additive that she thought would significantly change automobile gas mileage. She had three groups of 15 test standard eight-cylinder engines that simulated normal driving conditions. Each group received a different gasoline formulation (A1, A2, and A3) and was run for several hours. Simulated mileage for the A1 group was 35.60, 34.50, 36.20, 33.10, 36.10, 34.80, 33.90, 34.70, 35.20, 35.80, 36.60, 35.10, 34.90, 36.00, and 34.10. Mileage for A2 was 36.80, 35.30, 37.00, 32.90, 36.80, 35.60, 35.10, 35.80, 36.90, 36.60, 36.80, 36.60, 35.80, 36.30, and 36.00. Mileage for the A3 group was 37.79, 36.29, 38.01, 33.80, 37.79, 36.58, 36.03, 36.79, 37.89, 37.57, 37.79, 37.59, 36.78, 37.29, and 37.01.

Your job is to investigate the mileage numbers in an effort to provide evidence in support of her contention that the groups would have significantly different gas mileage. Write the *null* and *alternative* hypotheses. If you find a difference, therefore rejecting the null, you must identify the groups contributing to the significant F statistic with *post hoc* analysis. Can you provide evidence in support of the chemical engineer's contention that her formulas will significantly alter gas mileage for these test engines?

Answer: You have scale data and three group means to compare; therefore, the logical choice for analysis is ANOVA. You check the three mileage distributions for normality, and they are determined to be normal by the Kolmogorov-Smirnov test. You next write the null hypothesis as H_0: $\mu_1 = \mu_2 = \mu_3$ and the alternative hypothesis as H_A: One or more of the three gasoline formulations result in unequal gas mileages. If H_{01} is rejected, then the following null hypotheses should be tested: H_{02}: $\mu_1 = \mu_2$, H_{03}: $\mu_1 = \mu_3$, H_{04}: $\mu_2 = \mu_3$.

The ANOVA test computes an F value of 12.499, which is significant. You next do a Scheffe post hoc multiple comparisons test and find significant differences for A1 and A3, and between A2 and A3. We can say that we now have evidence in support of the engineer's idea that there are differences in two of the three mileage comparisons. Comparisons between A1 and A2 were found not to be significantly different.

Hypothesis Test Summary

	Null Hypothesis	Test	Sig.	Decision
1	The distribution of Miles per Gallon (1) is normal with mean 35.107 and standard deviation 0.97.	One-Sample Kolmogorov-Smirnov Test	.200[1,2]	Retain the null hypothesis.
2	The distribution of Miles per Gallon (2) is normal with mean 36.020 and standard deviation 1.06.	One-Sample Kolmogorov-Smirnov Test	.200[1,2]	Retain the null hypothesis.
3	The distribution of Miles per Gallon (3) is normal with mean 37.000 and standard deviation 1.08.	One-Sample Kolmogorov-Smirnov Test	.200[1,2]	Retain the null hypothesis.

Asymptotic significances are displayed. The significance level is .05.

Descriptives

Miles per Gallon (1)

	N	Mean	Std. Deviation	Std. Error	95% Confidence Interval for Mean		Minimum	Maximum
					Lower Bound	Upper Bound		
Formula A1	15	35.1067	.96767	.24985	34.5708	35.6425	33.10	36.60
Formula A2	15	36.0200	1.05844	.27329	35.4339	36.6061	32.90	37.00
Formula A3	15	37.0000	1.08218	.27942	36.4007	37.5993	33.80	38.01
Total	45	36.0422	1.27996	.19081	35.6577	36.4268	32.90	38.01

ANOVA

Miles per Gallon (1)

	Sum of Squares	df	Mean Square	F	Sig.
Between Groups	26.896	2	13.448	12.499	.000
Within Groups	45.189	42	1.076		
Total	72.085	44			

Multiple Comparisons

Dependent Variable: Miles per Gallon (1)

Scheffe

A1 and A3 significantly different

(I) Gas Formulas	(J) Gas Formulas	Mean Difference (I-J)	Std. Error	Sig.	95% Confidence Interval	
					Lower Bound	Upper Bound
Formula A1	Formula A2	A2 and A3 Sig. .876	.37876	.066	-1.8745	.0478
	Formula A3		.37876	.000	-2.8545	-.9322
Formula A2	Formula A1	.91333	.37876	.066	.0478	1.8746
	Formula A3	-.98000*	.37876	.045	-1.9412	-.0188
Formula A3	Formula A1	1.89333*	.37876	.000	.9322	2.8545
	Formula A2	.98000*	.37876	.045	.0188	1.9412

*. The mean difference is significant at the 0.05 level.

15.3 Bacteria counts were taken at the four Southern California beaches of Santa Monica, Malibu, Zuma, and Ventura. The researcher's idea was that the different beaches would yield significantly different bacteria counts. The Santa Monica beach count was 16.2, 12.0, 16.4, 15.5, 16.5, 22.0, and 23.0. The Malibu count was 18.3, 18.2, 18.3, 17.4, 18.4, 24.1, and 25.2. The Zuma count was 17.2, 17.3, 17.2, 16.4, 17.3, 23.0, and 24.3. The Ventura count was 20.2, 20.9, 21.1, 20.3, 20.2, 26.1, and 28.4. Check the distributions for normality—just by looking, you would suspect that they don't approximate the normal curve.

Select the correct testing approach based on your normality findings, and write the null and alternative hypotheses. If you find significant differences in the bacteria counts at the four beaches, do additional work to identify the specific beaches that contribute to the overall finding. What is the answer to the researcher's idea that the beaches have different bacteria counts?

Answer: The four distributions of bacteria counts at the beaches are found to be nonnormal. Therefore, the one-way ANOVA should *not* be used, and the Kruskal-Wallis test for nonparametric data is the test of choice. The null is that the median ranks for the beaches are the same. The alternative hypothesis is that one or more of the median ranks are different. The Kruskal-Wallis test shows that the bacteria counts were not the same for the four beaches. The null of equality is rejected, with a significance level of .036. This finding requires that you double click on **Hypothesis Statement** in the Output Viewer and then request the pairwise comparison. This reveals that the overall significance is obtained from the rather large difference between the Santa Monica and Ventura average ranks of 8.57 and 20.71, respectively. We can say that the researcher's idea that the beaches had different bacteria counts is only supported in one of the five comparisons.

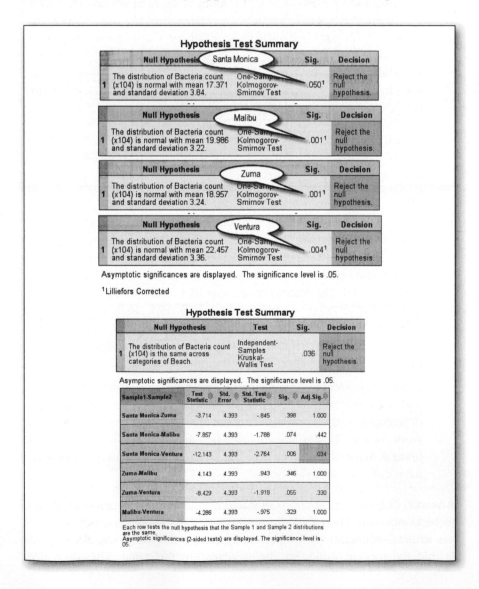

CHAPTER 16: TWO-WAY (FACTORIAL) ANOVA △

16.1 A corn farmer is interested in reducing the number of days it takes for his corn to silk. He has decided to set up a controlled experiment that manipulates the *nominal* variable "fertilizer," having two categories: 1 = *limestone* and 2 = *nitrogen*. Another *nominal* variable is "soil type," with two categories: 1 = *silt* and 2 = *peat*. The dependent variable is *scale* and is the "number of days until the corn begins to silk." The data are given below; note that they must be entered into Data View in one continuous string of 40 cases. Once the data are entered, you must look for significant differences between the four study groups. You will also look for any interaction effects between fertilizer and soil and any influence they may have on the number of days to the showing of silk. Write the null and alternative hypotheses, select and conduct the correct test(s), and interpret the results.

	fertilizer	soil	silk				
1	1	1	54	11	2	1	65
2	1	1	56	12	2	1	64
3	1	1	60	13	2	1	61
4	1	1	65	14	2	1	64
5	1	1	67	15	2	1	65
6	1	1	67	16	2	1	64
7	1	1	63	17	2	1	59
8	1	1	62	18	2	1	65
9	1	1	65	19	2	1	58
10	1	1	59	20	2	1	65

21	1	2	55	31	2	2	50
22	1	2	54	32	2	2	49
23	1	2	57	33	2	2	55
24	1	2	52	34	2	2	53
25	1	2	57	35	2	2	57
26	1	2	60	36	2	2	51
27	1	2	64	37	2	2	55
28	1	2	58	38	2	2	57
29	1	2	60	39	2	2	52
30	1	2	63	40	2	2	53

Answer: Since we have two independent *nominal* variables (each having two levels) and a dependent variable measured at the *scale* level, we should think of a two-way ANOVA. The first thing we do is check to see if the values of the dependent variable approximate the normal distribution. The Kolmogorov-Smirnov test confirms that it is normally distributed.

Hypothesis Test Summary

	Null Hypothesis	Test	Sig.	Decision
1	The distribution of Days to silking is normal with mean 59.000 and standard deviation 5.14.	One-Sample Kolmogorov-Smirnov Test	.066[1]	Retain the null hypothesis.

Asymptotic significances are displayed. The significance level is .05.

[1] Lilliefors Corrected

The next table tells us that the error variances of the dependent variable are equal across groups since we fail to reject the null of equality (*Sig.* is .275). This adds to our confidence that our findings can be taken seriously, since this is one of the requirements for our test.

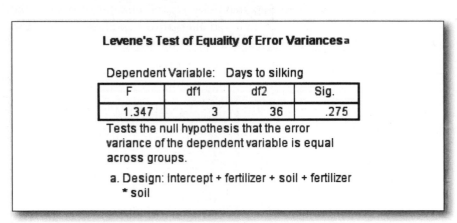

Levene's Test of Equality of Error Variances[a]

Dependent Variable: Days to silking

F	df1	df2	Sig.
1.347	3	36	.275

Tests the null hypothesis that the error variance of the dependent variable is equal across groups.

a. Design: Intercept + fertilizer + soil + fertilizer * soil

We next write the null and research hypotheses for both *main effects*:

Null for fertilizer type is H_{01}: $\mu_{nitrogen} = \mu_{limestone}$; Alternative is H_{A1}: $\mu_{nitrogen} \neq \mu_{limestone}$.

Null for soil type is H_{02}: $\mu_{peat} = \mu_{silt}$; Alternative is H_{A2}: $\mu_{peat} \neq \mu_{silt}$.

Proceed with the two-way ANOVA, and make sure you request *Descriptive Statistics*, as they are especially useful in learning what is happening with your data. You are interested in interpreting the mean values in the table in a way that will inform you about the *main* effects of soil type (peat and silt) and fertilizer (nitrogen and limestone) on the number of days it takes corn to silk. The interaction effects can also be interpreted from this table, but the explanation goes beyond the scope of this book. The values of 59.9 and 58.1 answer the question about the main effect of fertilizer (for significance, look at the next table. For the main effect of soil type, look for a significant difference between 62.4 and 55.6. For the interaction effect, look at the next table').

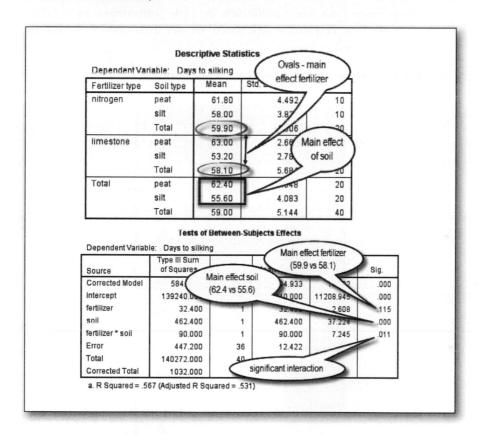

Descriptive Statistics

Dependent Variable: Days to silking

Fertilizer type	Soil type	Mean	Std.	
nitrogen	peat	61.80	4.492	10
	silt	58.00	3.87	10
	Total	59.90	.06	20
limestone	peat	63.00	2.66	
	silt	53.20	2.78	
	Total	58.10	5.68	20
Total	peat	62.40	.048	20
	silt	55.60	4.083	20
	Total	59.00	5.144	40

Ovals - main effect fertilizer

Main effect of soil

Tests of Between-Subjects Effects

Dependent Variable: Days to silking

Source	Type III Sum of Squares				Sig.
Corrected Model	584.		4.933		.000
Intercept	139240.0		0.000	11208.94	.000
fertilizer	32.400	1	32.400	2.608	.115
soil	462.400	1	462.400	37.22	.000
fertilizer * soil	90.000	1	90.000	7.245	.011
Error	447.200	36	12.422		
Total	140272.000	40			
Corrected Total	1032.000				

a. R Squared = .567 (Adjusted R Squared = .531)

Main effect fertilizer (59.9 vs 58.1)

Main effect soil (62.4 vs 55.6)

significant interaction

We fail to reject the null for fertilizer and reject the null for soil type; therefore, we have evidence that the soil type does affect the days to silking. Silt soil significantly reduces the time to silk, which is a positive outcome for this corn farmer. There is an interaction effect, but we would need more work to be more specific than this.

16.2 A psychologist had the idea that different types of music and room temperature would influence performance on simple math tasks. She had two independent variables measured at the nominal level: (1) "music type," hard rock and classical, and (2) "room temperature," comfortable and hot. The dependent variable was a series of minimally challenging mathematical problems that were scored on a 0 to 100 scale. She randomly selected 24 students and then once again randomly assigned them to one of four groups. The data that resulted from her experiment are presented in the following table. Your task is to select the correct test, write the null and alternative hypotheses, and then interpret the results. Was there any significance on task performance as a result of music type or room temperature, or did these two variables act together to cause change?

	roomtemp	music	math
1	1	1	95
2	1	1	100
3	1	1	85
4	1	1	75
5	1	1	95
6	1	1	87
7	2	1	76
8	2	1	76
9	2	1	65
10	2	1	100
11	2	1	54
12	2	1	78
13	1	2	58
14	1	2	76
15	1	2	95
16	1	2	56
17	1	2	79
18	1	2	100
19	2	2	65
20	2	2	73
21	2	2	82
22	2	2	65
23	2	2	97
24	2	2	76

Answer: Since we have scale data for the dependent variable and multiple independent nominal variables (two or more), we select the two-way ANOVA procedure. The Kolmogorov-Smirnov test indicates that the data are normally distributed. Levene's Test for Equality of Error Variances indicates that we can continue with the analysis.

Null for music type is H_{01}: $\mu_{\text{hard rock}} = \mu_{\text{classical}}$; Alternative is H_{A1}: $\mu_{\text{hard rock}} \neq \mu_{\text{classical}}$.

Null for room temperature is H_{02}: $\mu_{comfortable} = \mu_{hot}$; Alternative is H_{A2}: $\mu_{comfortable} \neq \mu_{hot}$.

The *Descriptive Statistics* table informs us that we will look for significance between 82.17 and 76.83 to either confirm or deny a statistically significant *main effect* of music type. Any main effect from room temperature will be identified by comparing 83.42 and 75.58. Interaction will be indicated in the *Tests of Between-Subjects Effects* table. Results indicate that no significance was found for any comparison, and the nulls remain in force. There was no evidence generated to support the psychologist's idea that math performance would be influenced by room temperature and/or type of music.

Hypothesis Test Summary

	Null Hypothesis	Test	Sig.	Decision
1	The distribution of Basic math task is normal with mean 79.500 and standard deviation 14.43.	One-Sample Kolmogorov-Smirnov Test	.171[1]	Retain the null hypothesis.

Asymptotic significances are displayed. The significance level is .05.

[1]Lilliefors Corrected

Descriptive Statistics

Dependent Variable: Basic math task

Ovals - main effect music

Music type	Room temperature	Mean	Std. De	
hard rock	comfortable	89.50	9.028	6
	hot	74.83	15.34	6
	Total	82.17	14. 7	12
classical	comfortable	77.33		
	hot	76.33		
	Total	76.83		
Total	comfortable	83.42	13.108	12
	hot	75.58	13.180	12
	Total	79.50	14.431	24

Main effect temperature

Levene's Test of Equality of Error Variances a

Dependent Variable: Basic math task

F	df1	df2	Sig.
.732	3	20	.545

Tests the null hypothesis that the error variance of the dependent variable is equal across groups.

a. Design: Intercept + music + roomtemp + music * roomtemp

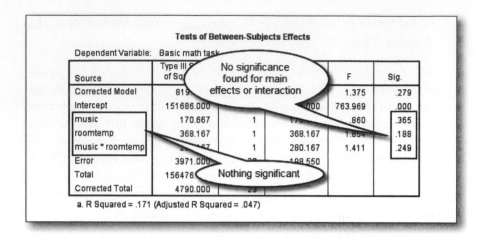

Tests of Between-Subjects Effects

Dependent Variable: Basic math task

Source	Type III Sum of Squares	df	Mean Square	F	Sig.
Corrected Model	819.			1.375	.279
Intercept	151686.000		.000	763.969	.000
music	170.667	1	17	.860	.365
roomtemp	368.167	1	368.167	1.854	.188
music * roomtemp	2 .67	1	280.167	1.411	.249
Error	3971.000		198.550		
Total	156476.				
Corrected Total	4790.000	23			

a. R Squared = .171 (Adjusted R Squared = .047)

No significance found for main effects or interaction

Nothing significant

16.3   The inspector general for a large state's motor vehicle department decided to collect some data on recent driving tests. The idea was to see if scores on the driving test (dependent *scale* variable) were significantly different for male and female (*nominal* independent variable) instructors. He also wanted to know if the time of day the test was given might also influence the scores. He first randomly picked two instructors and then collected data on recent tests they had administered. Time of day that the test was given was categorized as either early morning or late afternoon (the second nominal independent variable). He decided to randomly select six morning and six afternoon tests for each of his picked instructors. In the end, he had four unique groups consisting of six test takers each. You must write the null and alternative hypotheses and then select the correct test, interpret the results, and answer the inspector's questions. The data are given here.

Answer: We note that there are two nominal independent variables and one dependent scale variable. It is the perfect choice for a two-way ANOVA procedure. We first

	gender	timeday	test
1	1	1	87
2	1	1	76
3	1	1	93
4	1	1	89
5	1	1	74
6	1	1	87
7	2	1	73
8	2	1	71
9	2	1	81
10	2	1	69
11	2	1	75
12	2	1	63
13	1	2	63
14	1	2	51
15	1	2	52
16	1	2	61
17	1	2	52
18	1	2	51
19	2	2	89
20	2	2	91
21	2	2	89
22	2	2	83
23	2	2	77
24	2	2	76

check the values of the dependent variable (scores on a test) for normality, and the Kolmogorov-Smirnov confirms its normality.

Let's first write the null and alternative hypotheses:

Null for instructor's gender is H_{01}: $\mu_{Susan} = \mu_{Tom}$; Alternative is H_{A1}: $\mu_{Susan} \neq \mu_{Tom}$.

Null for time of day is H_{02}: $\mu_{morning} = \mu_{afternoon}$; Alternative is H_{A2}: $\mu_{morning} \neq \mu_{afternoon}$.

The *Descriptive Statistics* table informs us that we will look for significance between 69.67 and 78.08 to either confirm or deny a statistically significant *main effect* of gender. Any main effect from time of day will be identified by comparing 78.17 and 69.58. Interaction will be indicated in the *Tests of Between-Subjects Effects* table. Results indicate that there was significance found for the main effects of both gender and time of day. There was also significance identified for the interaction between gender and time of day. We reject both of the nulls; the identification of the specific interaction effect would require material not covered in this book. We now have statistical evidence to support the inspector general's idea that the factors of gender of the instructor and time of day the test was given influence the driving test scores.

Hypothesis Test Summary

	Null Hypothesis	Test	Sig.	Decision
1	The distribution of Scores on driving test is normal with mean 73.875 and standard deviation 13.65.	One-Sample Kolmogorov-Smirnov Test	.200[1,2]	Retain the null hypothesis.

Asymptotic significances are displayed. The significance level is .05.

[1] Lilliefors Corrected

[2] This is a lower bound of the true significance.

Levene's Test of Equality of Error Variances[a]

Dependent Variable: Scores on driving test

F	df1	df2	Sig.
.496	3	20	.689

Tests the null hypothesis that the error variance of the dependent variable is equal across groups.

a. Design: Intercept + gender + timeday + gender * timeday

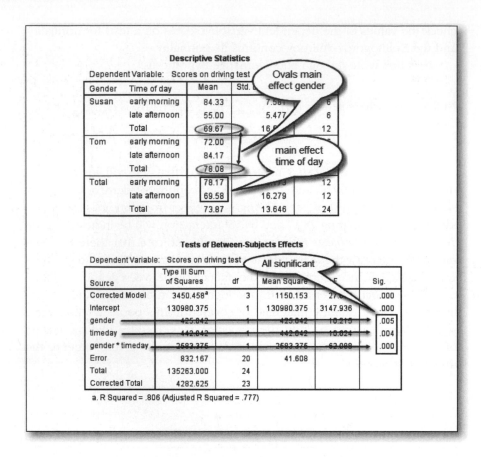

Descriptive Statistics

Dependent Variable: Scores on driving test

Gender	Time of day	Mean	Std.	
Susan	early morning	84.33	7.58	6
	late afternoon	55.00	5.477	6
	Total	69.67	16.	12
Tom	early morning	72.00		
	late afternoon	84.17		
	Total	78.08		
Total	early morning	78.17		12
	late afternoon	69.58	16.279	12
	Total	73.87	13.646	24

Ovals main effect gender

main effect time of day

Tests of Between-Subjects Effects

Dependent Variable: Scores on driving test

All significant

Source	Type III Sum of Squares	df	Mean Square	F	Sig.
Corrected Model	3450.458[a]	3	1150.153	27.	.000
Intercept	130980.375	1	130980.375	3147.936	.000
gender	425.042	1	425.042	10.215	.005
timeday	442.042	1	442.042	10.624	.004
gender * timeday	2583.375	1	2583.375	62.089	.000
Error	832.167	20	41.608		
Total	135263.000	24			
Corrected Total	4282.625	23			

a. R Squared = .806 (Adjusted R Squared = .777)

△ CHAPTER 17: ONE-WAY ANOVA REPEATED MEASURES TEST AND FRIEDMAN TEST

17.1 The high school track team coach recorded the time taken for the 100-yard dash by eight team members for three consecutive track meets during the regular season. His past experience informed him that they would improve their times throughout the season as they grew stronger and smarter. He had the idea that their level of improvement would qualify for statistical significance. Can you help the coach write the null and alternative hypotheses, select the correct test(s), interpret the analysis, and then answer his question? The data follow:

	runtime1	runtime2	runtime3
1	11.80	11.20	10.70
2	11.90	11.70	10.96
3	11.30	10.80	10.50
4	11.20	10.97	10.65
5	11.10	10.89	10.00
6	10.90	10.70	10.45
7	10.60	10.50	10.31
8	11.00	10.90	10.69

Answer: The run times were measured at the scale level; therefore, you can look for differences between the means. There are three measurements taken on the same individuals over a period of several weeks. The repeated measures ANOVA would be the first choice. We must first check for normality with the Kolmogorov-Smirnov test—the three run times are all found to approximate the normal curve.

We write the alternative and null hypotheses as follows. The alternative hypothesis is H_{A1}: One or more of the mean run times are unequal. The null hypothesis states the opposite and is written as H_{01}: $\mu_1 = \mu_2 = \mu_3$.

In this example, H_{01} states that there are no differences between the means of the run times. The track team coach would prefer to reject the null hypothesis, which would provide statistical evidence for the idea that the run times are significantly different.

If there is evidence of overall significance, leading to the rejection of the null hypothesis (H_{01}), the researcher would most likely wish to identify which of the three groups are different and which are equal. The following null and alternative hypotheses will facilitate that task. If the first null hypothesis (H_{01}) is rejected, then we may test the following additional null hypotheses: H_{02}: $\mu_1 = \mu_2$, H_{03}: $\mu_1 = \mu_3$, H_{04}: $\mu_2 = \mu_3$. The alternative hypotheses for these new null hypotheses are H_{A2}: $\mu_1 \neq \mu_2$, H_{A3}: $\mu_1 \neq \mu_3$, H_{A4}: $\mu_2 \neq \mu_3$.

Hypothesis Test Summary

	Null Hypothesis	Test	Sig.	Decision
1	The distribution of Run time 1 (100yrd dash) is normal with mean 11.225 and standard deviation 0.44.	One-Sample Kolmogorov-Smirnov Test	.200[1,2]	Retain the null hypothesis.
2	The distribution of Run time 2 (100 yrd dash is normal with mean 10.958 and standard deviation 0.36.	One-Sample Kolmogorov-Smirnov Test	.200[1,2]	Retain the null hypothesis.
3	The distribution of Run time 3 (100 yrd dash) is normal with mean 10.532 and standard deviation 0.29	One-Sample Kolmogorov-Smirnov Test	.200[1,2]	Retain the null hypothesis.

Asymptotic significances are displayed. The significance level is .05.

Mauchly's Test of Sphericity[a]

Measure: MEASURE_1

Within Subjects Effect	Mauchly's W	Approx. Chi-Square	df	Sig.	Epsilon[b]		
					Greenhouse-Geisser	Huynh-Feldt	Lower-bound
time	.587	3.198	2	.202	.708	.835	.500

Tests the null hypothesis that the error covariance matrix of the orthonormalized transformed dependent variables is proportional to an identity matrix.

a. Design: Intercept
   Within Subjects Design: time

Tests of Within-Subjects Effects

Measure: MEASURE_1

Source		Type III Sum of Squares	df	Mean Square	F	Sig.
time	Sphericity Assumed	1.951	2	.976	27.197	.000
	Greenhouse-Geisser	1.951	1.415	1.379	27.197	.000
	Huynh-Feldt	1.951	1.669	1.169	27.197	.000
	Lower-bound	1.951	1.000	1.951	27.197	.001
Error(time)	Sphericity Assumed	.502	14	.036		
	Greenhouse-Geisser	.502	9.907	.051		
	Huynh-Feldt	.502	11.684	.043		
	Lower-bound	.502	7.000	.072		

Pairwise Comparisons

Measure: MEASURE_1

(I) time	(J) time	Mean Difference (I-J)	Std. Error	Sig.[b]	95% Confidence Interval for Difference[b]	
					Lower Bound	Upper Bound
1	2	.267*	.065	.004	.114	.421
	3	.693*	.119	.001	.411	.974
2	1	-.267*	.065	.004	-.421	-.114
	3	.425*	.093	.003	.206	.644
3	1	-.693*	.119	.001	-.974	-.411
	2	-.425*	.093	.003	-.644	-.206

Based on estimated marginal means

*. The mean difference is significant at the .05 level.

17.2 A farm manager was interested in studying several first-time strawberry pickers over a period of 4 weeks. He felt that there was a significant difference in the number of pints picked per hour from one week to the next. Can you help him write the null and alternative hypotheses, input the data, select the correct tests, interpret the results, and answer his question concerning significant changes in the number of pints picked? The data are shown here.

Answer: You have *scale* data (number of pints picked), and therefore you can select a test that compares the means. The paired-samples *t* test

	time1	time2	time3	time4
1	10	11	13	16
2	9	10	12	15
3	10	11	13	13
4	8	9	10	11
5	11	12	12	14
6	8	9	11	12
7	11	13	15	17
8	10	11	12	14
9	11	12	14	16
10	8	8	10	13

won't work since you have more than two time periods. You should use the repeated measures ANOVA since you are measuring the same individuals at four different times. You can first check for normality of the distributions of the average number of pints picked per hour for each of the 4 weeks. The Kolmogorov-Smirnov test indicates normal distributions for all 4 picking weeks, so you proceed with the ANOVA test.

The alternative and null hypotheses for overall significance and all possible comparisons are as follows. The alternative hypothesis is H_{A1}: One or more of the mean numbers of pints picked over the four time periods are unequal. The null hypothesis states the opposite and is written as H_{01}: $\mu_1 = \mu_2 = \mu_3 = \mu_4$.

In this example, H_{01} states that there are no differences between the mean numbers of pints picked over the four time periods. The farm manager would prefer to reject the null hypothesis, which would provide statistical evidence for the idea that the numbers of pints picked are significantly different.

If there is evidence of overall significance, leading to the rejection of the null hypothesis (H_{01}), the farm manager would most likely wish to identify which of the four time periods are different and which are equal. The following null and alternative hypotheses will facilitate that task.

If the first null hypothesis (H_{01}) is rejected, then we may test the following additional null hypotheses: H_{02}: $\mu_1 = \mu_2$, H_{03}: $\mu_1 = \mu_3$, H_{04}: $\mu_1 = \mu_4$, H_{05}: $\mu_2 = \mu_3$, H_{06}: $\mu_2 = \mu_4$, H_{07}: $\mu_3 = \mu_4$. The alternative hypotheses for these new null hypotheses are H_{A2}: $\mu_1 \neq \mu_2$, H_{A3}: $\mu_1 \neq \mu_3$, H_{A4}: $\mu_1 \neq \mu_4$, H_{A5}: $\mu_2 \neq \mu_3$, H_{A6}: $\mu_2 \neq \mu_4$, H_{A7}: $\mu_3 \neq \mu_4$.

The second table in the output, *Descriptive Statistics*, clearly indicates a steady increase in the mean number of pints picked from Week 1 to Week 4. The production of these 10 workers is definitely on the increase, but now we want to answer the question of statistical significance. We first examine Mauchly's Test for Sphericity and see the borderline significance value of .054 (remember we don't want to reject the null for this test), but we make the decision to proceed. As you will see in the next table, there are tests other than Mauchly's that can guide our use of ANOVA. Also, the ANOVA has the ability to tolerate minor deviations. The following table, *Tests of Within-Subjects Effects*, confirms our decision to proceed as the tests that do not assume sphericity found significance for difference between the four times. The final table, *Pairwise Comparisons*, identifies significant differences between all possible time comparisons. All of the null hypotheses were rejected, thus we have evidence to support the alternatives. We can definitely inform the farm manager that there is a statistically significant increase (look at the direction of the means) in strawberry production for this group of workers.

Hypothesis Test Summary

	Null Hypothesis	Test	Sig.	Decision
1	The distribution of pints of strawberries per/hour (3 wk) is normal with mean 12.200 and standard deviation 1.62.	One-Sample Kolmogorov-Smirnov Test	.200[1,2]	Retain the null hypothesis.
2	The distribution of pints of strawberries per/hour (2 wk) is normal with mean 10.600 and standard deviation 1.58.	One-Sample Kolmogorov-Smirnov Test	.200[1,2]	Retain the null hypothesis.
3	The distribution of pints of strawberries per/hour (1 wk) is normal with mean 9.600 and standard deviation 1.26.	One-Sample Kolmogorov-Smirnov Test	.200[1,2]	Retain the null hypothesis.
4	The distribution of pints of strawberries per/hour (4 wk) is normal with mean 14.100 and standard deviation 1.91.	One-Sample Kolmogorov-Smirnov Test	.200[1,2]	Retain the null hypothesis.

Asymptotic significances are displayed. The significance level is .05.

Descriptive Statistics

	Mean	Std. Deviation	N
pints of strawberries per/hour (1 wk)	9.60	1.265	10
pints of strawberries per/hour (2 wk)	10.60	1.578	10
pints of strawberries per/hour (3 wk)	12.20	1.619	10
pints of strawberries per/hour (4 wk)	14.10	1.912	10

Mauchly's Test of Sphericity[a]

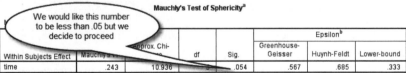

We would like this number to be less than .05 but we decide to proceed

					Epsilon[b]		
Within Subjects Effect	Mauchly's W	Approx. Chi-Square	df	Sig.	Greenhouse-Geisser	Huynh-Feldt	Lower-bound
time	.243	10.936	5	.054	.567	.685	.333

Tests the null hypothesis that the error covariance matrix of the orthonormalized transformed dependent variables is proportional to an identity matrix.

a. Design: Intercept
   Within Subjects Design: time

Tests of Within-Subjects Effects

Measure: MEASURE_1

Source		Type III Sum of Squares	df	Mean Square	F	Sig.
time	Sphericity Assumed	116.075	3	38.692	82.420	.000
	Greenhouse-Geisser	116.075	1.701	68.245	82.420	.000
	Huynh-Feldt					.000
	Lower-bound			6.075	82.420	.000
Error(time)	Sphericity			.469		
	Greenhouse			.828		
	Huynh-Feldt			.685		
	Lower-bound	12.675	9.000	1.408		

All tests indicate significance, even the ones that don't assume sphericity, so our decision to proceed was correct

Pairwise Comparisons

Measure: MEASU   Week 1

(I) time	(J) time	Mean Difference (I-J)	Std. Error	Sig.[b]	95% Confidence Interval for Difference[b] Lower Bound	95% Confidence Interval for Difference[b] Upper Bound
1	2	-1.000*	.149	.000	-1.337	-.663
	3	-2.600*	.267	.000	All significant	
	4	Week 3 .401		.000		-3.592
2	1	1.000*	.149	.000	.663	1.337
	3	-1.600*	.221	.000	-2.100	-1.100
	4	-3.500*	.401	.000	-4.408	-2.592
3	1	Week 4 .267		.000	1.997	3.203
	2	1.600*	.221	.000	1.100	2.100
	4	-1.900*	.314	.000	-2.611	-1.189
4	1	4.500*	.401	.000	3.592	5.408
	2	3.500*	.401	.000	2.592	4.408
	3	1.900*	.314	.000	1.189	2.611

Based on estimated marginal means

*. The mean difference is significant at the .05 level.

17.3 For this exercise, you must open the SPSS sample file called *bankloan. sav*. A bank president is interested in comparing the last three variables: "preddef1," "preddef2," and "preddef3." These three variables were three different models created to predict whether a bank customer would default on a bank loan. Since they were all created from the same basic information, we can treat them as the same object and analyze them using some type of repeated measures method. Can you help the bank president in determining if the three models are the same? Also, write the null and alternative hypotheses that you will be testing.

Answer: For the repeated measures for scale data, we might use the ANOVA approach, but the Kolmogorov-Smirnov test shows nonnormality for all variables.

Let's write the alternative and null hypotheses for this design. The alternative hypothesis is H_{A1}: One or more of the three prediction models resulted in unequal default predictions. The null hypothesis states the opposite and is written as H_{01}: $\mu_1 = \mu_2 = \mu_3$.

In this example, H_{01} states that there are no differences between the prediction models. The bank president would prefer to reject the null hypothesis, which would provide statistical evidence for the idea that the prediction models are significantly different.

If there is evidence of overall significance, leading to the rejection of the null hypothesis (H_{01}), the bank president would most likely wish to identify which of the three models are different and which are equal. The following null and alternative hypotheses will facilitate that task. If the first null hypothesis (H_{01}) is rejected, then we may test the following additional null hypotheses: H_{02}: $\mu_1 = \mu_2$, H_{03}: $\mu_1 = \mu_3$, H_{04}: $\mu_1 = \mu_3$. The alternative hypotheses for these new null hypotheses are H_{A2}: $\mu_1 \neq \mu_2$, H_{A3}: $\mu_1 \neq \mu_3$, H_{A4}: $\mu_2 \neq \mu_3$.

To test the hypotheses, we decide to use the Related-Samples Freidman's Test, which indicates no statistical difference between the ranks for the three model types. The above null hypotheses are not rejected, and all the models are equal. We double click on the SPSS output to obtain the graph showing the ranks for the three different default models just to add some visual evidence to our conclusion of nonsignificance. You can confidently inform the bank president that there is no difference in the three default models.

Hypothesis Test Summary

	Null Hypothesis	Test	Sig.	Decision
1	The distribution of Predicted default, model 1 is normal with mean 0.259 and standard deviation 0.26.	One-Sample Kolmogorov-Smirnov Test	.000[1]	Reject the null hypothesis.
2	The distribution of Predicted default, model 2 is normal with mean 0.258 and standard deviation 0.27.	One-Sample Kolmogorov-Smirnov Test	.000[1]	Reject the null hypothesis.
3	The distribution of Predicted default, model 3 is normal with mean 0.259 and standard deviation 0.17.	One-Sample Kolmogorov-Smirnov Test	.000[1]	Reject the null hypothesis.

Asymptotic significances are displayed. The significance level is .05.

[1] Lilliefors Corrected

Hypothesis Test Summary

	Null Hypothesis	Test	Sig.	Decision
1	The distributions of Predicted default, model 1, Predicted default, model 2 and Predicted default, model 3 are the same.	Related-Samples Friedman's Two-Way Analysis of Variance by Ranks	.000	Reject the null hypothesis.

Asymptotic significances are displayed. The significance level is .05.

Related-Samples Fridman's Two-Way Analysis of Variance by Ranks

CHAPTER 18: ANALYSIS OF COVARIANCE   Δ

18.1 A metallurgist has designed a way of increasing the strength of steel. She has discovered a chemical that is added to samples of molten metal during the manufacturing process that have already been measured for strength. These pre-additive values are recorded as the variable called "pre-add" in the data table shown here. She believes that the "preadd" values may influence the "postadd" measure of the steel's strength. She is looking for significant differences in strength for the four different manufacturing methods. If differences are found, she wishes to identify which ones contribute to the overall significance. Can you help her select the correct statistical procedure? She also needs help in writing the null and alternative hypotheses.

	method	preadd	postadd
1	1	5.00	12.65
2	1	5.10	12.85
3	1	4.24	11.18
4	1	4.47	12.29
5	1	5.29	12.73
6	1	4.90	12.08
7	2	5.20	13.23
8	2	5.20	12.81
9	2	4.69	12.88
10	2	4.58	12.53
11	2	5.20	13.23
12	2	4.69	12.85
13	3	4.36	12.29
14	3	5.20	13.56
15	3	4.47	11.53
16	3	5.48	13.60
17	3	4.58	12.45
18	3	4.90	12.92
19	4	6.00	14.00
20	4	5.20	12.96
21	4	5.66	13.96
22	4	5.57	13.71
23	4	4.36	11.70
24	4	5.57	13.56

Answer: You might initially think of the one-way ANOVA as you have two *scale* variables and one *nominal* and you are looking for differences in groups. However, the scientist stated that she suspected that the "preadd" values may have an undue influence on the "postadd" values. This suggests that she needs some method to control for any influence of the "preadd" variable—the ANCOVA seems to fit this requirement. Let's begin by checking the distributions for normality and then the homogeneity of regression slopes. The Kolmogorov-Smirnov test finds both variables normally distributed. The initial run of the ANCOVA procedure is to check for the homogeneity of regression slopes, which indicates that we could proceed with the main test.

Before starting the analysis, we write the null and alternative hypotheses as follows. We state the null hypothesis as H_{01}: $\mu_A = \mu_B = \mu_C = \mu_D$. We state the alternative hypothesis as H_{A1}: One or more of the four groups have mean steel strengths that are not equal.

If the null hypothesis (H_{01}) is rejected, the researcher then wishes to identify which of the four groups are different and which are equal. The following additional null hypotheses will facilitate that task: H_{02}: $\mu_A = \mu_B$, H_{03}: $\mu_A = \mu_C$, H_{04}: $\mu_A = \mu_D$, H_{05}: $\mu_B = \mu_C$, H_{06}: $\mu_B = \mu_D$, H_{07}: $\mu_C = \mu_D$.

The alternative hypotheses for these new null hypotheses are H_{A2}: $\mu_A \neq \mu_B$, H_{A3}: $\mu_A \neq \mu_C$, H_{A4}: $\mu_A \neq \mu_D$, H_{A5}: $\mu_B \neq \mu_C$, H_{A6}: $\mu_B \neq \mu_D$, H_{A7}: $\mu_C \neq \mu_D$.

Running the analysis, we see the *Descriptive Statistics* table showing small differences between the four methods used in the steel production process. It remains to be seen if these differences are significant. The results of the first ANCOVA, shown in the first *Tests Between-Subjects Effects* table, indicate no interaction (see *Sig.* of .331); thus, we have homogeneous regression slopes—we may proceed. Levene's test shows a *Sig.* value of .603, telling us that the error variances are equal across all groups, which adds to our confidence in our statistical method.

Our second *Tests Between-Subjects Effects* table shows an overall significance of .043. Thus, we next examine the *Pairwise Comparisons* table and find that A&B and A&C are significant. Thus, we have rejected the overall null and those for A&B and A&C. We can advise the scientist that the method used to manufacture these samples had a significant impact and that there was significant difference between methods A&B and A&C.

Hypothesis Test Summary

	Null Hypothesis	Test	Sig.	Decision
1	The distribution of strength pre additive is normal with mean 4.995 and standard deviation 0.47.	One-Sample Kolmogorov-Smirnov Test	.200[1,2]	Retain the null hypothesis.
2	The distribution of strength post additive is normal with mean 12.815 and standard deviation 0.74.	One-Sample Kolmogorov-Smirnov Test	.200[1,2]	Retain the null hypothesis.

Asymptotic significances are displayed. The significance level is .05.

Tests of Between-Subjects Effects

Dependent Variable: str...

Source				F	Sig.
Corrected M...				18.031	.000
Intercept				58.294	.000
method				.281	.315
preadd	5.135	1	5.135	58....	.000
method * preadd	.326	3	.109	1.230	.331
Error	1.411	16	.088		
Total	3954.000	24			
Corrected Total	12.543	23			

This is the value that informs us that we fail to reject the null of no interaction - we can say our regression slopes are homogeneous

a. R Squared = .887 (Adjusted R Squared = .838)

Descriptive Statistics

Dependent Variable: strength post additive

manufacturing method	Mean	Std. Deviation	N
A	12.2956	.61693	6
B	12.9205	.26930	6
C	12.7266	.80009	6
D	13.3177	.87500	6
Total	12.8151	.73848	24

Levene's Test of Equality of Error Variances [a]

Dependent Variable: strength post additive

F	df1	df2	Sig.
.632	3	20	.603

Tests the null hypothesis that the error variance of the dependent variable is equal across groups.

a. Design: Intercept + preadd + method

Tests of Between-Subjects Effects

Dependent Variable: strength post additive

Significance is found; we reject the null

Source	Type III Sum of Squares	df	Mean Square	F	Sig.
Corrected Model	10.806[a]	4		29.557	.000
Intercept	5.085	1	5.085	55.637	.000
preadd	7.558	1	7.558	82.687	.000
method	.902	3	.301	3.289	.043
Error	1.737	19	.091		
Total	3954.000	24			
Corrected Total	12.543	23			

a. R Squared = .862 (Adjusted R Squared = .832)

Pairwise Comparisons

Dependent Variable: strength post additive

Significant

(I) manufacturing method	(J) manufacturing method	Mean Difference (I-J)	Std. Error	Sig.[b]	Lower Bound	Upper Bound
A	B	-.496*	.175	.011		.130
	C	-.435*	.175	.022	-.801	-.070
	D	-.238	.195	.237	-.645	.170
B	A	.496*	.175	.011	.130	.863
	C	.061	.175	.731	-.305	.428
	D	.258	.189	.187	-.137	.654
C	A	.435*	.175	.022	.070	.801
	B	-.061	.175	.731	-.428	.305
	D	.197	.195	.324	-.211	.605
D	A	.238	.195	.237	-.170	.645
	B	-.258	.189	.187	-.654	.137
	C	-.197	.195	.324	-.605	.211

Pairwise comparisons identified A&B and A&C as significant

Based on estimated marginal means

*. The mean difference is significant at the .05 level.

18.2 A botanist measured the 3-day growth, in inches, of his marijuana plants at two different times (variables: "pregrowth" and "postgrowth") under four different growing conditions (variable: "peatsoil"). He felt that the initial growth rate influenced the second rate of growth. The scientist's main concern was the effect of soil type on growth

	peatsoil	pregrowth	postgrowth
1	4	1.40	2.20
2	4	1.41	2.22
3	4	1.26	2.10
4	4	1.30	2.18
5	4	1.45	2.21
6	4	1.38	2.16
7	3	1.43	2.24
8	3	1.43	2.21
9	3	1.34	2.22
10	3	1.32	2.20
11	3	1.43	2.24
12	3	1.34	2.22
13	2	1.28	2.18
14	2	1.43	2.26
15	2	1.30	2.12
16	2	1.48	2.27
17	2	1.32	2.19
18	2	1.38	2.22
19	1	1.56	2.29
20	1	1.43	2.23
21	1	1.51	2.29
22	1	1.49	2.27
23	1	1.28	2.14
24	1	1.49	2.26

rate during the second growth period. The problem was that he somehow wanted to statistically account for any differences in the second growth period that might be related to the first rate of growth. His ultimate quest was to identify any significant differences in the four samples that were grown in soils containing different percentages of peat. Select the correct statistical method, write the null and alternative hypotheses, do the analysis, interpret the results, and answer the botanist's questions.

Answer: The botanist has two *scale* variables and one *categorical* variable. One of the scale variables ("pregrowth") is thought to have an effect on the second ("postgrowth") that needs to be statistically controlled. The ANCOVA is the logical choice for analysis, but we first use the Kolmogorov-Smirnov to test for normality, and we find that both scale distributions pass. We next test for the homogeneity of regression slopes, and the test shows that we fail to reject the null of no interaction (*Sig.* = .332), which provides evidence of homogeneous regression slopes. The *Descriptive Statistics* table shows means that look pretty much alike—but we must test for significance. Levene's test informs us that we can confidently proceed with the ANCOVA since the null of equal variances across groups is not rejected (*Sig.* = .498).

We next write the null and alternative hypotheses as follows. We state the null hypothesis as H_{01}: $\mu_1 = \mu_2 = \mu_3 = \mu_4$. We state the alternative hypothesis as H_{A1}: One or more of the four groups have mean inches of growth that are not equal.

If the null hypothesis (H_{01}) is rejected, the researcher then wishes to identify which of the four groups are different and which are equal. The following additional null hypotheses will facilitate that task: H_{02}: $\mu_1 = \mu_2$, H_{03}: $\mu_1 = \mu_3$, H_{04}: $\mu_1 = \mu_4$, H_{05}: $\mu_2 = \mu_3$, H_{06}: $\mu_2 = \mu_4$, H_{07}: $\mu_3 = \mu_4$.

The alternative hypotheses for these new null hypotheses are H_{A2}: $\mu_1 \neq \mu_2$, H_{A3}: $\mu_1 \neq \mu_3$, H_{A4}: $\mu_1 \neq \mu_4$, H_{A5}: $\mu_2 \neq \mu_3$, H_{A6}: $\mu_2 \neq \mu_4$, H_{A7}: $\mu_3 \neq \mu_4$.

The main table of interest is *Tests of Between-Subjects Effects*, which shows whether the percentage of peat in the soil had a significant effect on the rate of growth during the second growth period. The significance level of .048 permits us to the reject the null (H_{01}: $\mu_1 = \mu_2 = \mu_3 = \mu_4$) of equality of all groups; therefore, we have evidence that soil type did affect the rate of growth of the marijuana plants. We can investigate further by an examination of the *Pairwise Comparisons* table, which indicates significant differences for the 12% & 16% and 14% & 16% peat groups. Therefore, we can also reject the null hypotheses H_{06}: $\mu_2 = \mu_4$ and H_{07}: $\mu_3 = \mu_4$

Hypothesis Test Summary

	Null Hypothesis	Test	Sig.	Decision
1	The distribution of growth first week is normal with mean 1.393 and standard deviation 0.08.	One-Sample Kolmogorov-Smirnov Test	.200[1,2]	Retain the null hypothesis.
2	The distribution of growth second week is normal with mean 2.214 and standard deviation 0.05.	One-Sample Kolmogorov-Smirnov Test	.200[1,2]	Retain the null hypothesis.

Asymptotic significances are displayed. The significance level is .05.

[1]Lilliefors Corrected

Tests of Between-Subjects Effects

Dependent Variable: growth

Fail to reject null of no interaction therefore the regression slopes are homogeneous

Source					
Corrected Model	.052	7		17.372	.000
Intercept	.117	1	.117	633	.000
peatsoil	.002	3	.001	1.2	.320
pregrowth	.025	1	.025	57.787	.000
peatsoil * pregrowth	.002	3	.001	1.229	.332
Error	.007	16	.000		
Total	117.706	24			
Corrected Total	.059	23			

a. R Squared = .884 (Adjusted R Squared = .833)

Descriptive Statistics

Dependent Variable: growth second week

soil peat content	Mean	Std. Deviation	N
10%	2.2472	.05931	6
12%	2.2080	.05510	6
14%	2.2224	.01811	6
16%	2.1786	.04471	6
Total	2.2140	.05073	24

Levene's Test of Equality of Error Variances[a]

Dependent Variable: growth second week

F	df1	df2	Sig.
.820	3	20	.498

Tests the null hypothesis that the error variance of the dependent variable is equal across groups.

a. Design: Intercept + pregrowth + peatsoil

Tests of Between-Subjects Effects

Dependent Variable:

Soil has a significant effect on growth rate

Source	of Squares	df	Square	F	Sig.
Corrected Model	.051[a]	4		28.451	.000
Intercept	.127	1	.127	712	.000
pregrowth	.036	1	.036	80.0	.000
peatsoil	.004	3	.001	3.175	.048
Error	.008	19	.000		
Total	117.706	24			
Corrected Total	.059	23			

a. R Squared = .857 (Adjusted R Squared = .827)

Pairwise Comparisons

Dependent Variable: growth second week

(I) soil peat content	(J) soil peat content	Mean Difference (I-J)	Std. Error	Sig.[b]	95% Confidence Interval for Difference[b] Lower Bound	Upper Bound
10%	12%	-.013	.013	.364		.016
	14%	-.017	.013	.212	*Significant*	.010
	16%	.017	.013	.211	-.011	.046
12%	10%	.013	.013	.364	-.016	.041
	14%	-.004	.012	.731		.021
	16%	.030[*]	.012	.024	*Significant*	.055
14%	10%	.017	.013	.212	.010	.044
	12%	.004	.012	.731	-.021	.030
	16%	.034[*]	.012	.011	.009	.060
16%	10%	-.017	.013	.211	-.046	.011
	12%	-.030[*]	.012	.024	-.055	-.004
	14%	-.034[*]	.012	.011	-.060	-.009

Based on estimated marginal means

*. The mean difference is significant at the .05 level.

18.3 An epidemiologist/psychologist was interested in studying the effects of early-childhood vaccinations and cognitive ability. He obtained records on randomly selected children who had received three levels of vaccinations during their first year of life. He randomly placed them in three groups defined by rates of vaccination ("vaccinated" is the nominal variable), where 1 = *high*, 2 = *low*, and 3 = *none*. The children had been tested for cognitive ability at 5 years of age ("precog" is the scale variable) and again at 10 years of age ("postcog" is another scale variable). The scientist's main reason for conducting the investigation was to search for any differential effects that the levels of vaccination might have on the children's cognitive ability. However, he was concerned about the potential effect that the "precog" scores might have on the "postcog" values. His major research question was whether the three levels of vaccination affected the children's cognitive ability at 10 years of age.

Can you help this scientist pick the appropriate statistical test that would offer a way to control for differences in the "precog" values? If you can, then write the null and alternative hypotheses, run the analysis, interpret the results, and answer his questions. The data are given in the table shown here.

	vaccinated	precog	postcog
1	1	50	49
2	1	63	55
3	1	45	46
4	1	78	69
5	1	54	53
6	1	81	80
7	2	43	44
8	2	65	66
9	2	43	42
10	2	75	73
11	2	42	40
12	2	80	74
13	3	34	45
14	3	35	43
15	3	67	87
16	3	41	55
17	3	78	82
18	3	53	65

Answer: The epidemiologist/psychologist has collected the data from prior observations. This makes it impossible to alter the research design in a way to control for differences in cognitive ability at age 5. He needs to consider using ANCOVA, which could statistically control for any earlier differences in cognitive ability. He has two *scale* variables and one *nominal*, which qualifies it for the ANCOVA approach. If the distributions of cognitive ability scores approximate the normal and if the regression slopes are homogeneous, then we can use the ANCOVA procedure. We run the Kolmogorov-Smirnov test and confirm that the "precog" and "postcog" values are normally distributed. The customized run of the ANCOVA analysis indicates that the null hypothesis of *no interaction* cannot be rejected; therefore, we have evidence that the regression slopes are homogeneous. We can proceed with the ANCOVA procedure.

We next write the null and alternative hypotheses as follows. We state the null hypothesis as H_{01}: $\mu_1 = \mu_2 = \mu_3$. We state the alternative hypothesis as H_{A1}: One or more of the three groups have mean cognitive abilities that are not equal.

If the null hypothesis (H_{01}) is rejected, the researcher then wishes to identify which of the three groups are different and which are equal. The following additional null hypotheses will facilitate that task: H_{02}: $\mu_1 = \mu_2$, H_{03}: $\mu_1 = \mu_3$, H_{04}: $\mu_2 = \mu_3$. The alternative hypotheses for these new null hypotheses are H_{A2}: $\mu_1 \neq \mu_2$, H_{A3}: $\mu_1 \neq \mu_3$, H_{A4}: $\mu_2 \neq \mu_3$.

The next SPSS output examined is the *Descriptive Statistics* table. Looking at the means gives us a hint that we might find significant differences in the cognitive ability of children and the level of vaccinations. The next table, showing the results of Levene's test, indicates that we can have more confidence in any significant levels that may be identified. This is because a significance level of .227 results in the decision *not* to reject the null of equal variances across all groups on values of the dependent variable.

The table *Tests of Between-Subjects Effects* indicates that the overall null hypothesis for vaccination (H_{01}: $\mu_1 = \mu_2 = \mu_3$) can be rejected. This finding gives us evidence that there is a difference in children's cognitive ability and their level of vaccine exposure. We examine the table titled *Pairwise Comparisons* and find significant differences in cognitive ability between groups of children in the *high* and *none* categories of vaccination. Also, we find significant differences for the *low* and *none* categories of vaccination levels. We reject H_{03}: $\mu_1 = \mu_3$ and H_{04}: $\mu_2 = \mu_3$. We can inform the epidemiologist/psychologist that, for this sample of children, the evidence supports the idea that vaccines negatively affect children's cognitive ability at age 10.

Hypothesis Test Summary

	Null Hypothesis	Test	Sig.	Decision
1	The distribution of cognitive ability age five is normal with mean 57.056 and standard deviation 16.48.	One-Sample Kolmogorov-Smirnov Test	.200[1,2]	Retain the null hypothesis.
2	The distribution of cognitive ability age 10 is normal with mean 59.333 and standard deviation 15.38.	One-Sample Kolmogorov-Smirnov Test	.200[1,2]	Retain the null hypothesis.

Asymptotic significances are displayed. The significance level is .05.

Tests of Between-Subjects Effects

Dependent Variable: cogniti

Source	T		ean Square	F	Sig.
Corrected Model	3		757.883	39.102	.000
Intercept	62.296	1		3.214	.098
vaccinated	20.494	2	10.2	.529	.602
precog	3488.658	1	3488.658	1	.000
vaccinated * precog	12.675	2	6.338	.327	.727
Error	232.584	12	19.382		
Total	67390.000	18			
Corrected Total	4022.000	17			

(callout: Fail to reject the null of no interaction)

a. R Squared = .942 (Adjusted R Squared = .918)

Descriptive Statistics

Dependent Variable: cognitive ability age 10

vaccination rate first year of birth	Mean	Std. Deviation	N
high	58.67	13.125	6
low	56.50	16.171	6
none	62.83	18.595	6
Total	59.33	15.381	18

Levene's Test of Equality of Error Variances a

Dependent Variable: cognitive ability age 10

F	df1	df2	Sig.
1.642	2	15	.227

Tests the null hypothesis that the error variance of the dependent variable is equal across groups.

a. Design: Intercept + precog + vaccinated

Tests of Between-Subjects Effects

Dependent Variable: cognitive ability age 10

Source	Type III Su of Squa				Sig.
Corrected Model	3776.741			71.862	.000
Intercept	53.377	1	53.37	3.047	.103
precog	3652.408	1	3652.408	20	.000
vaccinated	653.263	2	326.631	18.645	.000
Error	245.259	14	17.518		
Total	67390.000	18			
Corrected Total	4022.000	17			

(callout: Significant for the three vaccination categories)

a. R Squared = .939 (Adjusted R Squared = .926)

Pairwise Comparisons

Dependent Variable: cognitive ability age 10

(I) vaccination rate first year of birth	(J) vaccination rate first year of birth	Mean Difference (I-J)	Std. Error	Sig.[b]	95% Confidence Interval for	
					Lower	Upper Bound
high	low	-1.376	2.429	.580	-6.586	3.833
	none	-13.871*	2.508	.000		
low	high	1.376	2.429	.580		6.586
	none	-12.495*	2.454	.000	-17.75	
none	high	13.871*	2.508	.000		
	low	12.495*	2.454	.000	7.232	17.758

(callouts: Significant; Significant; Duplicates)

Based on estimated marginal means

*. The mean difference is significant at the .05 level.

CHAPTER 19: PEARSON'S CORRELATION △
AND SPEARMAN'S CORRELATION

19.1 Assume you have collected a random sample of first-year students at a local community college and given them a general survey that included a number of items. A series of questions results in self-esteem ratings, and part of their official record includes their IQ. You want to calculate a correlation coefficient for these two variables including a significance level and then chart the results and add the *Fit Line*. Select the correct correlation coefficient, write the null and alternate hypotheses, and interpret the results. A summary of the data is shown here.

	selfesteem	IQ
1	3.8	110
2	4.2	130
3	3.9	126
4	4.1	127
5	3.7	128
6	3.5	128
7	4.5	135
8	4.5	149
9	3.7	135
10	4.7	140
11	3.8	131
12	4.2	142

Answer: We first test both distributions using the Kolmogorov-Smirnov nonparametric one-sample test and determine that they are both normal. The null hypothesis is H_0: $\rho = 0$, and the alternative is H_A: $\rho \neq 0$.

We use Spearman's correlation and find a moderate to strong coefficient of .599 with a significance level of .04. We now have evidence that the population correlation is not equal to 0, and we may take our value seriously. The chart you build also shows a definite positive relationship between these two variables.

Hypothesis Test Summary

	Null Hypothesis	Test	Sig.	Decision
1	The distribution of scores is normal with mean 4.050 and standard deviation 0.38.	One-Sample Kolmogorov-Smirnov Test	.200[1,2]	Retain the null hypothesis.
2	The distribution of Intelligence Quotient is normal with mean 131.750 and standard deviation 9.78.	One-Sample Kolmogorov-Smirnov Test	.200[1,2]	Retain the null hypothesis.

Asymptotic significances are displayed. The significance level is .05.

Correlations

Correlation coefficient		scores	Intelligence Quotient
scores	Pearson Correlation	1	.599[*]
	Sig. (2-tailed)		.040
	N	12	12
Intelligence Quotient	Pearson Correlation	.599[*]	1
	Significance	.040	
	N	12	12

*. Correlation is significant at the 0.05 level (2-tailed).

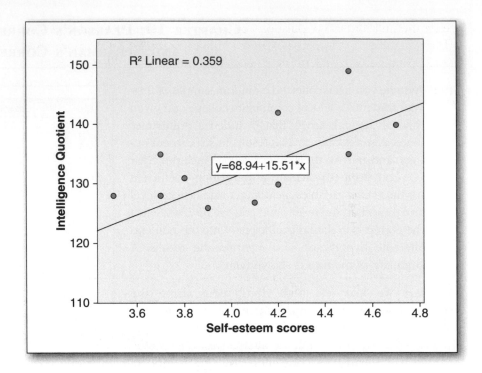

19.2 Let's say you live on a little used back road that leads to the ski slopes. Over the years, you have noticed that there seems to be a correlation between the number of inches of snowfall and traffic on your road. You collect some data and now wish to analyze them using correlation and a test of significance. You also wish to visualize the data on a graph that includes a *Fit Line*. Write the null and alternative hypotheses, calculate the coefficient and the significance level, and then build the graph. The data are shown here.

	snowfall	cars
1	12.7	23
2	13.0	16
3	6.0	10
4	23.0	32
5	10.0	12
6	20.0	28
7	15.0	17
8	24.0	32
9	16.0	18
10	11.0	10
11	14.0	14

Answer: You should first test both distributions using the Kolmogorov-Smirnov nonparametric one-sample test to see if they are normally distributed. The test provides evidence that both snowfall and number of cars are normally distributed, so you may proceed with the calculation of the Spearman correlation coefficient. The null hypothesis is H_0: $\rho = 0$, and the alternative is H_A: $\rho \neq 0$.

Spearman's correlation finds a strong positive correlation coefficient of .92, with a significance level of .000. We now have evidence that the population correlation is not equal to 0, and we may take the value of .92 seriously. You reject the null that the population correlation coefficient is equal to 0. The chart you build visually agrees with the finding by showing a strong positive relationship between inches of snowfall and car traffic on the back road.

Hypothesis Test Summary

	Null Hypothesis	Test	Sig.	Decision
1	The distribution of snowfall in inches is normal with mean 14.973 and standard deviation 5.51.	One-Sample Kolmogorov-Smirnov Test	.200[1,2]	Retain the null hypothesis.
2	The distribution of cars per hour is normal with mean 19.273 and standard deviation 8.27.	One-Sample Kolmogorov-Smirnov Test	.200[1,2]	Retain the null hypothesis.

Asymptotic significances are displayed. The significance level is .05.

Correlations

		High correlation ~~ll in~~	cars per hour
snowfall in inches	Pearson Correlation	1	.920**
	Sig. (2-tailed)		.000
	N		11
cars per hour	Pearson Correlati~~on~~ Significant	~~.000~~	1
	Sig. (2-tailed)	~~.000~~	
	N	11	11

**. Correlation is significant at the 0.01 level (2-tailed).

R^2 Linear = 0.846

$y = -1.41 + 1.38 * x$

19.3 Assume you own a furniture store and you decided to record a random sample of rainy days and the number of patrons on those days. Calculate the correlation coefficient, build a graph, and test the numbers for significance. The data are shown here.

	rainfall	storepat
1	1.20	446.00
2	2.95	235.00
3	5.12	123.00
4	6.20	72.00
5	4.10	174.00
6	2.21	347.00
7	4.46	156.00
8	5.49	97.00
9	2.49	293.00
10	6.48	46.00
11	1.45	393.00
12	3.49	197.00

Answer: We test the rainfall and number of patrons for normality using *the Kolmogorov-Smirnov* one-sample test and find that they are both normally distributed. We may proceed with the calculation of the Spearman correlation coefficient. The null hypothesis is H_0: $\rho = 0$, and the alternative is H_A: $\rho \neq 0$.

Spearman's correlation finds a strong negative correlation coefficient of −0.979, with a significance level of .000. The null hypothesis is rejected. We have statistical evidence that the population correlation coefficient is not equal to 0, and we may take the value of −0.979 seriously. The bivariate chart agrees with the correlation calculation and shows a strong negative relationship between inches of rain and the number of patrons coming to the store. More rain equals fewer patrons.

Hypothesis Test Summary

	Null Hypothesis	Test	Sig.	Decision
1	The distribution of inches of rain is normal with mean 3.803 and standard deviation 1.79.	One-Sample Kolmogorov-Smirnov Test	.200[1,2]	Retain the null hypothesis.
2	The distribution of number of people is normal with mean 214.917 and standard deviation 129.98.	One-Sample Kolmogorov-Smirnov Test	.200[1,2]	Retain the null hypothesis.

Asymptotic significances are displayed. The significance level is .05.

[1]Lilliefors Corrected

Correlations

		inches of rain	number of people
inches of rain	Pearson Correlation	1	-.979[**]
	Sig. (2-tailed)		.000
	N	12	12
number of people	Pearson Correlation		1
	Sig. (2-tailed)		
	N	12	12

Strong negative correlation

Significant

**. Correlation is significant at the 0.01 level (2-tailed).

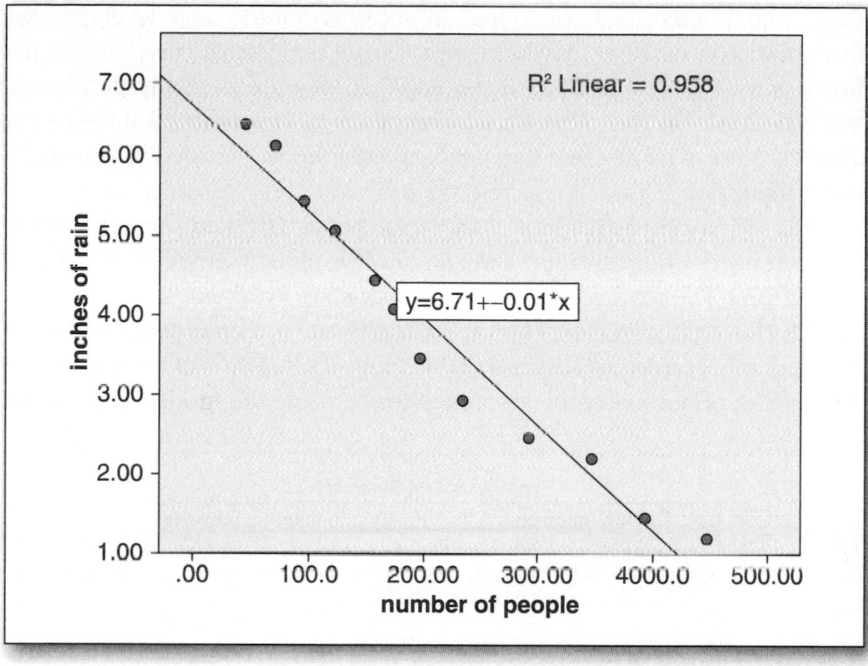
R² Linear = 0.958
y=6.71+-0.01*x
inches of rain
7.00
6.00
5.00
4.00
3.00
2.00
1.00
.00
100.0
200.00
300.00
4000.0
500.00
number of people

CHAPTER 20: SINGLE LINEAR REGRESSION △

20.1 Can you help the manager of a senior citizen center at the local library determine if there was any merit to her idea that the patron's age and the number of books checked out were related? Her thought was that as an individual got older, more books would be checked out. She would like to be able to predict the number of books that would be checked out by looking at a person's age. The manager is especially interested in the number of books checked out for those 65 years of age. She selected a random sample of 24 senior patrons and collected details of the age and the number of books checked out during a 4-week period. If you wish to help, select the correct statistical approach, write the null and alternative hypothesis, conduct the analysis, and interpret the results. Her data are shown here.

	age	books
1	62	2
2	67	6
3	65	4
4	70	10
5	66	7
6	63	4
7	67	6
8	65	4
9	63	2
10	68	9
11	65	5
12	69	7
13	68	7
14	62	3
15	64	4
16	70	9
17	63	2
18	68	6
19	64	3
20	65	5
21	69	8
22	64	3
23	62	2
24	68	7

Answer: Correlation and regression seems to be the best approach to answer the manager's questions. The manager has two variables ("age" and "number of books") that are measured at the scale level, and the first question is whether they both approximate the normal curve. We do the Kolmogorov-Smirnov test and determine that they are normally distributed. We will check on the other data assumptions as we proceed through the analysis. Since we have one independent variable, we decided to use single linear regression.

The dependent variable is "number of books checked out"—for statistics, you need *estimates* and *model fit*, and for plots, you will need *ZPRED* and *ZRESID*. The results of both plots are satisfactory in that the *ZPRED* values are close to the diagonal and the *ZRESID* values are scattered uniformly on the graph. The *Model Summary* table indicates that 86% of the variance in the number of books can be accounted for by using the model. The *ANOVA*

Hypothesis Test Summary

	Null Hypothesis	Test	Sig.	Decision
1	The distribution of age is normal with mean 65.708 and standard deviation 2.61.	One-Sample Kolmogorov-Smirnov Test	.183[1]	Retain the null hypothesis.
2	The distribution of number of checked out is normal with mean 5.208 and standard deviation 2.43.	One-Sample Kolmogorov-Smirnov Test	.182[1]	Retain the null hypothesis.

Asymptotic significances are displayed. The significance level is .05.

Normal P-P Plot of Regression Standardized Residual
Dependent Variable: number of checked out

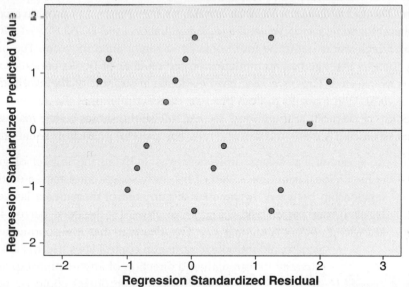

Scatterplot
Dependent Variable: number of checked out

Model Summary[b]

Model	R	R Square	Adjusted R Square	Std. Error of the Estimate
1	.941[a]	.885	.880	.841

a. Predictors: (Constant), age

b. Dependent Variable: number of checked out

ANOVA[a]

Model		Sum of Squares	df	Mean Square	F	Sig.
1	Regression	120.381	1	120.381	170.015	.000[b]
	Residual	15.577	22	.708		
	Total	135.958	23			

a. Dependent Variable: number of checked out

b. Predictors: (Constant), age

Coefficients[a]

Model		Unstandardized Coefficients		Standardized Coefficients	t	Sig.
		B	Std. Error	Beta		
1	(Constant)	-52.337	4.417		-11.850	.000
	age	.876	.067	.941	13.039	.000

a. Dependent Variable: number of checked out

	age	books	pred_books
1	62	2	1
2	67	6	6
3	65	4	4

table shows an *F* value of 170.015 that is significant at .000, providing additional evidence that our model has merit. The final table, titled *Coefficients*, provides the intercept and slope needed to write the regression equation. The *Transform/Compute Variable* feature can then be used to provide a prediction for someone 65 years of age. The equation used is −52.876 + (.876 * 65), which works out to be four books for a senior aged 65 years. The null hypothesis that age has no influence on the number of books checked out can be rejected. Thus, we now have evidence in support of *the alternative hypothesis* (the manager's idea) that age directly influences the number of books checked out at the library. We can advise the senior center manager that her idea has merit and that she now has a useful prediction equation.

20.2   An economist at a large university was studying the impact of gun crime on local economies. Part of his study sought information on the relationship between the number of gun control measures a lawyer/ legislator introduced and his score awarded by the state bar on his knowledge of constitutional law. His idea was that low-scoring lawyers would introduce more gun control laws. He wished to quantify the strength and direction of any relationship and also see if the number of laws introduced could be predicted by knowing the legislator's constitutional law rating. One specific value he wished to predict was the number of laws introduced by the average score of 76, a value not directly observed in the data.

	const_score	gun_control
1	98	1
2	86	2
3	74	3
4	63	4
5	51	6
6	97	1
7	85	2
8	77	5
9	65	6
10	53	8
11	94	2
12	83	2
13	74	4
14	69	4
15	55	6
16	97	2
17	84	4
18	71	4
19	64	5
20	57	8
21	99	1
22	82	3
23	75	4
24	63	4

His research has thus far shown that gun control laws have a negative impact on local economies. The researcher selected a random sample of lawyers elected to office and then compiled public information on the two variables of interest ("gun control" and "state bar rating"). As a consulting statistician, your task is to select the correct statistical method, write the null and alternative hypotheses, do the analysis, and interpret the results. His data are shown here.

Answer: It appeared that since the economist had two scale variables and was interested in correlation-type questions, single linear regression might be a good place to start. The first check is to use Kolmogorov-Smirnov to test for normality, which indicates both distributions are normal. To satisfy additional data assumptions, you will need to generate plots for error residuals and for homoscedasticity. In the *Statistics* window, you must request estimates, model fit, and also *ZPRED* and *ZRESID* for the plots. The plots look good since the values are aligned fairly well with the diagonal of the P-P plot and are uniformly scattered on the residual plot.

According to the *Model Summary*, 80% of the variance in the number of gun laws is accounted for by the level of understanding of constitutional law by the legislator. In the *ANOVA* table, we find an *F* value of 91.068, which is significant, indicating that our model is sound.

The *Coefficients* table gives us the intercept and slope, which we need to write the prediction equation. We use the *Transform/Compute Variable* feature to generate predicted values, which can then be compared with the observed data. The equation used to create a new predicted variable when $x = 76$ is $12.929 + (-.121x)$. If we plug in the average of 76, the value of interest, we get a predicted number of four gun laws.

The null hypothesis that score on constitutional law rating has no influence on the number of laws introduced is rejected. There is evidence that supports the alternative hypothesis that score on constitutional law influences the number of laws introduced.

Hypothesis Test Summary

	Null Hypothesis	Test	Sig.	Decision
1	The distribution of Score Constitution is normal with mean 75.667 and standard deviation 15.02.	One-Sample Kolmogorov-Smirnov Test	.200[1,2]	Retain the null hypothesis.
2	The distribution of Gun control measures is normal with mean 3.792 and standard deviation 2.02.	One-Sample Kolmogorov-Smirnov Test	.200[1,2]	Retain the null hypothesis.

Asymptotic significances are displayed. The significance level is .05.

Scatterplot
Dependent Variable: Gun control measures

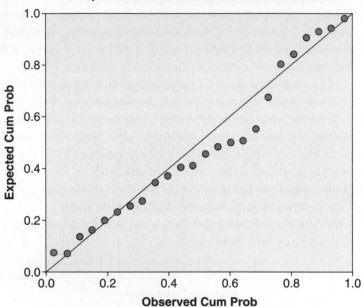

Normal P-P Plot of Regression Standardized Residual
Dependent Variable: Gun control measures

Model Summary[b]

Model	R	R Square	Adjusted R Square	Std. Error of the Estimate
1	.897[a]	.805	.797	.912

a. Predictors: (Constant), Score Constitution

b. Dependent Variable: Gun control measures

ANOVA[a]

Model		Sum of Squares	df	Mean Square	F	Sig.
1	Regression	75.677	1	75.677	91.068	.000[b]
	Residual	18.282	22	.831		
	Total	93.958	23			

a. Dependent Variable: Gun control measures

b. Predictors: (Constant), Score Constitution

Coefficients[a]

Model		Unstandardized Coefficients		Standardized Coefficients	t	Sig.
		B	Std. Error	Beta		
1	(Constant)	12.929	.975		13.255	.000
	Score Constitution	-.121	.013	-.897	-9.543	.000

a. Dependent Variable: Gun control measures

20.3 A deacon at St. Joseph the Worker Church had the theory that attendance at formal church services was a good indicator of the number of hours an individual volunteered. He randomly selected 12 individual volunteers and collected the required information. The deacon wanted to measure the strength and direction of any association. He also wanted a method whereby he might predict the number of hours volunteered by a person who attends church on average four times per month. Since you are an active volunteer and a student of statistics, he asked for your help. You have to select the appropriate statistical technique, write the null and alternative hypotheses, do the analysis, and interpret the results. The deacon's data are shown here.

	churchattend	hrsvolunteer
1	10	16
2	0	9
3	2	4
4	3	6
5	5	10
6	9	11
7	10	16
8	2	2
9	7	5
10	8	10
11	3	7
12	6	10

Answer: The deacon had two scale variables ("number of times of church attendance" and "hours and volunteer work") and wanted to measure their relationship. It was a basic correlation problem, but to make the prediction, we need to apply linear regression. We select single linear regression since there is only one independent variable, the number of times the parishioner attends formal services per month. Using the Kolmogorov-Smirnov test, we determine that both the independent and the dependent variables are normal. We also use SPSS to calculate values that will tell us whether the other assumptions required of the regression model are met. In the *Statistics* window for regression, we click **Estimates**, **Model Fit**, and also **ZPRED** and **ZRESID**. We generate plots for error residuals and homoscedasticity information. The plots indicate that the data are okay for regression. The scatterplot could be better if it showed a more uniform distribution, but it is acceptable for this small sample.

The *Model Summary* shows that 74% of the variance in the number of volunteer hours is accounted for by the number of formal church services attended each month. In the *ANOVA* table, we find an F value of 29.096, which is significant (.000), which tells us that we can have confidence in our model.

The *Coefficients* table gives us the intercept and slope, which are required to make the predictions desired by the deacon. We use the *Transform/Compute Variable* feature to generate predicted values, which can then be compared with the observed data. The equation used to create a new predicted variable when $x = 4$ is $1.396 + (1.257 * 4)$. If we plug in the average church attendance of four times into the equation, we get approximately 6 hours of volunteer work per month.

The null hypothesis that the level of church attendance has no relationship with the number of volunteer hours is rejected. The deacon now has evidence that supports the alternative hypothesis that there is a relationship between his two variables. He also has at his disposal a prediction equation that can be used to predict the number of volunteer hours if church attendance figures can be obtained.

Hypothesis Test Summary

	Null Hypothesis	Test	Sig.	Decision
1	The distribution of Times per month going to church services is normal with mean 5.917 and standard deviation 2.97.	One-Sample Kolmogorov-Smirnov Test	.200[1,2]	Retain the null hypothesis.
2	The distribution of hours per month doing volunteer work is normal with mean 8.833 and standard deviation 4.34.	One-Sample Kolmogorov-Smirnov Test	.200[1,2]	Retain the null hypothesis.

Asymptotic significances are displayed. The significance level is .05.

[1]Lilliefors Corrected

[2]This is a lower bound of the true significance.

Model Summary[b]

Model	R	R Square	Adjusted R Square	Std. Error of the Estimate
1	.859[a]	.738	.711	2.335

a. Predictors: (Constant), Times per month going to church services

b. Dependent Variable: hours per month doing volunteer work

ANOVA[a]

Model		Sum of Squares	df	Mean Square	F	Sig.
1	Regression	153.156	1	153.156	28.096	.000[b]
	Residual	54.511	10	5.451		
	Total	207.667	11			

a. Dependent Variable: hours per month doing volunteer work

b. Predictors: (Constant), Times per month going to church services

Coefficients[a]

Model		Unstandardized Coefficients		Standardized Coefficients	t	Sig.
		B	Std. Error	Beta		
1	(Constant)	1.396	1.557		.896	.391
	Times per month going to church services	1.257	.237	.859	5.301	.000

a. Dependent Variable: hours per month doing volunteer work

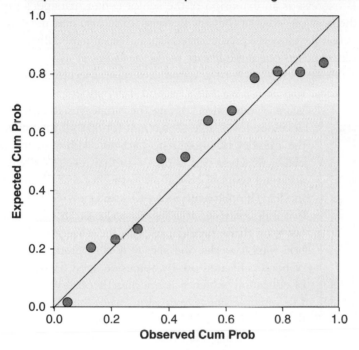

Normal P-P Plot of Regression Standardized Residual
Dependent Variable: hours per month doing volunteer work

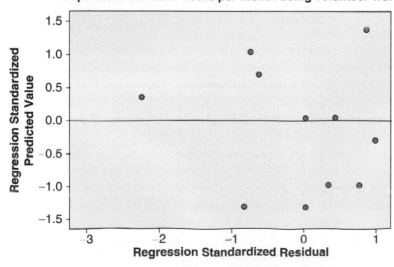

Scatterplot
Dependent Variable: hours per month doing volunteer work

△   Chapter 21: Multiple Linear Regression

21.1 This exercise is an extension of the senior center manager's problem in the previous chapter (Review Exercise 20.1). You may recall that the manager developed a prediction equation that estimated the number of books checked out at the library using the "patrons' age" as the *single independent* variable. For the current exercise, used to illustrate multiple linear regression, we add a second independent variable—"total years of education." Using the single variable, the model developed was able to account for 86% of the variance in the number of books checked out. Although the senior center manager was happy with that result, she wishes to add total years of education in the hope of improving her model. The manager wants to use a new equation (using two independent variables) to make predictions and then compare those predictions with the observed data to see how well it works. She also wishes to predict the number of books checked out by someone aged 63 with 16 years of education, which was not directly observed in her data. Use multiple linear regression, write the null and alternative hypotheses, conduct the analysis, write the prediction equations, make the predictions, and interpret the results. Her data are shown here.

	age	education	books
1	62	12	2
2	67	16	6
3	65	14	4
4	70	22	10
5	66	18	7
6	63	14	4
7	67	16	6
8	65	18	4
9	63	12	2
10	68	22	9
11	65	16	5
12	69	18	7
13	68	16	7
14	62	14	3
15	64	14	4
16	70	18	9
17	63	12	2
18	68	16	6

Answer: If we use multiple linear regression, we must first check the three variables for normality. The Kolmogorov-Smirnov test confirms that they are normally distributed, and we may proceed. We basically follow the same steps taken in single regression. However, we now move both independent variables ("age" and "education") to the *Independent Variables* box. In the *Statistics* window, we check **Estimates** and **Model Fit**. In the *Plot* window, we move *ZPRED* to the *Y* box and *ZRESID* to the *X* box, while checking the box next to *Normal probability plot*. This takes care of the initial output; later there will be more for the prediction part of this exercise.

The *ZPRED* values are close to the diagonal, and the *ZRESID* values are scattered uniformly but with a slight negative skew. The skew is not surprising with the small sample size, and we decide to proceed with the regression analysis. The *Model Summary* table indicates that 93.8% of the variance in the number of books can be accounted for by our model. The *ANOVA* table shows an *F* value of 48.882 (significance level is .000), indicating that our model is statistically sound.

As with our single regression, the *Coefficients* table is needed for the prediction part of this exercise. This table provides the intercept and slopes needed to write the multiple regression equation. As before, the *Transform/Compute Variable* function is used to provide the required predictions. The equation used for all cases is −35.443 + (.876 * age) + (.374 * education). The first four cases of this analysis are shown in the transformed Data View below. The results are very encouraging in that the predicted values for all cases are very close to the observed. The specific case that the manger requested was when someone aged 63 and with 16 years of education was entered into the equation—this resulted in a prediction of four books being checked out in 1 month.

The null hypothesis that age and education are not related to the number of books checked out can be rejected. Thus, we have evidence to support the alternative hypothesis that age and education are related to the number of books checked out at the library. We can advise the senior center manager that her idea is sound and that she has a useful prediction equation.

Hypothesis Test Summary

	Null Hypothesis	Test	Sig.	Decision
1	The distribution of age in years is normal with mean 65.833 and standard deviation 2.66.	One-Sample Kolmogorov-Smirnov Test	.200[1,2]	Retain the null hypothesis.
2	The distribution of total years is normal with mean 16.000 and standard deviation 2.99.	One-Sample Kolmogorov-Smirnov Test	.200[1,2]	Retain the null hypothesis.
3	The distribution of checked out per month is normal with mean 5.389 and standard deviation 2.48.	One-Sample Kolmogorov-Smirnov Test	.200[1,2]	Retain the null hypothesis.

Asymptotic significances are displayed. The significance level is .05.

Model Summary[b]

Model	R	R Square	Adjusted R Square	Std. Error of the Estimate
1	.968[a]	.938	.929	.659

a. Predictors: (Constant), total years, age in years

b. Dependent Variable: checked out per month

ANOVA[a]

Model		Sum of Squares	df	Mean Square	F	Sig.
1	Regression	97.764	2	48.882	112.559	.000[b]
	Residual	6.514	15	.434		
	Total	104.278	17			

a. Dependent Variable: checked out per month

b. Predictors: (Constant), total years, age in years

Coefficients[a]

Model		Unstandardized Coefficients		Standardized Coefficients	t	Sig.
		B	Std. Error	Beta		
1	(Constant)	-35.443	5.495		-6.450	.000
	age in years	.529	.100	.569	5.314	.000
	total years	.374	.089	.451	4.215	.001

a. Dependent Variable: checked out per month

Normal P-P Plot of Regression Standardized Residual
Dependent Variable: checked out per month

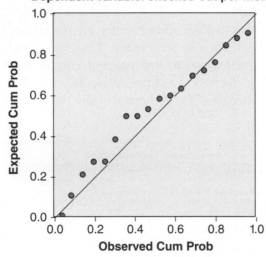

Scatterplot
Dependent Variable: checked out per month

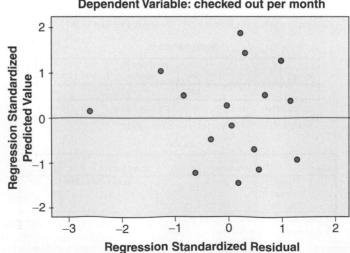

	age	education	books	predbooks
1	62	12	2	2
2	67	16	6	6
3	65	14	4	4
4	70	22	10	10

21.2 This problem is based on the single regression you did in the previous chapter. We just added another variable called "freedom index" to turn it into an example of multiple regression. You now have two independent variables ("constitutional law score" and "freedom index") and one dependent variable that counts the "number of measures introduced" by the legislator.

The political consultant wants to determine if the scores on knowledge of constitutional law and score of the freedom index are related to the number of gun control laws introduced. He also wishes to extend any findings into the realm of prediction by using it to estimate the number of measures introduced by a legislator rated average on both these independent variables. He also wishes to use the equation to predict for his data, which will permit him to examine the equation's performance when the predicted values are directly compared with the observed values. Use multiple linear regression, write the null and alternative hypotheses, conduct the analysis, write the prediction equations, make the predictions, and interpret the results. His data are shown here.

	const_score	gun_control	freeindex
1	98	1	28
2	86	2	23
3	74	3	26
4	63	4	24
5	51	6	14
6	97	1	25
7	85	2	21
8	77	5	23
9	65	6	18
10	53	8	12
11	94	2	21
12	83	2	26
13	74	4	15
14	69	4	19
15	55	6	10
16	97	2	21
17	84	4	21
18	71	4	18
19	64	5	17
20	57	8	18
21	99	1	30
22	82	3	23
23	75	4	14
24	63	4	28

Answer: Checking the three variables using the Kolmogorov-Smirnov test, we find three normal distributions. The *Model Summary* indicates that 83.7% of the variance in the number of gun control measures introduced can be accounted for by the two independent variables. The *ANOVA* table reports an *F* value of 53.865, which is significant at .000. This indicates that our overall model is significant and that the amount of variance accounted for (83.7%) can be taken seriously. The normal P-P plot of standardized residuals could be better, but it is within tolerance when taking into consideration our sample size. The plot for homogeneity of variances is fine, so we have additional confidence of our model.

We next turn to the *Coefficients* table for the values needed for our prediction equation. The intercept is 13.243, the slope for "constitutional law score" is −0.101, and the slope for "freedom index" is −0.088. We insert these values into the *Compute Variable* window as 13.243 + (−.101 * const_score) + (−.088 * freeindex). Once **OK** is clicked, the new variable (named "predgunlaw") is automatically inserted into your database. The predicted values can now be directly compared, which gives you immediate insight into the value of the equation. Next, we can insert the value of particular interest to the political consultant, which is the value of the dependent variable for a legislator rated average on both the constitutional law score and freedom index measures. We do a descriptive analysis and find 75.67 for constitutional law score and 20.63 for freedom index. Plugging these into the equation, we can either do it with a hand calculator or use SPSS to arrive at four gun control measures that are predicted to be introduced by an average-rated legislator.

The null hypothesis that the *constitutional law score* and *freedom index* are not related to the *number of gun control laws introduced* is rejected. There is evidence in support of the alternative hypothesis that these variables are related. The political consultant has a useful prediction equation that can predict the number of gun control measures that will be introduced if the legislator's score on constitutional law and freedom index are known.

Hypothesis Test Summary

	Null Hypothesis	Test	Sig.	Decision
1	The distribution of State BA rating on knowledge of constitutional law is normal with mean 75.667 and standard deviation 15.02.	One-Sample Kolmogorov-Smirnov Test	.200[1,2]	Retain the null hypothesis.
2	The distribution of Gun control measures is normal with mean 3.792 and standard deviation 2.02.	One-Sample Kolmogorov-Smirnov Test	.200[1,2]	Retain the null hypothesis.
3	The distribution of Freedom index (0 to 30) is normal with mean 20.625 and standard deviation 5.30.	One-Sample Kolmogorov-Smirnov Test	.200[1,2]	Retain the null hypothesis.

Asymptotic significances are displayed. The significance level is .05.

Model Summary[b]

Model	R	R Square	Adjusted R Square	Std. Error of the Estimate
1	.915[a]	.837	.821	.854

a. Predictors: (Constant), Freedom index (0 to 30), State BA rating on knowledge of constitutional law

b. Dependent Variable: Gun control measures

ANOVA[a]

Model		Sum of Squares	df	Mean Square	F	Sig.
1	Regression	78.631	2	39.315	53.865	.000[b]
	Residual	15.328	21	.730		
	Total	93.958	23			

a. Dependent Variable: Gun control measures

b. Predictors: (Constant), Freedom index (0 to 30), State BA rating on knowledge of constitutional law

Coefficients[a]

Model		Unstandardized Coefficients		Standardized Coefficients	t	Sig.
		B	Std. Error	Beta		
1	(Constant)	13.243	.927		14.280	.000
	State BA rating on knowledge of constitutional law	-.101	.015	-.750	-6.545	.000
	Freedom index (0 to 30)	-.088	.044	-.231	-2.012	.057

a. Dependent Variable: Gun control measures

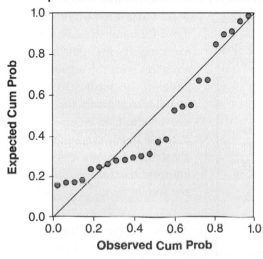

Normal P-P Plot of Regression Standardized Residual
Dependent Variable: Gun control measures

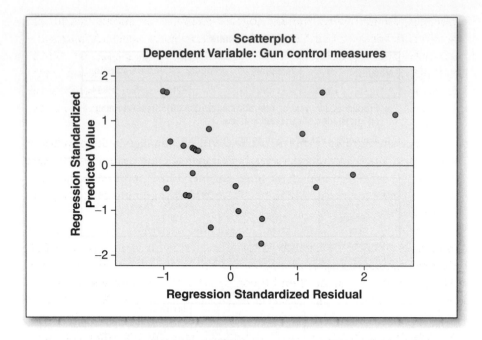

Scatterplot
Dependent Variable: Gun control measures

21.3 As we have done in the previous two exercises, we bring forward from the previous chapter a single linear regression problem and add an additional variable. In that exercise, you had one independent variable, which was "the number of times an individual attended church during a month." For this current exercise, you will add another independent variable, which is "the number of times one prays in a day." The deacon of the church wants to see if the earlier prediction equation could be improved by adding this additional variable. As before, he wants to compare the performance of the new equation with the actual observed values. In addition, he wishes to predict the number of volunteer hours for those rated as average on the two independent variables. Use multiple linear regression, write the null and alternative hypotheses, do the analysis, and interpret the results. The deacon's new data are shown here.

Answer: Use multiple linear regression since you have two independent variables. Check all data, independent and dependent

	churchattend	pray	hrsvolunteer
1	10	6	16
2	6	5	9
3	2	1	4
4	3	2	6
5	5	5	10
6	9	6	11
7	10	7	16
8	2	2	2
9	7	3	5
10	8	6	10
11	3	2	7
12	6	4	10

variables, for normality. The Kolmogorov-Smirnov test shows that all are normally distributed. In the regression statistics window, click *Estimates* and *Model Fit*, and in the *Plots* window, move *ZPRED* to the *Y* box and *ZRESID* to the *X* box. These will generate output that further informs you whether regression is appropriate for your data. We find that the P-P plot is just acceptable and the scatterplot excellent; therefore, we proceed with our regression analysis.

The *Model Summary* shows that 79.6% of the variance in the number of volunteer hours is accounted for by the number of church services attended each month and the number of times an individual prays each day. The *ANOVA* table shows an *F* value of 17.54, which is significant (.001), and informs us that the overall model is good.

The *Coefficients* table shows an intercept of 0.874 and a slope for "church attendance per month" of 0.478 and for "prayers per day" of 1.257. We next use the *Transform/Compute Variable* feature to calculate the predicted values, which can then be compared with the observed data. The equation inserted in the *Numeric Expression* window is .874 + (.478 * churchattend) + (1.257 * pray). This gives us the predictions, which can be compared with the actual observations. Next, we see what the equation predicts for the average values of six ("church attendance per month") and four ("prayers per day"). The equation now becomes .874 + (.478 * 6) + (1.257 * 4) and predicts that such a person would be expected to volunteer for 9 hours per month.

The null hypothesis that the level of church attendance and number of times one prays daily has no relationship to the number of volunteer hours is rejected. The deacon now has evidence that supports the alternative hypothesis that there is a relationship between these variables.

Hypothesis Test Summary

	Null Hypothesis	Test	Sig.	Decision
1	The distribution of Times per month going to church services is normal with mean 5.917 and standard deviation 2.97.	One-Sample Kolmogorov-Smirnov Test	.200[1,2]	Retain the null hypothesis.
2	The distribution of self reported prayer times per day is normal with mean 4.083 and standard deviation 2.02	One-Sample Kolmogorov-Smirnov Test	.200[1,2]	Retain the null hypothesis.
3	The distribution of hours per month doing volunteer work is normal with mean 8.833 and standard deviation 4.34.	One-Sample Kolmogorov-Smirnov Test	.200[1,2]	Retain the null hypothesis.

Asymptotic significances are displayed. The significance level is .05.

Model Summary[b]

Model	R	R Square	Adjusted R Square	Std. Error of the Estimate
1	.892[a]	.796	.750	2.171

a. Predictors: (Constant), self reported prayer times per day, Times per month going to church services

b. Dependent Variable: hours per month doing volunteer work

ANOVA[a]

Model		Sum of Squares	df	Mean Square	F	Sig.
1	Regression	165.266	2	82.633	17.540	.001[b]
	Residual	42.401	9	4.711		
	Total	207.667	11			

a. Dependent Variable: hours per month doing volunteer work

b. Predictors: (Constant), self reported prayer times per day, Times per month going to church services

Coefficients[a]

Model		Unstandardized Coefficients		Standardized Coefficients		
		B	Std. Error	Beta	t	Sig.
1	(Constant)	.874	1.483		.589	.570
	Times per month going to church services	.478	.534	.327	.896	.394
	self reported prayer times per day	1.257	.784	.584	1.603	.143

a. Dependent Variable: hours per month doing volunteer work

Normal P-P Plot of Regression Standardized Residual
Dependent Variable: hours per month doing volunteer work

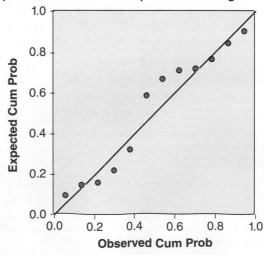

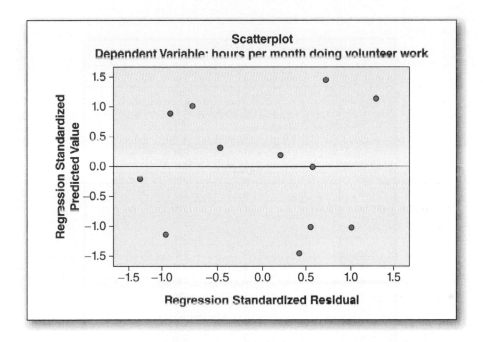

CHAPTER 22: LOGISTIC REGRESSION △

22.1 A major in the Air Force wanted to find a way to predict whether a particular airman would be promoted to sergeant within 4 years of enlisting in the military. He had data on many characteristics of individuals prior to enlistment. He chose three variables that he thought might be useful in determining whether they would get the early promotion. They are listed in the table showing the SPSS Variable View below. He selected a random sample of 30 individuals and compiled the required information. Your task is to develop a prediction equation that might assist the major in efforts to predict early promotion for his young airman. Write the research question(s) and the null and alternative hypotheses. The major's variable information and data are as follows:

	Name	Type	Width	De...	Label	Values	Missing		Align	Measure	Role
1	sports	Numeric	8	0	HS contact sports	{0, no}..	N	0=no 1=yes	ter	Nominal	Input
2	hunt	Numeric	8	0	hunting license	{0, no}..	None	0=no 1=yes	r	Nominal	Input
3	test	Numeric	8	0	induction test	None	None		Center	Scale	Input
4	sgt1	Numeric	8	0	4 years to sergeant	{0, no}..	None	8	Center	Scale	Input

	sports	hunt	test	sgt1
1	0	1	210	0
2	0	0	210	0
3	0	1	280	1
4	0	0	210	0
5	0	1	280	0
6	1	0	210	0
7	1	1	290	1
8	0	0	280	0
9	1	0	210	1
10	0	0	220	0
11	1	1	270	1
12	1	0	210	0
13	1	1	290	1
14	1	0	290	0
15	1	0	210	0
16	0	0	220	0
17	1	1	280	1
18	0	0	220	0
19	0	0	290	0
20	0	0	220	0
21	0	1	260	0
22	0	0	210	0
23	0	1	250	0
24	1	1	270	1
25	1	0	290	0
26	0	0	270	0
27	1	1	220	0
28	0	0	270	0
29	1	0	220	0
30	0	1	280	0

Answer: You can identify this problem as requiring a regression method since we are attempting to make a prediction based on observed data. Since we have a discrete dependent variable ("promoted to sergeant in 4 years") with two categories (*yes* or *no*), we should immediately think of binary logistic regression. Levels of measurement for our independent variables can be either discrete or continuous. Okay, now we have the selected statistical approach—we next write the research questions and the null and alternative hypotheses. *Questions:* (1) Can we predict whether an airman will be promoted to sergeant within 4 years given the pre-enlistment data? (2) If we can accurately predict the outcome, then which of the variables are the most important? The null hypothesis is the opposite of the research questions and would state that knowledge of the pre-enlistment variables would be of no help in predicting early promotion to sergeant.

First, let's check the data for multicollinearity. You want lower correlations between independent variables and higher correlations between the independent and dependent variables. The pattern observed is acceptable, and we make the decision to proceed.

The tables under *Block 0: Beginning Block* show the prediction of the dependent variable with none of our selected variables in the equation at 76.7% (*Classification Table*). The *Variables in the Equation* table shows significance under these same conditions. The results shown in the *Omnibus Tests of Model Coefficients* under *Block 1: Method = Enter* are much more interesting, and since significance is .001, we now have evidence that we have a good fit. The *Model Summary* indicates that between 43% and 64.9% of the variance in our dependent variable can be accounted for by the model. The *Hosmer-Limeshow* test provides additional support for the model since we fail to reject the null. The chi-square value of 5.677 at 8 degrees of freedom and a significance level of .683 is evidence that the model is worthwhile.

The new *Classification Table* indicates that the predictive power is 90%, an increase of 13.3% from the original *Classification Table*. This percentage increase gives us more confidence that our equation will have the ability to actually make predictions. The final table, *Variables in the Equation*, adds more evidence that our model can be useful in predicting the binary outcome. The most desirous result would be for all variables to be significant. However, the one variable ("test") showing *no significance* at .340 could be easily removed from the equation with no negative impact. Try it, and see what happens!

We can now answer the research question and say that we can write the prediction equation. The two variables that are significant are "sports activity in high school" and "whether the airman had a hunting license in high school." The performance on the induction test does not have a positive effect on the equation's performance in predicting early promotion to sergeant.

Correlations

			HS contact sports	hunting license	induction test	4 years to sergeant
Spearman's rho	HS contact sports	Correlation Coefficient	1.000	.110	.091	.472[**]
		Sig. (2-tailed)	.	.563	.632	.008
		N	30	30	30	30
	hunting license	Correlation Coefficient	.110	1.000	.369[*]	.515[**]
		Sig. (2-tailed)	.563	.	.045	.004
		N	30	30	30	30
	induction test	Correlation Coefficient	.091	.369[*]	1.000	.325
		Sig. (2-tailed)	.632	.045	.	.080
		N	30	30	30	30
	4 years to sergeant	Correlation Coefficient	.472[**]	.515[**]	.325	1.000
		Sig. (2-tailed)	.008	.004	.080	.
		N	30	30	30	30

**. Correlation is significant at the 0.01 level (2-tailed).

*. Correlation is significant at the 0.05 level (2-tailed).

Classification Table[a,b]

Prediction with no idependent variable

	Observed		no	yes	Percentage Correct
Step 0	4 years to sergeant	no	23	0	100.0
		yes	7	0	.0
	Overall Percentage				76.7

a. Constant is included in the model.

Variables in the Equation

		B	S.E.	Wald	df	Sig.	Exp(B)
Step 0	Constant	-1.190	.432	7.594	1	.006	.304

Block 1: Method = Enter

Omnibus Tests of Model Coefficients

		Chi-square	df	Sig.
Step 1	Step	16.858	3	.001
	Block	16.858	3	.001
	Model	16.858	3	.001

Model Summary

Step	-2 Log likelihood	Cox & Snell R Square	Nagelkerke R Square
1	15.738[a]	.430	.649

a. Estimation terminated at iteration number 7 because parameter estimates changed by less than .001.

Hosmer and Lemeshow Test

Step	Chi-square	df	Sig.
1	5.677	8	.683

Classification Table[a]

Prediction with independent variables

	Observed		no	yes	Percentage Correct
Step 1	4 years to sergeant	no	22	1	95.7
		yes	2	5	71.4
	Overall Percentage				90.0

a. The cut value is .500

Variables in the Equation

		B	S.E.	Wald	df	Sig.	Exp(B)
Step 1[a]	sports(1)	3.345	1.514	4.883	1	.027	28.360
	hunt(1)	3.110	1.527	4.150	1	.042	22.426
	test	.021	.022	.910	1	.340	1.021
	Constant	-10.371	6.035	2.954	1	.086	.000

a. Variable(s) entered on step 1: sports, hunt, test.

22.2 A social scientist wanted to develop an equation that would predict whether a male student would be successful in getting a date for the senior prom. The scientist had access to many student records and took a random sample of 40 students. She choose four characteristics that she felt would predict whether a male would get a date or not—a binary outcome. These variables are shown below in the SPSS Variable View. The *Label* column shows the description of the variable. Your job is to select the correct statistical approach and then assist the social scientist in developing the equation. Write the research question(s) and the null and alternative hypotheses. The variable information and data are as follows:

	Name	Type	Width	Decimals	Label	Values	Missing	Columns	Align	Measure	Role
1	work	Numeric	8	0	have personal income	{0, no}...	None	8	Center	Nominal	Input
2	height	Numeric	8	0	taller than 5'8"	{0, no}...	None	8	Center	Nominal	Input
3	grade	Numeric	8	0	GPA >3.5	{0, no}	None	8	Center	Nominal	Input
4	activities	Numeric	8	0	3 or more	{0, no}...	None	8	Center	Nominal	Input
5	date	Numeric	8	0	date for prom	{0, no}...	None	8	Center	Nominal	Input

	work	height	grade	activities	date
1	0	1	0	0	0
2	0	0	0	1	0
3	0	1	1	0	1
4	0	0	0	0	0
5	0	1	0	0	0
6	1	0	0	0	0
7	1	1	0	0	1
8	0	0	1	0	0
9	1	0	0	1	1
10	0	0	0	0	0
11	1	1	1	0	1
12	0	0	0	0	0
13	1	0	0	1	1
14	0	0	1	0	0
15	1	0	0	0	0
16	0	0	0	0	0
17	1	1	1	0	1
18	0	0	0	0	0
19	0	0	0	1	0
20	0	0	0	0	0

21	1	0	1	0	0
22	0	0	0	0	0
23	0	1	0	0	0
24	1	1	1	0	1
25	0	1	0	0	0
26	0	0	1	0	0
27	0	1	0	1	0
28	0	0	0	0	0
29	0	1	0	0	0
30	1	0	0	0	0
31	0	1	0	0	0
32	0	1	0	0	0
33	1	0	0	0	0
34	0	0	0	0	0
35	1	1	1	0	0
36	0	1	1	0	1
37	1	0	0	1	0
38	0	0	1	0	0
39	1	1	0	0	0
40	0	0	0	1	0

Answer: The tip-off as to the type of statistical method needed is that you need a prediction equation—hence, regression is required. Since you have a discrete dependent variable that has only two possible outcomes (getting a date or not), you should select binary logistic regression. For this method, you can have a mix of discrete and continuous independent variables. There are really just two research questions or alternative hypotheses (H_A): (1) Can you predict whether the high school student will get a date for the prom? (2) If we can develop a reliable prediction equation, which variables are the most important? The null hypothesis (H_0) is the opposite and states that knowledge of the variables would not help in predicting whether someone gets a date or not.

Before using the logistic approach, we check the variables for multicollinearity. We generate a bivariate correlation table to see how the independent variables correlate with one another (we want low correlations) and then separately with the dependent variable (high). The pattern observed is passable, and we decide to proceed with the analysis.

The first *Classification Table* shows an 80% prediction with no variables. The *Variables in the Equation* table shows a Wald statistic of 12.3 and significance of .000 under these same conditions. The *Omnibus Tests of Model Coefficients* under *Block 1: Method = Enter* shows a significance of .000, which provides the first evidence of a good fit for our model. The *Model Summary* shows that between 39.5% and 62.4% of the variance in the dependent variable (getting a date or not) can be accounted for by the model. The Hosmer-Limeshow test provides additional support for the model since we fail to reject the null. The chi-square value of 3.775 at 6 degrees of freedom and a significance level of .707 informs us that the model is worthwhile.

The *Classification Table* indicates that the predictive power is now 82.5%, which represents an increase over the original table. *Variables in the Equation* adds more evidence that our model can be useful in predicting the binary outcome. In this table, we see the most desirous result in that all variables are found to be significant. Thus, we have evidence that all our variables do contribute to the model for predicting the outcome.

To answer our two research questions, we can say that we developed a prediction equation that made it possible to predict whether the high school student would get a date for the prom. Furthermore, we found that the variables had a positive significant impact on our equation.

Correlations

			have personal income	taller than 5'8"	GPA >3.5	3 or more	date for prom
Spearman's rho	have personal income	Correlation Coefficient	1.000	.043	.135	.076	.419[**]
		Sig. (2-tailed)	.	.793	.406	.642	.007
		N	40	40	40	40	40
	taller than 5'8"	Correlation Coefficient	.043	1.000	.183	-.242	.357[*]
		Sig. (2-tailed)	.793	.	.259	.133	.024
		N	40	40	40	40	40
	GPA >3.5	Correlation Coefficient	.135	.183	1.000	-.284	.392[*]
		Sig. (2-tailed)	.406	.259	.	.076	.012
		N	40	40	40	40	40
	3 or more	Correlation Coefficient	.076	-.242	-.284	1.000	.099
		Sig. (2-tailed)	.642	.133	.076	.	.545
		N	40	40	40	40	40
	date for prom	Correlation Coefficient	.419[**]	.357[*]	.392	.099	1.000
		Sig. (2-tailed)	.007	.024	.012	.545	.
		N	40	40	40	40	40

**. Correlation is significant at the 0.01 level (2-tailed).

*. Correlation is significant at the 0.05 level (2-tailed).

Block 0: Beginning Block

Classification Table[a,b]

			Predicted		
			date for prom		Percentage
	Observed		no	yes	Correct
Step 0	date for prom	no	32	0	100.0
		yes	8	0	.0
	Overall Percentage				80.0

a. Constant is included in the model.

Variables in the Equation

		B	S.E.	Wald	df	Sig.	Exp(B)
Step 0	Constant	-1.386	.395	12.300	1	.000	.250

Block 1: Method = Enter

Omnibus Tests of Model Coefficients

		Chi-square	df	Sig.
Step 1	Step	20.081	4	.000
	Block	20.081	4	.000
	Model	20.081	4	.000

Model Summary

Step	-2 Log likelihood	Cox & Snell R Square	Nagelkerke R Square
1	19.952[a]	.395	.624

a. Estimation terminated at iteration number 7 because parameter estimates changed by less than .001.

Hosmer and Lemeshow Test

Step	Chi-square	df	Sig.
1	3.775	6	.707

Classification Table[a]

			Predicted		
			date for prom		Percentage
	Observed		no	yes	Correct
Step 1	date for prom	no	30	2	93.8
		yes	5	3	37.5
	Overall Percentage				82.5

Variables in the Equation

		B	S.E.	Wald	df	Sig.	Exp(B)
Step 1[a]	work(1)	2.592	1.280	4.101	1	.043	13.353
	height(1)	3.112	1.498	4.318	1	.038	22.469
	grade(1)	3.084	1.535	4.038	1	.044	21.839
	activities(1)	3.542	1.779	3.963	1	.047	34.540
	Constant	6.424	2.149	8.939	1	.003	.002

a. Variable(s) entered on step 1: work, height, grade, activities.

22.3 For this review exercise, you will use the SPSS sample file titled *customer_dbase.sav*. You are a statistical consult with a contract to help a phone company executive develop a way to predict whether a customer would order the paging service. Based on prior experience, the executive feels that customers using voice mail ("voice"), caller ID ("callid") and electronic billing ("ebill") would also be inclined to utilize the paging service ("pager"). He is seeking statistical evidence and a written equation to support his intuitive feeling. He also wishes to utilize any equation that may result from the analysis to predict for future customers. Select the appropriate statistical method, open the database, select the variables, do the analysis, and then interpret the results.

Answer: You will need to develop a prediction equation utilizing the method that accommodates categorical data. The executive wants to predict whether someone will purchase the paging service or not—a categorical outcome. And to be more specific, it is a binary outcome. The logical choice for analysis is binary logistic regression.

The two research questions or alternative hypotheses (H_A) are as follows: (1) Can you predict whether the customer will purchase the paging service? (2) If a reliable prediction equation can be developed, which of the selected variables are the most important? The null hypothesis (H_0) is the opposite and states that the variables would be of no use in predicting whether someone purchases the paging service.

We first check the selected variables for multicollinearity. We request a Spearman bivariate correlation table to see how the independent variables correlate with one another. We want these to have low correlation coefficients. The ideal would be for the independent variables to then have high coefficients with the dependent variable. The observed pattern is not perfect, but we believe it will not have a negative impact on the development of an equation, so we proceed with logistic regression.

The first *Classification Table* shows a 75.6% prediction with no variables. The *Variables in the Equation* table shows a Wald statistic of 12.3 and significance of .000 under these same conditions.

The *Omnibus Tests of Model Coefficients* under *Block 1: Method = Enter* shows a significance of .000, providing evidence that the model fit is good. The *Model Summary* shows that between 35.7% and 53.2% of the variance in the dependent variable (pager service purchase or not) can be accounted for by the model. The Hosmer-Limeshow test provides additional support for the model, since we fail to reject the null. The chi-square value of 6.624 at 4 degrees of freedom and a significance level of .157 informs us that the model is worthwhile.

The *Classification Table* indicates that the predictive power has increased to 85.1%, up from 75.6% in the original table. *Variables in the Equation* adds more evidence that our model can be useful in predicting the binary outcome. All the variables are found to contribute significantly to the model. This result gives us additional confidence in our predictive model.

To answer our two research questions, we can say that we developed a prediction equation that made it possible to predict whether an average customer would purchase the paging service or not. The second question was also answered in that we found that all the variables had a positive impact on our equation.

Correlations

			Voice mail	Caller ID	Electronic billing	Paging service
Spearman's rho	Voice mail	Correlation Coefficient	1.000	.391**	.368**	.606**
		Sig. (2-tailed)	.	.000	.000	.000
		N	5000	5000	5000	5000
	Caller ID	Correlation Coefficient	.391**	1.000	-.007	.385**
		Sig. (2-tailed)	.000	.	.635	.000
		N	5000	5000	5000	5000
	Electronic billing	Correlation Coefficient	.368**	-.007	1.000	.349**
		Sig. (2-tailed)	.000	.635	.	.000
		N	5000	5000	5000	5000
	Paging service	Correlation Coefficient	.606**	.385**	.349**	1.000
		Sig. (2-tailed)	.000	.000	.000	.
		N	5000	5000	5000	5000

**. Correlation is significant at the 0.01 level (2-tailed).

Classification Table[a,b]

			Predicted		
			Paging service		Percentage Correct
Observed			No	Yes	
Step 0	Paging service	No	3778	0	100.0
		Yes	1222	0	.0
	Overall Percentage				75.6

a. Constant is included in the model.

Variables in the Equation

		B	S.E.	Wald	df	Sig.	Exp(B)
Step 0	Constant	-1.129	.033	1176.318	1	.000	.323

Block 1: Method = Enter

Omnibus Tests of Model Coefficients

		Chi-square	df	Sig.
Step 1	Step	2209.234	3	.000
	Block	2209.234	3	.000
	Model	2209.234	3	.000

Model Summary

Step	-2 Log likelihood	Cox & Snell R Square	Nagelkerke R Square
1	3351.755[a]	.357	.532

a. Estimation terminated at iteration number 6 because parameter estimates changed by less than .001.

Hosmer and Lemeshow Test

Step	Chi-square	df	Sig.
1	6.624	4	.157

Classification Table[a]

			Predicted		
			Paging service		Percentage Correct
Observed			No	Yes	
Step 1	Paging service	No	3424	354	90.6
		Yes	393	829	67.8
Overall Percentage					85.1

Variables in the Equation

		B	S.E.	Wald	df	Sig.	Exp(B)
Step 1[a]	voice(1)	2.317	.092	634.093	1	.000	10.147
	callid(1)	1.728	.103	281.421	1	.000	5.631
	ebill(1)	1.421	.096	220.270	1	.000	4.141
	Constant	-3.875	.110	1250.924	1	.000	.021

a. Variable(s) entered on step 1: voice, callid, ebill.

Chapter 23: Factor Analysis △

23.1 For this exercise, you will use the SPSS sample file called *customer_dbase.sav*. This database is composed of 5,000 cases and 132 variables. You will select the first 10 *scale* variables and search for underlying *latent* variables (*factors*) within these variables. The idea is that you must explore the data in an attempt to reduce the 10 variables into smaller *clusters*, referred to as *components* in *principle component factor analysis*. Write the null and alternative hypotheses, open the database, select the correct statistical approach, and search for any underlying latent factors in the first 10 scale variables.

Answer: Before getting into the actual analysis, we write the null and alternative hypotheses. The null hypothesis (H_0) is that there are no underlying structures and that all the variables load equally. The alternative hypothesis (H_A) is that there are one or more underlying components (factors) and the variables do not load equally.

Go to *Analyze/Dimension reduction/Factor*, and then select and move the 10 variables ("Age in years" through "Spouse years of education") for analysis; then open the *Descriptives* window and check **Univariate**, **Initial**, **Coefficients**, **Significance**, and **KMO**. In *Extraction*, click **Unrotated** and **Scree**. You have now generated all the output you will need to determine if you have evidence in support of the alternative hypothesis. Next, we interpret each generated table to learn more about the 10 variables.

The table titled *Correlation Matrix* is not shown, but there are a number of variables that have correlations below the recommended .3, so we decide to proceed. Remember that high correlations are best when doing factor analysis. Two additional tests presented in the next table give us evidence supportive of efforts to uncover the underlying factors. The KMO shows a value of .698, which is good since any value greater than .6 is considered satisfactory. Bartlett's test shows significance at .000, which is also a positive indicator that any identified factors can be taken seriously.

The table titled *Communalities* gives the proportion of variability in the original variable that is accounted for by the high-loading factors. The value of .934 means that 93.4% of the variance in the variable "Debt" is accounted for by the high-loading factors, which are those with eigenvalues >1. The high-loading factors are numbers 1, 2, and 3 in the next table.

The *Total Variance Explained* table shows the eigenvalues for the 10 new factors. Look at the column called *Initial Eigenvalues*, and notice the value of 4.382 for the first factor. It means that 43.82% of the total variance is accounted for by this first factor. Factor 2 shows an eigenvalue of 1.696, meaning that it accounts for 16.96% of the total variance for all variables. This percentage is not related to the variance of the first component; therefore, the two taken together (43.82 + 16.96) can be said to account for 60.78% of the variance for all variables.

The scree plot shows three factors with eigenvalues greater than 1.0. These three factors account for 73.317% of the variance for all variables. We can say that the scree plot provides additional support for a two- or three-factor solution to our factor analysis problem.

The final table is called *Component Matrix*, and it shows the factor-loading values for factors with eigenvalues of 1.0 or more. These values are interpreted by any correlation coefficient. Zero indicates no loading, while minus values, such as −.174 for Component 2, indicate that as the particular variable score increases the component score decreases. Those values with a plus sign, such as .194 for Component 1, indicate that as the variable score increases, so does the component score.

The *Component Matrix* can be very useful when the data analyst is attempting to name the newly discovered latent variable (factor). For instance, look at Factor 1, where we see the highest loadings on "debt" and "income." This is suggestive of name dealing with financial conditions within a particular household. Factor 2 suggests a name associated with education. Factor 3 should probably be dropped.

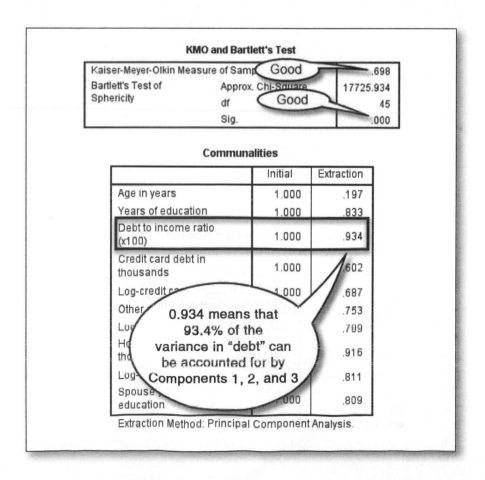

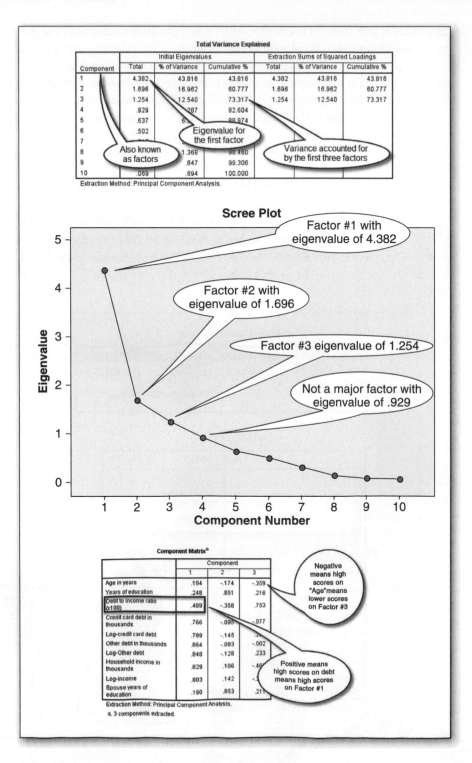

23.2 This review exercise uses the same SPSS sample file database as in the previous example (*customer_dbase.sav*). However, this time you will examine eight scale variables dealing with pet ownership (#29 through #36). You are to look for underlying latent variables that would permit the reduction of these eight variables. Write the null and alternative hypotheses, select the correct procedures, examine the initial statistics, and interpret the results.

Answer: The null hypothesis (H_0) is that there are no underlying structures and that all the variables load equally. The alternative hypothesis (H_A) is that there are one or more underlying components (factors) and the variables do not load equally.

Open *customer_dbase.sav*, and go to *Analyze/Dimension reduction/Factor*; then select and move the eight variables. In the *Descriptives* window, check **Univariate**, **Initial**, **Coefficients**, **Significance**, and **KMO**. In *Extraction*, click **Unrotated** and **Scree**. Click **OK**, and you generate several tables that are ready for your interpretation.

After looking at the very low correlation coefficients in the initial *Correlation Matrix*, you see that there is no hope for a successful solution of this factor analysis problem. However, in this example, we proceed to explain the following tables, which will point out the results of this initial data screening. Just keep in mind that you would be justified to just stop at this point. We could say that the initial analysis indicates that the null cannot be rejected and there is no evidence to support the alternative.

You may have noticed that the KMO and Bartlett's test are not reported, another indication that there was no underlying factor for these eight variables. The *Total Variance Explained* table indicates very little difference in the eigenvalues for all variables. The scree plot is also atypical for a successful solution. A scree plot showing THE underlying components (factors) would drop to some elbow point and then level off with steadily diminishing eigenvalues. The table for *Component Matrix* clearly shows that there are no unique factor loadings and that all factors basically measure the same quantity as the original variable.

Correlation Matrix[a]

		Number of pets owned	Number of cats owned	Number of dogs owned	Number of birds owned	Number of reptiles owned	Number of small animals owned	Number of saltwater fish owned	Number of freshwater fish owned
Correlation	Number of pets owned	1.000	.252	.225	.129	.104	.187	.134	.899
	Number of cats owned	.252	1.000	.011	-.013	-.004	-.006	-.018	.007
	Number of dogs owned	.225	.011	1.000	.002	.017	.020	.007	-.015
	Number of birds owned	.129	-.013	.002	1.000	-.006	-.008	.001	-.011
	Number of reptiles owned	.104	-.004	.017	-.006	1.000	-.021	-.003	.011
	Number of small animals owned	.187	-.006	.020	-.008	-.021	1.000	.001	.022
	Number of saltwater fish owned	.134	-.018	.007	.001	-.003	.001	1.000	-.003
	Number of freshwater fish owned	.899	.007	-.015	-.011	.011	.022	-.003	1.000

a. This matrix is not positive definite.

Total Variance Explained

Component	Initial Eigenvalues			Extraction Sums of Squared Loadings		
	Total	% of Variance	Cumulative %	Total	% of Variance	Cumulative %
1	2.004	25.044	25.044	2.004	25.044	25.044
2	1.024	12.803	37.847	1.024	12.803	37.847
3	1.021	12.757	50.604	1.021	12.757	50.604
4	1.017	12.709	63.313	1.017	12.709	63.313
5	1.002	12.531	75.845	1.002	12.531	75.845
6	.989	12.357	88.202			
7	.944	11.798	100.000			
8	.000	.000	100.000			

Extraction Method: Principal Component Analysis.

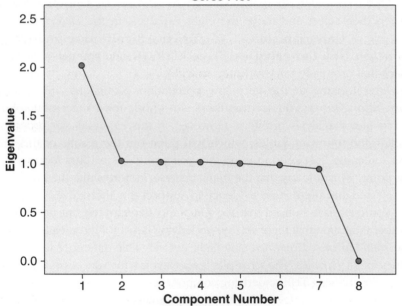

Scree Plot

Component Matrix[a]

	Component				
	1	2	3	4	5
Number of pets owned	1.000	.004	.009	-.012	.015
Number of cats owned	.254	-.532	-.117	.393	.288
Number of dogs owned	.220	.063	.617	.601	.080
Number of birds owned	.114	.295	.246	-.294	.836
Number of reptiles owned	.112	-.416	.588	-.236	-.290
Number of small animals owned	.206	.461	-.237	.508	-.129
Number of saltwater fish owned	.128	.514	.336	-.161	-.318
Number of freshwater fish owned	.898	-.033	-.225	-.273	-.113

Extraction Method: Principal Component Analysis.

a. 5 components extracted.

23.3 For this review exercise, you will use the SPSS sample file titled *telco. sav*. This database has 1,000 cases and 22 variables measured at the *scale* level. You will select the first 16 of the scale variables (up to *wireten/wireless over tenure*, #25) and attempt to identify any underlying factor(s) that would permit data reduction. State the null and alternative hypotheses, select the statistical method, and proceed with the analysis.

Answer: We begin by writing the null and alternative hypotheses. The null hypothesis (H_0) is that there are no underlying structures and that all the variables load equally. The alternative hypothesis (H_A) is that there are one or more underlying components (factors) and the variables do not load equally.

Go to *Analyze/Dimension reduction/Factor*, and then select and move the 16 variables for analysis; then open the *Descriptives* window, and check **Univariate**, **Initial**, **Coefficients**, **Significance**, and **KMO**. In *Extraction*, click **Unrotated** and **Scree**. Click **OK**, and you have all the output needed to answer the questions regarding the underlying factors.

The *Correlation Matrix* does show many correlations exceeding the recommended .3 level, so we proceed with the analysis of the output. The KMO value of .654 and Bartlett's significance of .00 are also evidence that factor analysis was worthwhile.

The *Communalities* table indicated the proportion of variability in the original variable that was accounted for by the high-loading factors (1, 2, 3, and 4). These high-loading *factors* are numbers 1, 2, 3, and 4 in the next table.

The *Total Variance Explained* table shows the eigenvalues for the new factors. Look at the column called *Initial Eigenvalues*, and notice the value of 5.226 for the first factor. This value represents the fact that 37.327% of the total variance is accounted for by this first factor. The first four major factors (those with eigenvalues >1.0) account for 76.365% of the total variance for all variables.

The scree plot shows four factors with eigenvalues greater than 1.0. These four factors account for 76.365% of the variance for all variables. We can say that the scree plot provides additional support for a two-, three-, or four-factor solution.

The *Component Matrix* shows the factor-loading values for the four factors having eigenvalues of 1.0 or more. We interpret these values, as we do any correlation coefficient. We can readily see strong correlations for Factors 1 and 2, which would argue for a two- or three-factor solution.

Looking at Factor 1, it appears that the high loadings are associated with convenience of uninterrupted communication—perhaps, you could name this Factor "convenient communication." Factor 2 loaded more on equipment-related issues and might be named "reliability."

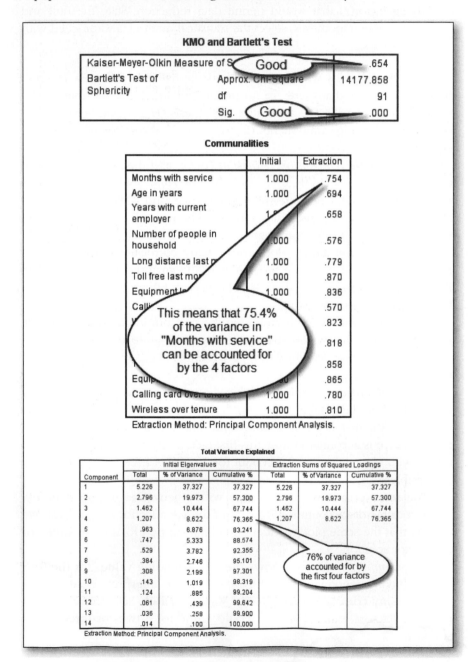

KMO and Bartlett's Test

Kaiser-Meyer-Olkin Measure of S... Good		.654
Bartlett's Test of Sphericity	Approx. Chi-Square	14177.858
	df	91
	Sig. Good	.000

Communalities

	Initial	Extraction
Months with service	1.000	.754
Age in years	1.000	.694
Years with current employer	1...	.658
Number of people in household	.000	.576
Long distance last ...	1.000	.779
Toll free last mo...	1.000	.870
Equipment ...	1.000	.836
Cal...		.570
...		.823
...		.818
...		.858
Equip...		.865
Calling card over tenure	1.000	.780
Wireless over tenure	1.000	.810

This means that 75.4% of the variance in "Months with service" can be accounted for by the 4 factors

Extraction Method: Principal Component Analysis.

Total Variance Explained

Component	Initial Eigenvalues			Extraction Sums of Squared Loadings		
	Total	% of Variance	Cumulative %	Total	% of Variance	Cumulative %
1	5.226	37.327	37.327	5.226	37.327	37.327
2	2.796	19.973	57.300	2.796	19.973	57.300
3	1.462	10.444	67.744	1.462	10.444	67.744
4	1.207	8.622	76.365	1.207	8.622	76.365
5	.963	6.876	83.241			
6	.747	5.333	88.574			
7	.529	3.782	92.355			
8	.384	2.746	95.101			
9	.308	2.199	97.301			
10	.143	1.019	98.319			
11	.124	.885	99.204			
12	.061	.439	99.642			
13	.036	.258	99.900			
14	.014	.100	100.000			

76% of variance accounted for by the first four factors

Extraction Method: Principal Component Analysis.

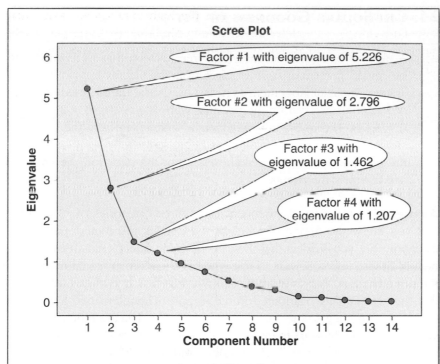

Scree Plot

Factor #1 with eigenvalue of 5.226

Factor #2 with eigenvalue of 2.796

Factor #3 with eigenvalue of 1.462

Factor #4 with eigenvalue of 1.207

Component Number

Component Matrix[a]

	Component			
	1	2	3	4
Months with service	.807	-.257	.187	.046
Age in years	.564		.006	-.492
Years with current employer	.606	-.38		-.368
Number of people in household	-.064			
Long distance last month	.773			
Toll free last month	.543	.24		.032
Equipment last month	.151	.794	.415	-.099
Calling card last month	.658	-.069	.024	.362
Wireless last month	.445	.773	-.140	-.087
Long distance over tenure	.799	-.326	.251	.105
Toll free over tenure	.723	.123	-.566	-.012
Equipment over tenure	.437	.664	.476	-.074
Calling card over tenure	.789	-.199	.150	.309
Wireless over tenure	.603	.652	-.095	-.113

"Months with service" loads strongly on Factor #1

Extraction Method: Principal Component Analysis.

a. 4 components extracted.

△   Chapter 24: Chi-Square Goodness of Fit

24.1 A medical biologist was studying three bacteria that were known to be equally present in samples taken from the healthy human digestive system. We shall label them as A, B, and C. The transformed numbers of bacteria observed by the scientist were A = 3,256, B = 2,996, and C = 3,179. The question is whether the difference from the expected shown in these values qualifies as being statistically significant. Write the null and alternative hypotheses, input the data, run the analysis, and interpret the results.

Answer: You are given the frequencies of three categories (A, B, and C), and you know the frequencies should be equally distributed among the categories but you suspect that they are not. You select a chi-square goodness-of-fit test, and you want to test whether there are an equal number of the three bacteria dispersed in the sample. The null hypothesis (H_0) states that they are equally distributed in the sample. The alternative hypothesis (H_A) states that they are not equally dispersed. The data are inputted as two variables, "bacteria" and "frequency." The test results indicate a chi-square value of 11.347 at 2 degrees of freedom and a significance level of .003. You reject the null, and you now have evidence that the bacteria are not equally distributed in this sample.

	bacteria	frequency
1	1	3256
2	2	2996
3	3	3179

bacteria

	Observed N	Expected N	Residual
A	3256	3143.7	112.3
B	2996	3143.7	-147.7
C	3179	3143.7	35.3
Total	9431		

Reject the null

Test Statistics

	bacteria
Chi-Square	11.347[a]
df	2
Asymp. Sig.	.003

24.2 For this problem, you will open the SPSS *sample file telco.sav* and use the first variable, named "region" (labeled *Geographic indicator*). The variable "region" represents five different zones in which the 1,000 cases reside. The researcher believes that the cases are not equally distributed among the five zones. You are to write the null and alternative hypotheses and then study the variable "region" in an attempt to develop evidence in support of the researchers' hypothesis.

Answer: You are concerned with checking to see how many people live in each of the five different regions. These are the five categories that contain the frequencies, which makes your solution sound like a chi-square problem. The null hypothesis (H_0) is that the cases are equally dispersed among the five geographic regions. The alternative hypothesis (H_A), which is the researcher's idea, is that the people are not equally dispersed among the five regions. Open the database, and then use the nonparametric one-sample test approach in the Fields tab; remember to customize the test and specify *chi-square test*. The results indicate that you fail to reject the null as you have a *Sig.* of .695. There is now statistical evidence that the cases are in fact equally dispersed among the five regions.

Hypothesis Test Summary

	Null Hypothesis	Test	Sig.	Decision
1	The categories of Geographic indicator occur with equal probabilities.	One-Sample Chi-Square Test	.695	Retain the null hypothesis.

Asymptotic significances are displayed. The significance level is .05.

24.3 A high school principal was concerned that several of her teachers were awarding "A" grades at vastly different rates—she had heard many complaints from students and parents. She compiled the data on the teachers and decided to see if there was in fact a statistically significant difference. Her first look at the grades gave her the idea that there was indeed a difference. Can you write the null and alternative hypotheses, select the appropriate statistical test, conduct the analysis, and then interpret the results? The data are as follows—the teacher's name and the number of "A" grades are given in parentheses: Thomas (6), Maryann (10), Marta (12), Berta (8), and Alex (10).

Answer: We have five categories (the five teachers) and frequencies (number of "A" grades awarded by each of the five teachers), and we wish to determine if the numbers of "A" grades are equal in the five categories. This is a one-sample goodness-of-fit chi-square test. The principal believes that the teachers do not award an equal number of "A" grades; therefore, this becomes the alternative hypothesis. The alternative hypothesis (H_A) is that the grades are not equally dispersed among the five teachers, and the null hypothesis (H_0) is that the grades are equally dispersed among the five teachers. We input the data in two columns—one for *teacher* and one for *frequency*—then use the *weighted measure* function found under *Data* on the Main Menu. Finally, we use the nonparametric one-sample test to obtain the results. A chi-square value of 2.61 at 4 degrees of freedom and *significance* of .688 informs us that we must fail to reject the null hypothesis. Therefore, we conclude that any grading differences are simply attributable to chance—the principal should not be overly concerned with the complaining students and parents.

	teacher	frequency
1	1	6
2	2	10
3	3	12
4	4	8
5	5	10

teacher name

	Observed N	Expected N	Residual
Thomas	6	9.2	-3.2
Maryann	10	9.2	.8
Marta	12	9.2	2.8
Berta	8	9.2	-1.2
Alex	10	9.2	.8
Total	46		

Fail to reject null

Test Statistics

	teacher name
Chi-Square	2.261[a]
df	4
Asymp. Sig.	.688

CHAPTER 25: CHI-SQUARE TEST OF INDEPENDENCE △

25.1 A community activist believed that there was a relationship between membership in the police SWAT Team and prior military experience. He collected data from several police departments in an effort to support his belief. He found that there were 57 members of the SWAT team with prior military experience and 13 members with no prior military service. There were also 358 police personnel who had military experience but were not members of SWAT and another 413 with no military experience and not members of SWAT. You must write the null and alternative hypotheses, select the correct statistical method, do the analysis, and interpret the results.

Answer: The activist believes that individuals having prior military experience tend to seek out and join the police department's SWAT team. He is basically saying that there is a relationship between military experience and SWAT team membership—this is the alternative hypothesis (H_A). The null hypothesis (H_0) is that there is no relationship between prior military experience and SWAT team membership. We have frequency data and four categories; therefore, a logical choice for analysis is the chi-square test of independence. Input the data as three variables—"military experience," "SWAT," and "numbers"—in each of the four categories (mil + swat), (no mil + swat), (mil + no swat), and (no mil + no swat). Run the data using Analyze/Descriptive/Crosstabs, then click **Statistics** and **Options**. Interpretation of the chi-square value of 31.442 at 1 degrees of freedom and *significance* of .000 informs us that we must reject the null hypothesis of independence. The community activist has statistical evidence to support his idea that there is a relationship between membership in the SWAT team and prior military experience.

prior military * swat member Crosstabulation			swat member		
			yes	no	Total
prior military	yes	Count	57	358	415
		Expected Count	34.5	380.5	415.0
	no	Count	13	413	426
		Expected Count	35.5	390.5	426.0
Total		Count	70	771	841
		Expected Count	70.0	771.0	841.0

Chi-Square Tests

	Value	df	Asymp. Sig. (2-sided)	Exact Sig. (2-sided)	Exact Sig. (1-sided)
Pearson Chi-Square	31.442[a]	1	.000		
Continuity Correction[b]	30.058	1	.000		
Likelihood Ratio	33.631	1	.000		
Fisher's Exact Test				.000	.000
Linear-by-Linear Association	31.405	1	.000		
N of Valid Cases	841				

a. 0 cells (0.0%) have expected count less than 5. The minimum expected count is 34.54.

b. Computed only for a 2x2 table

25.2 For this exercise, you will open the SPSS sample file *bankloan.sav* and determine if there is a relationship between gender and the size of their hometown for these 5,000 bank customers. The bank official conducting the research believes that "size of hometown" is definitely related to "gender." Your task is to assist the bank official in uncovering evidence in support of his belief. Write the null and alternative hypotheses, select the appropriate statistical method, conduct the analysis, and interpret the results.

Answer: You have categorical data consisting of the two variables: "size of hometown," which has five categories, and two categories for "gender." The sample consists of 5,000 cases, which must be subdivided into gender and size of hometown. Once this is done, you must determine the expected number in each category and then determine if the difference is significant.

Begin by writing the null and alternative hypotheses, which will serve as a guide for your work. The null hypothesis is H_0: Gender is independent of (not related to) the size of one's hometown. Another way to state the same thing is that the numbers of males and females coming from hometowns of different sizes are the same. The alternative hypothesis is H_A: Gender and the size of hometown are related. That is, the numbers of females and males coming from different-sized hometowns are not the same. Remember that the alternative hypothesis is the bank official's belief and you are attempting to develop evidence to support his belief.

You select the chi-square test of independence and use *Analyze/ Descriptive/Crosstabs* to generate the *Crosstabulation* table and request a chi-square statistic and expected values for each of the 10 categories. The calculated chi-square of 3.021 at 4 degrees of freedom has the significance level of .554 and informs you that the null cannot be rejected. You must inform the bank official that there is no statistical evidence of a difference in the numbers of females and males originating from different-sized hometowns—they are equal. You hope he doesn't get angry, as this is really important to him. You decide not to tell him until tomorrow and go home and have a single malt scotch.

Size of hometown * Gender Crosstabulation

			Gender		
			Male	Female	Total
Size of hometown	> 250,000	Count	713	717	1430
		Expected Count	700.7	729.3	1430.0
	50,000-249,999	Count	523	532	1055
		Expected Count	516.9	538.1	1055.0
	10,000-49,999	Count	425	471	896
		Expected Count	439.0	457.0	896.0
	2,500-9,999	Count	432	429	861
		Expected Count	421.9	439.1	861.0
	< 2,500	Count	356	400	756
		Expected Count	370.4	385.6	756.0
Total		Count	2449	2549	4998
		Expected Count	2449.0	2549.0	4998.0

Chi-Square Tests

	Value	df	Asymp. Sig. (2-sided)
Pearson Chi-Square	3.021[a]	4	.554
Likelihood Ratio	3.022	4	.554
Linear-by-Linear Association	.960	1	.327
N of Valid Cases	4998		

a. 0 cells (0.0%) have expected count less than 5. The minimum expected count is 370.44.

25.3 A nutritionist was developing a healthy-eating educational program and was seeking evidence to support her belief that males and females do not consume the same amount of vegetables. She conducted a survey that categorized people by gender and whether they consumed low, medium, or high amounts of vegetables. The numbers for males were low = 29, medium = 21, and high = 16. The numbers for females were low = 21, medium = 25, and high = 33. Write the null and alternative hypotheses, select the correct test, do the analysis (include percentages for all categories), and interpret the results.

Answer: You have categorized the data with simple counts in each category. You wish to determine if the counts in the categories are equal or unequal. The investigator suspects that they are unequal; therefore, you seek evidence to support this contention. Let's begin by writing the null and alternative hypotheses. The null is H_0: Gender is independent of (not related to) the amount of vegetables consumed. You could also say that males and females consume the same amount of vegetables. The alternative hypothesis is H_A: Gender and the amount of vegetables consumed are related. Or females and males consume different quantities of vegetables.

We choose chi-square analysis since we have frequency data in unique categories. The data are inputted as three variables—"gender," "vegetable consumption," and "frequency"—in each of the six categories (male + low veg = 29), (male + med veg = 22), (male + high veg = 16), (female + low veg = 21), (female + med veg = 25), and (female + high veg = 33). Use Analyze/Descriptive./Crosstabs, then click **Statistics** for chi-square test and the Options tab to request all categorized percentages.

Interpretation of the chi-square value of 6.412 at 2 degrees of freedom and *significance* of .041 tells us that we can reject the null hypothesis of independence. The nutritionist has statistical evidence to support her idea that there is a relationship between gender and quantity of vegetables consumed. She may now proceed with the development of her educational program better informed about her target audience.

gender * daily vegetable consumption Crosstabulation

| | | | daily vegetable consumption | | | |
			low	medium	high	Total
gender	female	Count	21	25	33	79
		Expected Count	27.2	25.1	26.7	79.0
		% within gender	26.6%	31.6%	41.8%	100.0%
		% within daily vegetable consumption	42.0%	54.3%	67.3%	54.5%
		% of Total	14.5%	17.2%	22.8%	54.5%
	male	Count	29	21	16	66
		Expected Count	22.8	20.9	22.3	66.0
		% within gender	43.9%	31.8%	24.2%	100.0%
		% within daily vegetable consumption	58.0%	45.7%	32.7%	45.5%
		% of Total	20.0%	14.5%	11.0%	45.5%
Total		Count	50	46	49	145
		Expected Count	50.0	46.0	49.0	145.0
		% within gender	34.5%	31.7%	33.8%	100.0%
		% within daily vegetable consumption	100.0%	100.0%	100.0%	100.0%
		% of Total	34.5%	31.7%	33.8%	100.0%

Chi-Square Tests

	Value	df	Asymp. Sig. (2-sided)
Pearson Chi-Square	6.412[a]	2	.041
Likelihood Ratio	6.489	2	.039
Linear-by-Linear Association	6.366	1	.012
N of Valid Cases	145		

a. 0 cells (0.0%) have expected count less than 5. The minimum expected count is 20.94.

INDEX